QUICK GUIDES:

....

For a helpful way to review for comprehension, see the Quick Guides located throughout the text and listed in the following contents:

QUICK GUIDE 4   LISTENING TO A CLIENT SYSTEM

As you listen to a client system, ask yourself the following questions:

- What brings the client system here today?
- How does the client system describe the situations, and what meaning do these situations have for them?
- What will life look like when the situation is better?
- What strengths and talents do they have?
- What are their expectations of me?
- What do they want to happen in their work with me?
- What can we accomplish together?

Julie Birkenmaier is an Associate Professor in the School of Social Work at Saint Louis University. Dr. Birkenmaier's practice experience includes community organizing, community development, and nonprofit administration. Her research and writing focuses on community development, financial capability, financial credit, and asset development. Along with Marla Berg-Weger, she also co-authored *The Practicum Companion for Social Work: Integrating Class and Field Work* (3rd edition).

Marla Berg-Weger is a Professor in the School of Social Work at Saint Louis University. Dr. Berg-Weger's social work practice experience includes public social welfare services, domestic violence services, mental health, medical social work, and gerontological social work. Her research and writing focuses on gerontological social work and social work practice. She is the author of *Social Work and Social Welfare: An Invitation* (3rd edition). Along with Julie Birkenmaier, she co-authored *The Practicum Companion for Social Work: Integrating Class and Field Work* (3rd edition). She is the past president of the Association of Gerontology in Social Work and currently serves as the Chair of the *Journal of Gerontological Social Work* Editorial Board Executive Committee and is a fellow in the Gerontological Society of America.

Martha P. Dewees, Associate Professor Emerita of the University of Vermont, made her way into social work through counseling and then focused on mental health work at the state psychiatric facility. Her inclusion on the faculty at the University of Vermont provided the impetus for looking at social work practice through the lenses of human rights, social justice, strengths, and social construction.

www.routledgesw.com

Alice A. Lieberman, The University of Kansas, Series Editor

An authentic breakthrough in social work education . . .

New Directions in Social Work is an innovative, integrated series of texts, web site, and interactive case studies for generalist courses in the Social Work curriculum at both undergraduate and graduate levels. Instructors will find everything they need to build a comprehensive course that allows students to meet course outcomes, with these unique features:

- All texts, interactive cases, and test materials are **linked to the 2008 CSWE Policy and Accreditation Standards (EPAS)**.

- **One Web portal with easy access** for instructors and students from any computer—no codes, no CDs, no restrictions. Go to www.routledgesw.com and discover.

- **The Series is flexible and can be easily adapted for use in online distance-learning courses as well as hybrid and bricks-and-mortar courses.**

- Each Text and the Web site can be used **individually** or as an **entire Series** to meet the needs of any social work program.

TITLES IN THE SERIES

The Practice of Generalist Social Work

Third Edition
by Julie Birkenmaier and Marla Berg-Weger, Saint Louis University, and Martha P. Dewees, University of Vermont

To access the innovative digital materials integral to this text and completely FREE to your students, go to www.routledgesw.com/practice

In this book and companion custom website you will find:

- Complete coverage of the range of social work generalist practice within the framework of planned change, encompassing engagement, assessment, intervention, and evaluation and termination—for work with individuals, families, groups, organizations, and communities. This edition features expanded coverage of practice with individuals, families, groups, organizations, and communities.

- Consistent and in-depth use of key theoretical perspectives and case examples to demonstrate essential knowledge, values, and skills for generalist social work practice. But the text does not overwhelm the student reader with a plethora of nuances of intervention and other skills that will occur in a variety of practice settings and roles. *Instead, this book presents clearly the core competencies for general social work practice.*

- Six *unique*, in-depth, interactive, easy-to-access cases, which students can easily reach from *any* computer, provide a "learning by doing" format unavailable with any other text(s). Your students will have an advantage unlike any other they will experience in their social work education. One case, Brickville, is brand new to this book. Go to www.routledgesw.com/cases to see each of these cases on the free website.

- In addition, *four* streaming videos relate to competencies and skills discussed in the book at the three client system levels—individuals and families, groups, and communities. The videos depict social workers demonstrating skills discussed in the chapters, and offer instructors numerous possibilities for classroom instruction. In the video for the new case, "Brickville," a social worker combines individual and family practice skills with multicultural community engagement as he works with an African American family about to be displaced by redevelopment and facing multiple stressors. Go to http://routledgesw.com//sanchez/ engage/video, http://routledgesw.com//

riverton/engage/video, and http://routledgesw.com//washburn/engage/video, and http://routledgesw.com//brickville/engage/video to see *each* of these videos that are included within *each* of the web-based cases.

- At least 10 exercises at the end of each chapter provide you with the means to insure that your students can *demonstrate their mastery* of the theoretical frameworks, skills, and core competencies of generalist social work practice as presented *not just* in the text, but in the free web-based cases as well. Instructors can choose from among the approximately 5 exercises that relate to relevant practice issues, and 5 relate specifically to one of the on-line cases.

- A wealth of instructor-only resources also available at www.routledgesw.com/practice provide: full-text readings that link to the concepts presented in each of the chapters; a complete bank of objective and essay-type test items, all linked to current CSWE EPAS standards; PowerPoint presentations to help students master key concepts; a sample syllabus; annotated links to a treasure trove of social work assets on the Internet and teaching tips on how to use them in your practice sequence of courses.

- A clear focus on generalist social work practice, informed by the authors' decades of real-world practice experience, at *all* levels of engagement and intervention.

ADAPTING THE THIRD EDITION TO YOUR COURSE NEEDS

CUSTOM OPTIONS FOR THE TEXT ARE AVAILABLE: if an instructor or social work program wishes to assign *only a portion* of the text for a micro-level practice course, or a course focused on group practice, they may do so. Smaller units of the text are available in modular formats from the publisher. Please contact orders@taylorandfrancis.com if you would like to pursue this option.

The Practice of Generalist Social Work

Third Edition

Julie Birkenmaier
Saint Louis University

Marla Berg-Weger
Saint Louis University

Martha P. Dewees
University of Vermont

Routledge
Taylor & Francis Group

NEW YORK AND LONDON

Third edition published 2014
by Routledge
711 Third Avenue, New York, NY 10017

and by Routledge
2 Park Square, Milton Park, Abingdon, Oxon OX14 4RN

Routledge is an imprint of the Taylor & Francis Group, an informa business

Second edition published 2011 by Routledge

Trademark notice: Product or corporate names may be trademarks or registered trademarks, and are used only for identification and explanation without intent to infringe.

Library of Congress Cataloging in Publication Data
The Library of Congress has cataloged the one-volume edition as follows:
Dewees, Martha.
  [Contemporary social work practice]
  The practice of generalist social work / by Julie Birkenmaier, Marla Berg-Weger, and Martha Dewees. – [Third edition].
    pages cm. – (New directions in social work)
  Includes bibliographical references and index.
  1. Social service. 2. Social service–United States.   I. Birkenmaier, Julie.   II. Berg–Weger, Marla, 1956-   III. Title.
  HV40.D534 2014
  361.3'20973--dc23
  2013020892

ISBN: 978–0–415–73174–4 (pbk)
ISBN: 978–1–315–84963–8 (ebk)

Typeset in Stone Serif
by RefineCatch Limited, Bungay, Suffolk, UK

BRIEF CONTENTS

DETAILED CONTENTS

PREFACE

MAJOR CHANGES TO THE THIRD EDITION

Like the previous editions, this new edition of *The Practice of Generalist Social Work* provides detailed coverage of the knowledge, skills, values, competencies, and practice behaviors needed for contemporary generalist social work practice. Using a strengths-based perspective, students are given a comprehensive overview of the major areas relevant for social work practice, including: theoretical frameworks; values and ethics; expanded coverage of communication skills for all client systems; and extensive coverage of practice with all client systems through all phases of the change process. *The Practice of Generalist Social Work* offers a comprehensive discussion of practice with individuals, families, groups, communities, and organizations within the concepts of planned change, encompassing engagement, assessment, intervention, evaluation, and termination. Students have the opportunity to learn about generalist practice through in-depth case studies, examples, and exercises integrated throughout the text.

This edition expands greatly on the previous edition to provide all the material necessary and relevant for a two or three course sequence. This third edition provides additional contemporary case studies and applications of theory and knowledge for all client system levels. New exhibits provide in-depth information relevant to practice, summarize pertinent facts from the chapter, and give practical examples of the application of key chapter content. The Quick Guides provide students with brief guidelines for practice and sample documents used in practice. These expanded resources contain up-to-date individual, family, group, community and organizational guidance for the beginning practitioner. New and expanded end-of-chapter exercises, and online supplemental material, including videos, podcasts, and other web-based resources with teaching tips give the instructor additional avenues to challenge students to integrate and expand on chapter content.

For the new editions of all five books in the New Directions in Social Work series, each addressing a foundational course in the social work curriculum, the publisher has created a brand-new, distinctive teaching strategy that revolves around the print book but offers much more than the traditional text experience. Quick Guides within the text offer students guidance for their field experiences. Book-specific websites are accessible through the series website, www.routledgesw.com, and offer a variety of features to support your integration of the many facets of an education in social work.

At www.routledgesw.com/, you will find a wealth of resources to help you create a dynamic, experiential introduction to social work for your students:

- Companion readings linked to key concepts in each chapter, along with questions to encourage further thought and discussion.

- Six interactive fictional cases (three new for this edition) with accompanying exercises that bring to life the concepts covered in the book, readings, and classroom discussions.

- A bank of exam questions (both objective and open-ended).

- PowerPoint presentations, which can serve as a starting point for class discussions.

- Sample syllabi demonstrating how the text and website, when used together through the course, satisfy the 2008 Council on Social Work Educational Policy and Accreditation Standards (EPAS).

- Quick Guides from the books offered online for students to copy and take into the field for guidance.

- Annotated links to a treasure trove of articles and other readings, videos, podcasts, and internet sites.

ORGANIZATION OF THE BOOK

The following paragraphs serve to briefly introduce each of the chapters included in this book with emphasis on the updated content. All chapters have updated and expanded end-of-chapter exercises that use online resources.

Chapter 1

Understanding Social Work Practice provides an overview of social work practice by grounding students in the purpose of social work, social work competencies, types of client grouping, and the practice framework of engagement, assessment, intervention, termination, and evaluation. A discussion of the ethics that guide social work practice, licensure of social work, client populations that social workers work with, and the tensions in social work provides students with real-world information about the profession. Students are also introduced to major theoretical perspectives for social work practice, including the ecosystems, social justice, human rights, strengths, and postmodern perspectives. In this third edition, Chapter 1 features five new exhibits and two new quick guides to highlight key content and summarize material, including a summary of social work approaches.

Chapter 2

In contrast to a straightforward overview of values and ethics, **Applying Values and Ethics to Practice** provides a brief history of social work ethics and the NASW *Code of Ethics* (2008), then contrasts the *Code of Ethics* with the International Federation of Social Workers' Ethical Statement, and also discusses the limits of ethical codes. A discussion of the intersection of ethics and the law gives students information about the interplay between the two, followed by a discussion of ethical dilemmas and processes for resolving them. Extensive discussion about common practice dilemmas gives students exposure to situations that they may encounter in practice, followed by an emphasis on risk management. Expanded coverage of ethics violations and state sanctions round-out the discussion. New content in Chapter 2 includes expanded coverage of ethics violations and state sanctions.

Chapter 3

Individual Engagement: Relationship Skills for Practice at All Levels provides students with the characteristics of core relationships qualities, as well as a description of the specific skills for dialogue with clients at all system levels, including coverage of common communication pitfalls. As the helping relationship includes the dimension of power, the chapter provides extensive coverage of sources of power within relationships, and provides guidance on the use of power through a case study of "Jasmine and the Social Worker." Practical questions guide students toward active listening. Students are also provided with strategies and skills for promoting social justice and human rights within helping relationships. New content in this third edition includes an exhibit on nonverbal behavior guidelines, a quick guide that helps students discover their listening skills, and content about using children as translators.

Chapter 4

Social Work Practice with Individuals: Assessment and Planning includes a focus on the assessment and planning process within the global environment in which practicing social workers live and practice. The chapter begins with a discussion of the history of assessment and moves to an overview of theoretical approaches to social work practice, both classic and contemporary (strengths, narrative, and solution-focused). The application of evidence-based practice approaches is highlighted. The need for practice knowledge and behaviors in the area of diversity within the assessment and planning phases emphasizes the need for cultural competence. The chapter concludes with a discussion of the relevant skills and practice behaviors in the assessment and planning phases of the social work intervention process, including skills needed for strengths-based, narrative, and solution-focused approaches, documentation, and self-care for the social worker. This edition offers more content on narrative and solution-focused approaches, documentation, self-care, and suicide risk assessment with vulnerable populations with more examples on applications of knowledge and theory.

Chapter 5

Social Work Practice with Individuals: Intervention, Termination, and Evaluation introduces students to key areas of social work practice that will impact virtually every dimension of their professional lives. With an emphasis on theoretical perspectives, students learn to apply various intervention, termination, and evaluation practice behaviors. Traditional and contemporary social work roles are highlighted and discussed. Documentation and record-keeping for social work interventions is explained. Interventions with individuals are also framed within an empowerment practice approach. Framed within theoretical perspectives for understanding diversity, students are offered an overview of the skills required to be a culturally competent social work practitioner. New features in Chapter 5 include additional content on cognitive behavioral treatment and expanded content on motivational interviewing, documentation, and empowerment.

Chapter 6

Social Work Practice with Families: Engagement, Assessment, and Planning The chapter begins with a history of social work practice with families, grounded within a systems framework. Theoretical perspectives, including narrative and solution-focused, are discussed within the context of the engagement and assessment phases of interventions with families with emphasis on empowerment. Students encounter a broad range of family constellations as they read about contemporary family social work. Practice behaviors and skills are presented for achieving engagement and assessment with families and documentation strategies are included. This newest version of Chapter 6 offers more content on documentation, empowerment, and more in-depth discussion about solution-focused and narrative assessment and planning.

Chapter 7

Social Work Practice with Families: Intervention, Termination, and Evaluation conceptualizes generalist social work practice interventions with families. Continuing with the theoretical perspectives discussed in Chapter 6, this chapter develops interventions with families using strengths and empowerment, narrative, and solution-focused approaches. Skills and practice behaviors for intervening, terminating, evaluating, and documenting family-focused interventions are discussed in detail. New to this edition is more in-depth content on empowerment and resiliency, and extended exploration of narrative and solution-focused family interventions.

Chapter 8

Social Work Practice with Groups: Engagement, Assessment, and Planning provides students with up-to-date perspectives on social work practice with groups. The chapter opens with an overview of the role of groups within our communities and profession followed by a historical and contemporary perspective on the use of groups for change. The dimensions of group practice are presented within the framework of theoretical perspectives (i.e., narrative and solution-focused). Planning for group interventions, including the engagement and assessment of group members, is emphasized from a practice behaviors perspective along with the importance of cultural competence in the group setting. With this edition, Chapter 8 now includes expanded coverage on cultural competence in group work.

Chapter 9

Social Work Practice with Groups: Intervention, Termination, and Evaluation Developing and implementing interventions with various types of groups is the emphasis of this chapter. Continuing the framing of skills and techniques within theoretical perspectives, the use of evidence-based interventions with groups is introduced using the strengths, narrative, and solution-focused frameworks. Models for group intervention are described along with an in-depth examination of the roles, skills, and practice behaviors required for carrying out a group-level intervention. Termination and evaluation of group interventions are also covered. New to Chapter 9 is additional content on narrative group work, social worker roles, group member behaviors, and evaluation.

Chapter 10

Social Work Practice with Communities: Engagement, Assessment, and Planning introduces students to the concept of community. The chapter defines and discusses types and functions of communities. Students learn about various theoretical perspectives, including contemporary perspectives for community practice. Engagement and assessment concepts, including community-based analysis, evidence-based practice, and community needs assessments, are extensively discussed. Examples of types of needs assessments, surveys used in needs assessments, and needs assessment summaries provide additional practice guidance. Community practice skills are thoroughly covered, as are the implications of global interdependence for community practice in the United States. This edition contains expanded content on needs assessments, including types, examples, and surveys used to collect needs assessment data.

Chapter 11

Social Work Practice with Communities: Intervention, Termination, and Evaluation builds on the engagement and assessment content of Chapter 10 to present strategies and techniques for community practice. Using the insights gained about practice at the individual, family, and group levels, this chapter expands the students' awareness of social work practice with communities through a discussion of today's trends and skills for intervention, including community social and economic development, and community organizing. Included in this discussion is coverage of international community practice. Examples of public and private efforts to promote evidence-based community practice assist students in applying the material. Additional guidance on advocacy efforts and asset based development are presented. Students also learn the knowledge and skills needed for termination and evaluation of community practice. The third edition offers a host of examples of contemporary community interventions, with a special focus on community development and community organizing examples, as well as quick guides that offer students concrete tools to use in community interventions, termination, and evaluation.

Chapter 12

Social Work Practice with Organizations: Engagement, Assessment, and Planning covers a challenging client system for beginning practitioners—the organization. Students learn a wealth of practical and theoretical aspects of organizations, including a discussion about the purpose and structure of organizations, power relations within organizations, and social work within host organizational settings. The chapter provides discussion about the elements of an internal assessment of organizations, to include organizational culture, and external assessments as well. Material about organizational policy advocacy and nonprofit partnerships help guide practice. The many new, recent examples of organizational engagement and assessment provide students with contemporary illustrations of key content in Chapter 12. Three new quick guides offer handy tools to assist students in their efforts to contribute to organizational engagement and assessment work.

Chapter 13

Social Work Practice with Organizations: Intervention, Termination, and Evaluation uses the foundation built in Chapter 12 to discuss approaches, perspectives, and models for intervening with organizations. This chapter provides extensive coverage of the relationship between theoretical perspectives and organizational change, as well as a practical framework for thinking about generating change and the needed knowledge for a social work generalist in this endeavor. Termination and evaluation of change efforts within organizations, including a discussion about the role of the generalist practitioner in this process, help students see their potential role in a change effort with organizations. Content about the challenges of implementing organizational change, and persuasion skills to assist in these efforts, provide direction for the practitioner. In this edition, Chapter 13 has expanded content that includes examples of intervention at the organizational level that includes developing and refining new programming, as well as the associated challenges.

INTERACTIVE CASES

The website www.routledgesw.com/cases presents six unique, in-depth, interactive, fictional cases with dynamic characters and real-life situations. Three of them—the RAINN, Hudson City, and Brickville cases—are entirely new to this edition of the series. Your students can easily access the cases from any computer. The cases provide a "learning by doing" format unavailable with any other book, and the experience will be unlike any other your students will experience in their social work training.

Each of the interactive cases uses text, graphics, and video to help students learn about engagement, assessment, intervention, and evaluation and termination at multiple levels of social work practice. The "My Notebook" feature allows students to take and save notes, type in written responses to tasks, and share their work with classmates and instructors by e-mail. Through these interactive cases, you can integrate the readings and classroom discussions:

The Sanchez Family: Systems, Strengths, and Stressors   The 10 individuals in this extended Latino family have numerous strengths but are faced with a variety of challenges. Students will have the opportunity to experience the phases of the social work intervention, grapple with ethical dilemmas, and identify strategies for addressing issues of diversity.

Riverton: A Community Conundrum   Riverton is a small Midwest city in which the social worker lives and works. The social worker identifies an issue that presents her community with a challenge. Students and instructors can work together to develop strategies for engaging, assessing, and intervening with the citizens of the social worker's neighborhood.

Carla Washburn: Loss, Aging, and Social Support   Students will get to know Carla Washburn, an older African American woman who finds herself living alone after the loss of her grandson and in considerable pain from a recent accident. In this case, less complex than the Sanchez family case, students can apply their growing knowledge of gerontology and exercise the skills of culturally competent practice at the individual, family, and group levels.

RAINN   Based on the first online hotline for delivering sexual assault services, this interactive case includes a variety of exercises to enable students to gain knowledge and skills related to the provision of services to persons in crisis. With a focus on social work practice at all levels, exercises provide insight into program services and evaluation, interactions with volunteers and clients, and research.

Hudson City: An Urban Community Affected by Disaster   A natural disaster in the form of Hurricane Diane has hit Hudson City, a large metropolitan area on the northeastern coast of the United States. This interactive case will provide students with insights into the complexities of experiencing a disaster, including the phases of the human response to disaster and the social work role in responding to natural disasters.

Brickville   A real estate developer has big plans to redevelop Brickville, an area of a major metropolitan area that has suffered from generations of disinvestment and decay. The redevelopment plans have stirred major controversy among community residents, neighborhood service providers, politicians, faith communities, and invested outsiders. This case is a "community case" in which a "family case" is embedded; the case is multi-layered and detailed. Students will be challenged to think about two levels of client systems, and the ways in which they influence and are influenced by one another.

IN SUM

We have written this book with the purpose of providing you and your students with the information needed to learn the knowledge, skills, values, competencies, and practice behaviors that are required for a competent and effective generalist social work practice. The multiple options for supporting your teaching of this content are intended to help you address the diverse range of student learning styles and needs. The design of this text and the instructor support materials are aimed at optimizing the experiential options for learning about generalist practice. We hope this book and the support materials will be of help to you and your students as they embark on their journey toward social work practice.

ACKNOWLEDGEMENTS

We would like to thank the many colleagues who helped to make this book and previous editions possible. To Alice Lieberman, we are grateful for your innovation and vision that has resulted in this series and the web-based supplements that bring the material alive. We appreciate the camaraderie and support of the authors of the other books in this series—Rosemary Chapin, Anissa Rogers, Judy Kryzik, and Jerry Finn. A special thank you to Anissa Rogers and Shannon Cooper-Sadlo, and Andrea Seper whose creativity makes the exercises, test questions, and PowerPoint slides enticing and easy to use. We want to thank the group who participated in the production of the video vignettes: actors John Abram, Patti Rosenthal, Beverly Sporleder, Sabrina Tyuse, Kristi Sobbe, Myrtis Spencer, Phil Minden, Katie Terrell, and Shannon Cooper-Sadlo and videographers, Tom Meuser and Elizabeth Yaeger. A special thanks goes to Sue Tebb, for her careful review and invaluable feedback, and Andrea Seper, graduate student assistant, for her extensive assistance with this edition and supplemental materials. Thanks also goes to social work graduate student assistants, Michelle Siroko and Luxiaofei Li, for their help with research, writing and editing. We also want to thank:

Toni Johnson	University of Kansas
Lara Vanderhoof	Tabor College
Chrys C. Ramirez Barranti	Sacramento State University
Kameri Christy	University of Arkansas
Robin Bonifas	Arizona State University
Bill Milford	Thomas University
Martha Haley-Bowling	Ferrum College
Mary Clay Thomas	Mary Baldwin College
Armon Perry	University of Louisville

for their reviews of the book as it was evolving. Finally, we are most appreciative to the staff of Routledge for their support and encouragement for making this book a reality. It takes a village . . . .

ABOUT THE AUTHORS

Julie Birkenmaier is an Associate Professor in the School of Social Work at Saint Louis University, Missouri. Dr. Birkenmaier's practice experience includes community organizing, community development, and nonprofit administration. Her research and writing focuses on financial capability, financial credit, community development, and asset development. With colleagues, she co-edited *Financial Capability and Asset Development: Research, Education, Policy and Practice*. With Marla Berg-Weger, she also co-authored the textbook, *The Practicum Companion for Social Work: Integrating Class and Field Work* (3rd edition).

Marla Berg-Weger is a Professor in the School of Social Work at Saint Louis University, Missouri and Executive Director of the Geriatric Education Center. Dr. Berg-Weger holds social work degrees at the bachelor's, master's, and doctoral levels. Her social work practice experience includes public social welfare services, intimate partner violence services, mental health, medical social work, and gerontological social work. Her research and writing focuses on gerontological social work and social work practice. She is the author of *Social Work and Social Welfare: An Invitation* (3rd edition). With Julie Birkenmaier, she co-authored the textbook, *The Practicum Companion for Social Work: Integrating Class and Field Work* (3rd edition). She is the Past President of the Association of Gerontology in Social Work and currently serves as the Chair of the *Journal of Gerontological Social Work* Editorial Board Executive Committee and is a fellow in the Gerontological Society of America.

Martha P. Dewees, Associate Professor Emerita of the University of Vermont, made her way into social work through counseling and then focused on mental health work at the state psychiatric facility. Her inclusion on the faculty at the University of Vermont provided the impetus for looking at social work practice through the lenses of human rights, social justice, strengths, and social construction. These are reflected in her previous publications and in the first edition of this book.

Understanding Social Work Practice

Debbie, Joan, Marcy, and Kate felt as if the whole cosmos had just opened up for them. As a group they had already shared their experiences as women who had experienced intimate partner violence. Their social worker had explained how they could use their stories to help educate children in schools about violence at home. Their collective power felt liberating to them.

Chan, an 11-year-old orphaned child in Southeast Asia, hung on every word the community development worker uttered. She spoke of human rights that he had in his workplace. He had never imagined that he had any rights.

Jamie was very relieved after the hospital social worker provided information about potential rehab centers for his father, who was struggling with physical and mental health issues. Jamie's father could not be cared for at home, and Jamie now knew how his father would be cared for during his recovery.

At their annual meeting, the neighborhood association expressed gratitude for the work of the community social worker. Her efforts, along with her colleagues', had resulted in neighbors who were more connected to job opportunities, teens involved in working for a new community park, and additional city funding made available to the neighborhood by local elected officials.

Key Questions for Chapter 1

(1) How can I prepare to engage, assess, intervene, and evaluate with client systems (EPAS 2.1.10(a)–(d))?

(2) How do you define the practice of social work? (EPAS 2.1.1)

(3) How does theory relate to social work practice? (EPAS 2.1.7)

WELCOME TO THE WORLD OF SOCIAL work practice. This world is sometimes exhilarating, sometimes frustrating, and sometimes heartbreaking. It is nearly always challenging, offers a deep sense of purpose, and once you have entered it, you may find it impossible to imagine doing any other kind of work.

This chapter examines the professional competence that the social worker offers and the ways in which the profession's mission has taken shape. You will explore ways to think about social work practice, types of client groupings and processes, tensions in social work, a brief account of the way the profession has dealt with theory, and contemporary practice commitments. The brief vignettes you have just read describe only a few of the many types of social work practice and client systems.

PURPOSE OF SOCIAL WORK

What is social work trying to accomplish? The primary mission of social work, according to the National Association of Social Workers (NASW) *Code of Ethics* is "to enhance human well-being and help meet the basic human needs of all people, with particular attention to the needs and empowerment of people who are vulnerable, oppressed, and living in poverty" (2008). This mission can be narrowly defined as tied to a particular type of system (e.g., mental health, correction, and child welfare). Alternatively, the mission can be defined as tied to a specific method (e.g., behavioral or cognitive change), a particular problem-solving process or the elimination of a problem-focus (e.g., through the strengths perspective). The mission can also be defined as trying to achieve goals, such as adaptation, sobriety, or a clean criminal record. Although these endpoints are real examples of social work purpose, they are embedded in the deeper and broader mission. The purpose of social work is expressed generally and somewhat vaguely in the *Code of Ethics*, and therefore leaves much open to interpretation. Tensions, therefore, have arisen within the profession about the identity of the profession, some of which are discussed below.

PRACTICING SOCIAL WORK

Social work practice can be conceptualized in several different ways, including: (1) type of practice or range of practice settings; (2) set of activities; (3) set of roles; (4) set of competencies and practice behaviors; (5) types of client grouping; (6) practice framework; (7) a profession that is licensed by states; (8) by purpose (as mentioned above); and (9) by the tensions experienced in the profession. Self-knowledge is also a critical component of effective practice. These concepts are described below, and some are described in depth in this chapter.

Generalist social work practice, included in the curricula of undergraduate social work programs and in the foundation coursework in social work graduate programs, prepares students to work with a range of systems, from direct, one-to-one practice to group-level practice to local and global community development work and work in international settings. Advanced social work practice, taught in the concentration curriculum of graduate school, focuses on more specialized practice, such as social work within a medical setting, family therapy, and administration of social service organizations. The range of generalist and advanced practice settings includes psychiatric facilities, schools, community organizations, family service organizations, legislatures, correctional settings, and a host of others.

Social work practice is also framed as a set of activities, often associated with particular agencies and the functions they carry out in the community. Examples of these include: (1) advocating for policy change regarding the rights of older adults to services from an area agency on aging; (2) facilitating an empowerment group in a domestic violence program, such as the group in which Debbie, Joan, Marcy, and Kate belong; (3) mentoring students in a neighborhood school; (4) developing a psycho-educational group in a mental health agency; (5) supporting families in an emergency housing shelter; (6) implementing human rights policies in another culture by learning a new language and traveling to another part of the world to assist children like Chan; (7) advocating with public officials, companies, and corporations; (8) working with a group of adolescents; and (9) fundraising from government and foundation sources to assist in the empowerment of disadvantaged neighborhoods.

Another way to look at generalist social work practice emphasizes the role of the social worker and the relationship between social worker and client. In the helping process, the generalist social worker working in direct practice with individuals may take on the role of *case manager* (i.e., assisting clients to assess for, arrange, and coordinate needed goods and services), *counselor* (i.e., providing suggestions to assist clients to reach their goals), *broker* (i.e., referring clients to appropriate needed goods and services), *mediator* (i.e., assisting two parties to mutually resolve a dispute), *educator* (i.e., providing relevant information to client systems), and *client advocate* (i.e., working with or on behalf of a client to obtain goods and services). Some definitions of role relationships are driven largely by theoretical perspectives. These roles are discussed in great detail in Chapter 5. All social worker functions are associated with roles that clarify the nature of the interaction between a social worker and the client system. These roles are fluid, and can change from interaction to interaction, and even within interactions. The roles define responsibilities for both the client system and the social worker. Social work practice involves a wide variety of roles because social work is involved in the breadth of human experience.

Social work is a complex and multifaceted profession, with practitioners found in virtually every setting from health clinics to housing programs. Social workers are also engaged in a wide variety of activities. The idea of social work practice is defined by competencies and practice behaviors, client populations, practice frameworks, as a licensed profession, and by the purpose of social work. These various ways of defining and describing the profession are discussed in the following section.

© Lisa F. Young

SOCIAL WORK COMPETENCIES

Generalist social work practitioners bring a set of competencies and practice behaviors to serve client systems. These competencies, or "... measureable practice behaviors that comprise knowledge, values and skills" (Council on Social Work Education (CSWE), 2012, p. 3), are defined by the Council on Social Work Education's Educational Policy and Accreditation Standards (EPAS) to guide social work education, and therefore influence social work practice. The 10 competencies define practice elements, as well as address the practice structure of engagement, assessment, intervention, and termination and evaluation. Each competency consists of several descriptive practice behaviors that social workers should competently be able to demonstrate in practice settings. These competencies encompass knowledge, values, and skills, and are the attributes that social workers bring to the interaction with the client system of individuals, families, groups, organizations, and communities.

Knowledge, Values, and Skills

The trio of knowledge, values, and skills is the core of social work education and training. As an integrated whole, these elements can be considered the tools that a practitioner brings to the work, as they reflect personal aspects and attributes acquired by—and some scholars think inherent to—the social worker. Each is necessary for effective service delivery.

Knowledge Facts and research findings comprise professional knowledge along with broader topics such as intuition and cultural awareness. A practice-based example may include the need for social workers to know facts, histories, theories, and trends about human development, policy, research, and practice, which are often combined and labeled "biopsychosocial" knowledge. Knowledge of many theories of human behavior and the social environment is central to generalist social work practice. This knowledge is needed to understand normal (or expected) behavior for an adolescent under stress, or the likely dynamics in an agency when a policy change is initiated from the administration. This arena also assumes that the knowledge required is largely agreed upon and that it is attainable through study, discussion, research experimentation, and related activities. According to the EPAS competencies, social workers should also know the history of social work, the value base and ethical standards of the profession, and the history and current structures of social policies and services (CSWE, 2012). Knowledge may also include your experience of utilizing theory and ideas, and even the knowledge emanating from the first-hand experience of seasoned practitioners, or "practice wisdom." This form of practice knowledge may elude traditional empirical measurement, an ongoing challenge both in the field and in social work education. This practice knowledge is the integration of knowledge learned "on the job" with knowledge from other sources, such as prior education and continuing education. Practice wisdom is an invaluable part of social work practice.

Two other critical areas of knowledge are culture and spirituality. First, social workers must understand the broad ways in which a culture's structure and values may create or enhance privilege and power for certain groups in society, and to eliminate the influence of personal biases and values in working with diverse clients (Council on Social Work Education, 2008). Cultural competence refers to the "process by which individuals and systems respond respectfully and effectively to people of all cultures, languages, classes, races, ethnic backgrounds, religions, and other diversity factors in a manner that recognizes, affirms, and values the worth of individuals, families, and communities, and protects and preserves the dignity of each" ("Indicators for the Achievement of the NASW Standards for Cultural Competence in Social Work Practice," 2008). The notion of spirituality is also included in models of both human development and social work education. Spirituality is integrally connected with culture and therefore gives practitioners access to knowledge about important dimensions of their clients. For example, recognizing a Navajo child's spiritual affirmation of harmony may help a social worker understand the child's reluctance to engage in aggressive competition in school. In some areas of practice, such as end of life, a biopsychosocial-spiritual model is being utilized to serve the needs of clients as whole persons (Buck, Overcash, & McMillan, 2009).

Values   Social work has always strongly identified itself as a profession of values, which are strongly held beliefs about preferred conditions of life. The values of the social work profession articulate several key elements, such as (1) the inherent worth of people; (2) the need for open and honest communication to build relationships; and (3) respect for the unique characteristics of diverse populations (to name just a few of the values of the social work profession). Although the social work profession has clearly stated values, you may hear spirited discussion about the ways to implement these values in practice. Consideration of values and ethics will present many challenges to social workers, and will often look different in complicated practice contexts from the way they do in isolation. Through the professional organization, the National Association of Social Workers' (NASW), *Code of Ethics* (NASW, 2008) plays a significant role in sorting out complex situations in which values and ethical conflicts are at issue. While values, as beliefs, guide professional thinking about behavior and judgments about conduct, ethics are the rules, or prescriptions, for behavior that reflect those values. For example, one of the cardinal values held by the social work profession is that all human life is worthy. From that value arises the ethical principle that human life should be protected in social work practice, which means that social workers actively try to prevent any harm that might occur to another person. Ethical dilemmas arise when ethical principles conflict, such as the case when trying to prevent harm to one person puts another at risk. The National Association of Social Workers developed a *Code of Ethics* to clarify the principles of ethical practice. The role of values provides critical criteria for the ways in which the profession shapes itself and the professional rules of conduct.

Skills   The third element the social worker brings to the professional work is skills, or the implementation of the knowledge, theoretical perspectives, and values the social worker brings to her or his work with client systems. Social workers need a wide range of skills, from traditional communication skills and individual assessment skills, to skills in working with families, groups, communities, and organizations. The range of skills will be discussed in subsequent chapters of this book, and the skills mentioned in the EPAS core competencies (CSWE, 2012) will be emphasized.

EXHIBIT 1.1

The full NASW Code of Ethics is available in English and Spanish here:

http://www.socialworkers.org/pubs/code/code.asp

NASW has also published a series of publications that articulate standards for social work in various practice settings (e.g., school social work) and with various populations (e.g., family caregivers of older adults). See the listing of standards publications here:

http://www.socialworkers.org/practice/

Types of Client Groupings

Competencies to practice social work may be viewed within the context of client groupings; social workers work with individuals, couples, families of all types, groups, communities, and organizations, both domestically and internationally. These client groups often overlap, as in the need for group work within community practice, and the need to engage in community work to assist families. These groupings define the constellation of the service beneficiaries involved in the interaction. The more recent expansion of global awareness has resulted in the inclusion of international and global social work, extending the arena in which social work practice is both relevant and critical. Social work education frequently uses this

EXHIBIT 1.2

Types of client groupings

- *Individual work, family work, or casework:* This face-to-face focus, or individual work, family work, or casework, spans all fields of practice, populations, and settings. Casework may include working with an adolescent struggling with sexuality issues, a mother concerned with her child's development, an older adult who is facing an inability to care for himself, or a five-year-old who does not pay attention in school. The nature of the work will be heavily influenced by the agency's purpose, the practice perspective, the personal characteristics of the client (individual or family), and your skills and theoretical perspective. Cultural context, social and political influences, and current community concerns will also affect client relationships.
- *Group work:* Social workers frequently practice with groups. The focus of group work may be on helping the group members make individual changes, the group as a whole make changes, or for the group to make changes in the environment. Group-level social work practice is conceptually social, and embodies the relationship emphasis of the profession. Group work can be a powerful tool for change, and has similarities and differences in the many different practice settings and contexts within which it is implemented. Social work with groups is one of the most distinctive practices of the profession.
- *Community practice:* The method of practice known as community practice usually involves a common locality, such as a city neighborhood, small town, or rural area. Beyond locality, however, *community* also refers to a common concern, interest, or identification. For example, you might consider yourself a member of a hometown as well as a community of gay men or Jewish women or people of Irish descent. Social work practice has historical work in community practice, and aspires toward inclusive participation of community citizens in addressing their self-defined concerns.
- *International work:* An emerging focus of practice is international social work. International social work is a broad concept that can include many facets, including: 1) a worldview; 2) a domestic practice and/or action informed by knowledge of international issues; 3) participation in international professional associations; and 4) development and human rights (Healy & Link, 2012).

organizational scheme in the layout of its curriculum by specific course content or emphasis.

The social work profession increasingly integrates a global, transnational perspective into local issues, such as employment, homelessness and health. Such a perspective provides a full account of the challenges faced by client systems.

An organizational scheme based on client groupings contains a number of underlying assumptions rooted in culture about social work skill development. For example, many U.S. social workers agree that one-to-one practice work is the natural starting place for social work practice, and other types of practice refer back to one-to-one methods. However, in other cultures in which family or community is the major organizing structure or reference point, this assumption may not fit. For example, a social worker providing mental health services may be accustomed to seeing clients individually. However, it may be more culturally appropriate to work with a Latino client's entire family, rather than just the individual client, in order to best assist the client (Conan, 2012). Although the distinctions between types of groupings continue to be a useful way of thinking about social work, generalist practice emphasizes an integration of knowledge and skills across all system levels and sizes.

Practice Framework

The practice framework views social work practice as it relates to the progression of the work. The most commonly used practice framework describes the activities of the social worker and the client system as they proceed together through relatively standard phases. Although these phases are described here in a linear progression, the social worker and client frequently loop back and forth between phases as necessary. The phases described below are more fully discussed relative to client systems in Chapters 3–13.

Engagement   Building a relationship among the social worker, the client, and the client's environment is referred to as engagement. Successful engagement involves establishing a degree of trust and a sense that the work ahead will be helpful to the client system and professionally rewarding and satisfying to the social worker. Engaging a client system requires effective communication and engagement skills with individuals, as well as establishing significant and collaborative connections with the client system's environment and the relevant service systems that will impact your work and the client's goals. For example, your client has asked for your help in negotiating and advocating within her disabled child's school system. You carefully develop a relationship with her so that you may understand the issues and her experiences. You will also need to engage with the client's network (in this case, the school system), to learn about the system constraints and challenges, in order to effectively facilitate a more productive relationship between your client and the school.

Without a successful engagement, the effectiveness or helpfulness of the work that follows may be compromised or more challenging. Engagement may not be effective for a variety of reasons: for instance, the client may lose investment in a process due to a lack of significant meaningful connection to the work, or the critical network contacts in the environment may feel that their role is neither appreciated nor fully understood. In the situation just described, if the school staff thinks that you do not understand the difficulties of providing extensive services with shrinking resources, they may actually become less willing to work on improving their relationship with your client.

Assessment and Planning   Assessment encompasses recognition of the parameters of the practice situation and the way in which the participant is affected. The client's goals are central to the process of assessment and planning for an intervention. Through mutual exploration of the issues, the client and social worker decide the best ways to address the client's goals. Assessment focuses on the analysis of the major area for work; the aspects of the client environment that can offer support

A social worker engages in a child welfare assessment

© Lisa F. Young

for a solution; the client knowledge, skills, and values that can be applied to the situation; and ways in which the client can meet her or his goals.

The social worker's theoretical perspective guides the selection of assessment styles and planning approaches. The overall mission of the agency also plays an important, perhaps defining role in the activity. For example, the assessments conducted in foster care agencies may vary, but they are all likely to be focused on children and parenting, rather than vocational development or personal growth counseling.

Planning is an integral part of the assessment process. Developing the plan will require the client and social worker to assess or evaluate the options, the resources, the barriers, client preferences, and the agreed-upon goals, along with the established methods of achieving them. The plan should also include consideration of the less-than-obvious or unexpected outcomes of reaching a goal. For example, a social worker may assist a couple in adopting a child for whom they provided foster care. Another child in the family may have accepted the temporary nature of the foster care arrangement but suddenly feels threatened by the permanency of foster care turning into an adoption. The complexity of human emotion frequently appears in unexpected places and times in social work practice.

Intervention   The next stage in this sequence, intervention, refers to the action—the doing of the work that will enable the client and the practitioner to accomplish the goals decided upon in the assessment. Generalist practice interventions vary widely, from helping an older adult tenants' union organize a rent strike to helping a family receiving Medicaid benefits receive health care for an ill child. Intervention is the joint activity of the client system and the social worker. In some cases, the intervention may involve the social worker listening to and reflecting on the client's situation, helping the client think about the situation and her or his role in it, and facilitating an opportunity to create a different, more preferable situation. The intervention process is also influenced by the theoretical and/or practice perspective of the social worker. In some practice models, the social worker is very active in determining the client's best interests and initiating action to achieve a specific outcome. In other models, the planning and intervention is a more collaborative process with shared responsibility for actions, and together deciding on the best course. As with other aspects of practice, the type, level, and focus of the intervention vary widely.

Termination   Ending with client systems, or termination, is a long-standing area of focus for social workers. Many clients have experienced abrupt, sometimes violent or completely disconcerting endings to relationships or arrangements; therefore, the profession is committed to facilitating appropriate and effective termination with client systems. In general, the termination process includes reviewing the work and accomplishments, discussing the development of the working relationship, and planning the future to sustain the changes that have been achieved. To many social workers, termination is one of the most difficult as well as one of the most important aspects of the work. Those who want to focus more on the future sometimes call it "consolidation" or "graduation."

Evaluation   Although many social workers are pressed for time to complete other tasks, present and future clients benefit from evaluation to determine the effectiveness of the practice intervention. The social work *Code of Ethics* (NASW, 2008) mandates social workers engage in and utilize research to improve practice, which can include program evaluation and research on individual effectiveness, as well as using research findings to inform their practice. CSWE accreditation guidelines require social workers to "engage in research-informed practice and practice-informed research" (CSWE, 2012, EPAS, 2.1.6). At the same time, the profession has put increasing emphasis on evidence-based practice, defined as "an educational and practice paradigm that includes a series of predetermined steps aimed at helping practitioners and agency administrators identify, select, and implement efficacious interventions for clients" (Jensen & Howard, 2008). Use of evidence-based practice facilitates the integration of research findings, client values and preferences, practitioner knowledge and expertise, and other factors to make practice, policy, and research decisions.

Research on practice may involve evaluation tools that assess the progress of a program, and may provide encouragement for a reflective approach by individual workers on the quality of their work. Others see an important link between research activities and effective client advocacy. For example, if you want to advocate for persons who are homeless by demonstrating that existing services are inadequate or are not directed effectively, an understanding of the demographics of the homeless population, the number of homeless, and previous research findings about effective services for homeless populations (among other items) is imperative.

There are many kinds of evaluation, including quantitative, qualitative, subjective, objective, formative (during the work), summative (at the end of the work), self-report, reflection, and standardized tests. The prevailing trend within the various funding and accountability arrangements associated with social work practice is to require more evaluative activity both to demonstrate effectiveness and to justify continued or increased financial support for practice initiatives and policy programs.

Licensure of Social Work

Another way to conceptualize social work is as a helping profession that is licensed by states to protect the public. Social work, as a profession, is regulated within and by all 50 states, the District of Columbia, Puerto Rico, the Virgin Islands, and 10 Canadian provinces. Regulation protects the public by establishing: (1) the qualifications that a professional must possess; (2) a means of holding professionals accountable; and (3) a system for the public to make complaints against incompetent or unethical practitioners and have them investigated and adjudicated. Today, there are almost 400,000 licensed social workers practicing in the United States and Canada (Randall & DeAngelis, 2008).

The social work profession has four types of licensure available: BSW (usually upon graduation); MSW (one type upon graduation, and a second type, independent, after two years of supervised general experience); and clinical (after two years of

EXHIBIT 1.3 *Continuing Education*	A variety of on-line and face-to-face options are available for CE credit for social workers. Below are some examples of resources for social work CEs: From NASW: http://www.naswwebed.org/ From NASW California: http://www.socialworkweb.com/nasw/ Rutgers School of Social Work: http://ssw-web.rutgers.edu/ssw/ce/

supervised clinical experience). Most jurisdictions license social workers at two or more of these categories. While the requirements vary from jurisdiction to jurisdiction, the general requirements for licensure are a specific level of education, supervised experience by a social worker, and demonstration of knowledge and minimum competence by passing an exam, providing references, and demonstrating evidence of good moral character. Some locales require more than two years of supervised clinical experience, or proof of a minimum number of hours in a clinical field placement, or proof of specific clinical coursework. After licensure, many states require that licensed social workers complete ongoing professional continuing education units (CEUs) through professional workshops or conferences that provide skills training for new situations, new populations, and new ways of thinking about the work (Randall & DeAngelis, 2008).

Tensions in Social Work

Almost since its inception, the social work profession has been challenged by a set of tensions that have sometimes seemed to obscure the identity of the profession. The most significant tensions (discussed below) are those that strongly shape the identity of the social work profession and the way in which social workers practice. They are:

- Whether to promote a clinical or nonclinical approach to working with clients
- The extent to which social workers exercise social control or promote social change
- The extent to which social workers promote change or acceptance of their clients
- The struggle between encouraging clients to adjust to their circumstances or challenge their circumstances
- Whether social workers promote their expert position or share power with their clients
- The profession's adjustment to globalization.

Clinical and Nonclinical Approaches   A central tension involves the question of whether practitioners should focus primarily on clinical or nonclinical work. The word *clinical* has many meanings, and can represent a code for a medically based private practice model that involves such concepts as diagnosis and managed care. *Clinical* can be associated with cold, calculated, stiff, or impersonal interchanges that are driven by the expert and received by the patient.

In this book, *clinical work* encompasses social work with individuals, groups, and/or families that is not only direct practice, or direct work (face-to-face), but also designed to change behaviors, solve problems, or resolve emotional or psychological issues (Grant, 2008). For example, clinical work can include intervening individually with a young woman to address her self-harming behavior, facilitating a series of groups for children who have experienced the death of a parent, and assisting a family to redefine the communication patterns among the three generations of its members. Clinical work can extend into many practice arenas such as physical and mental health, substance abuse treatment, school social work, gerontological social work, and some child welfare work. Some states define clinical work explicitly and require a master's degree. For example, West Virginia has separate levels of licensure for new MSW graduates, MSWs with over two years experience, and MSW clinical social workers with over two years of experience (West Virginia Board of Social Work Examiners, n.d.). Additional or different credentials may be required to work in some areas (e.g., an addictions certificate for substance abuse work).

Nonclinical work usually implies that the work addresses the environment. Macro, or indirect, practice addresses social problems in community, organizations, institutional, and society systems. Nonclinical, or macro, practice social workers, achieve social change through neighborhood organizing, community planning, locality development, public education, policy development, administration, and social action. Nonclinical work can also include social work that is political or focused on social reform efforts. These efforts involve such activities as working for improved institutional responses or changing or supporting laws, policies, and social structures relating to various dimensions of diversity, such as class, gender, ability, and cultural ethnicity. This field of work, policy practice, focuses specifically on ensuring that policies are more responsive to client needs and rights. Nonclinical work focuses less on the internal dynamics of an individual's experience and more on opportunity and change in the environment. While tension can exist between clinical and nonclinical work, thoughtful and principled efforts can connect the approaches. For example, clinical social workers can identify and communicate client needs to administrators and other nonclinical social workers, so that policy practice efforts are appropriately channeled to the client needs that are the most significant. In another example, clinical social workers can be involved in and refer clients to neighborhood organizing efforts to make social connections with others concerned about similar topics. The following discussion considers some dimensions of this tension.

Developmental Socialization and Resocialization   The late Harry Specht, a social work policy educator, distinguished between "developmental socialization" and "resocialization" (Specht, 1990). Specht defined developmental socialization as the attempt, through providing support, information, and opportunities, to help people enhance their environments by making the most of their roles. Developmental socialization also involves confronting obstacles such as abuse or oppression that impede people's attempts to make the most of their roles. Developmental socialization was the natural domain of social work, and is basically nonclinical. In contrast, Specht defined resocialization as the attempt to help people with feelings and inner perceptions that relate primarily to the self. Specht purported that psychotherapeutic approaches associated with psychology and psychiatry, rather than social work, should deal with such issues. Significantly, Specht argued that social work had been seduced from its original mission by clinical psychotherapy, which many people characterize as a higher-status activity. As an advocate for an emphasis on the social environment, rather than inner psychological life, Specht called for social work to build its professional core in public, rather than private, services and institutions and to replace all its clinical training with adult education, community work, and group work. His last major co-written publication was tellingly called *Unfaithful Angels: How Social Work Has Abandoned Its Mission* (Haynes, 1998).

Many social workers are committed to the kind of work that Specht rejects as inappropriate. These practitioners view social work as providing useful perspectives for dealing with clinical, interactional, interpersonal, and sometimes intrapersonal issues. Many practitioners view these perspectives as appropriate tools in the realms of individual counseling, family intervention, and a host of other areas that might be called clinical, or therapeutic. This location of social work's appropriate domain has been a prevalent controversy since Mary Richmond's day. In more recent times, some social workers who are committed to environmental or structural intervention have seen the movement to license social workers as a negative continuation of the move into professionalism, which in this context usually means individual psychotherapy and, often, private-pay practice. Clinical practitioners, however, report values, ethical principles, and practices consistent with social work's mission to improve human well-being and promote social justice, and promoting wellness over pathology (Bradley, Maschi, O'Brien, Morgan, & Ward, 2012).

The perceived polarity between clinical and nonclinical practice is a challenge in contemporary social work. The *Code of Ethics* (NASW, 2008) provides some guidance by requiring social workers to engage in work that supports socially equitable allocations of opportunity, which by definition is environmental and political. When clinical practice focuses entirely on individual issues and ignores or excludes the nonclinical work of advocacy and power analysis inherent in the pursuit of social justice, it comes into conflict with the *Code*. The proximity of some clinical practice settings (such as mental health and substance abuse) to the health care delivery world and its requirements for individualized, de-contextualized labels of pathology, sometimes discourages or diverts social workers from entering into social justice pursuits.

Integrating Approaches for Clinical and Indirect (Macro) Practice The debate between an individual, clinical focus and a nonclinical, or macro, environmental focus can be reconciled through integrating approaches. For example, Gitterman and Germain (2008b) argue that social work professionals must be prepared to work with all types of client systems, as situations require. Many methods and skills are common across all client systems. The historical loyalties to both the individual and to the macro advocacy pursuits to match client needs with environmental resources are a strength of the profession. The Educational Policy and Accreditation Standards (EPAS) (CSWE, 2012) for social work education underscore the commitment of social work to work with all client systems, with competencies and practice behaviors that require mastery of knowledge, skills, and values across all client systems.

There are many contemporary and generally complementary efforts to connect clinical practice with social justice issues. One approach expands the definition of clinical work to include "case management, advocacy, teamwork, mediation, and prevention roles, as well as therapeutic and counseling roles" (Swenson, 1998, p. 527). The clinician also uses self-reflection to consider her or his privilege. This in turn addresses the worker's accountability to clients, clearly an important component of a social justice approach. Newer areas of contemporary practice, such as environmental social work, also propel social workers to connect individual work with work at a systems level (Gray, Coates, & Hetherington, 2012).

A second way to integrate approaches is through the use of theory, perspectives, and methods that guide social workers through working with client systems. For example, a generalist, empowerment perspective (discussed below) also bridges the clinical and indirect practice tension. Within this perspective, social work practice is both clinical and indirect, and is at the intersection of private troubles and public issues. According to the empowerment perspective, social work employs an integrated view of humans in the context of their physical and social environments. Social workers using the empowerment perspective seek to promote a mutually beneficial interaction between individuals and society (Parsons, 2008). Narrative approaches (discussed in more detail in Chapters 4 and 6) provide an example of the way in which the issue of clinical practice and nonclinical pursuit of social justice occurs. A narrative approach to working with clients focuses on empowerment, collaboration, and viewing problems in social context. A narrative-oriented practitioner addresses the societal injustices individuals and families encounter in the client intervention (Kelley, 2008). The issues that individuals and families bring to the work are always put into the contextual arena of the social and power relations in the client's experience. Although narrative approaches began in family therapy work, narrative approaches have also been applied to work at the individual, group, and community levels (Kelley, 2008).

The use of the concept of social construction also blurs the distinction between clinical and nonclinical work. Social construction emphasizes the power of agency, a person's ability to affect her or his own circumstances that human beings exercise in creating their social locations. The focus of the work involves helping the client to create a new identity that is more empowering (Dybicz, 2012). In this view, people shape their environments, which in turn (or recursively) influence them. For example, you are a member of a community, and as such, you respond to other members of the community. However, you also take a part in creating your community, a process that goes beyond simply reacting to various individuals. By focusing on the client as a participant in the creation of the environment, this approach blurs the distinction between clinical work with the person and nonclinical work with the environment. Finally, a highly integrative model bridges individual, clinical work and environmental practice concerns through the use of deconstruction (Vodde & Gallant, 2002). In this conceptualization, the social worker moves from helping the client conquer the internal ramifications of the problem (such as clinical depression) to helping the client connect with others who are experiencing and resisting the same kind of oppression. Georgia's story will help demonstrate many of these points about deconstruction.

> Georgia is a 25-year-old woman from the southern United States. She has come to a large city in the Midwest to see a different world and another part of the country. Very early in her stay, she met and fell in love with Tom, a native to the Midwest. Georgia and Tom developed a serious relationship, and Georgia moved into his apartment when she discovered that she was pregnant. From that time on, things did not go well for Georgia. Tom began to resent her interest in the coming birth, and at times he was verbally abusive, insulting her southern background and degrading every personal aspect he could find wrong with her. He then started to be physically rough with her when they had any difference of opinion. Eventually he began to shove her into the wall, slap her, and kick at her belly. Georgia was disillusioned and frightened both for her own safety and that of her child. She could not understand how she had failed Tom, or what she had done to become so disgusting to him, or how she had become so hard for him to be around. When Georgia finally believed she could no longer manage this situation, she contacted a local women's shelter. She was devastated. By this time she believed herself to be entirely worthless and thoroughly unlovable due to Tom's labeling her as ugly and stupid. Georgia was at a very low point and was fearful for her future.

Georgia's social worker, over time and with the use of her professional skills, helped her to see that Tom's battering was not a function of her personality or unworthiness but rather of his own impulses. Tom, not Georgia, was responsible for Tom's abusive behavior. With support, Georgia began to regain her sense of worth and resilience and to feel stronger about her own capacities. Gradually she felt less in need of intense work on her esteem. She began to explore how our society supports violence. She met with other women at the shelter and joined in their resistance to societal violence through advocating for education in the schools and providing personal testimonies to groups of women.

This work empowered Georgia by enabling her to take control of her own emotions while addressing an environmental issue in a meaningful manner. The scenario presented here demonstrates the integrated approaches described earlier that address both clinical issues and social justice concerns. The approach spans the inner psychological turmoil that Georgia experienced as it began to become more political and facilitates Georgia's role in influencing her environment. In this way the work has gone from an individual clinical focus to a nonclinical, integrated political focus without sacrificing either, as each supports the other.

Social Control and Social Change   Social work is part of the society that it tries to change. Some of the regulatory bodies in which social workers practice, such as child protection, criminal justice, and mental health, carry an authoritative sanction for social control. In some circumstances this control appears to be at odds with the commitment of the social work profession to social change, a process in which social workers attempt to alter basic social structures. The two functions—control and change—are not inherently irreconcilable, but the ways in which they play out in their respective practice arenas tend to make them appear incompatible at times.

Change and Acceptance   Another tension involves how much the goal of social work is to implement change—either in individual or environmental—or to help a client accept their status as "good enough." As you would expect, the particular circumstances and settings associated with each situation strongly influence the social worker's approach to this issue. In fact, although the pull between acceptance and change occupies a place in the historical tensions of the profession, it may prove to be so contextually influenced that it will never be fully put to rest as long as individuals continue to evaluate social contexts, based on their own idiosyncrasies, in different ways.

Adjustment and Challenge   The dilemma of helping people to adjust to their circumstances or helping people challenge their circumstances is another historical tension in social work. This question must be answered in context according to factors relating to social justice. For example, many practitioners might support a female client who feels angry, distressed, and overburdened in a marital relationship and struggles with the sociological realities of contemporary families. These realities might include the expectations of partners, employers, and society at large that most mothers, even those working full time outside the home, should assume more responsibility than fathers for child care and home life. In this situation the social worker would offer support to bolster the client's existing coping mechanisms rather than facilitating a change within the client.

The social worker (who could possibly see herself in the same scenario) may offer suggestions for child care respite, recreation, or self-care that would help mitigate the client's sense of injustice in the arrangement, but not address an overall change in a significant or structural way, while other clients and practitioners would respond differently. They may not be willing to wait until parenting and homemaking become more equitable but instead demand or at least work for substantial change, both within the marriage and in the larger society. The possible stances that either client or practitioners take are not polarized but can be thought of as situated on a continuum, meaning that both the client and social worker may wish for large-scale social change, and both may believe that they need personally to make peace with the current reality of an individual situation.

The question becomes, are their respective positions compatible enough so they can agree on the goal of the work? If the social worker has a strong position that the client needs to change the situation, and the client really wants to learn to accept the situation, setting mutually agreeable goals will be challenging. There may be situations in which the tension of adjustment or challenge involves a significant concern for personal safety of an adult. The response to such a situation is not clear. You may suggest that to overcome this relationship stress, the client can work through its internal dynamics in some interpersonal work rather than through overt activism. Others may suggest that practitioners need to respond to any available opening for services, even if a woman cannot or will not separate from her partner. This is a controversial issue, in part because of the differences of opinion about what is tolerable, and the larger or political ramifications of inequitable relationships.

Experts and Shared Power   The relationship between the ideas of expertise and shared power has recently created a tension within the social work profession. The history and development of both the educational and professional systems in U.S. culture have revolved around the idea of expertise, or expert knowledge. Social workers are educated and socialized in professional programs to become respected members of a profession in which there are others with the same or similar expertise. In contrast, the idea of shared power is relatively recent in professional culture and challenges the claims of expert power. In shared power, the individual is the expert on her or his life, culture, dreams, experience, and goals. This creates a mandate in the work for social work practitioners to assume power only over limited activities in which they are trained while the client retains the power to direct the work. For example, if you are working in a college setting as an advocate for an African American student who experienced discrimination in housing options on campus, you may claim expertise on the process of advocacy, but your client would retain control regarding those issues that require your focus and the goals to be met through the advocacy. In this way, the power is shared.

Shifting your views of your expertise and interest in sharing power may not come easily; the U.S. population is socialized to value expertise, thus, we may interpret our own value and contributions through the lens of expertise. Given the competition that can develop between complex specialization on the one hand and client empowerment on the other, this tension is likely to persist. Some clients may want social workers to be the expert on their lives and relationships in the same way people want their dentist to be an expert on dental care.

Like much of U.S. culture, the social work profession has been caught up in the status and perceived legitimacy of being scientific and thus tends to take on the same metaphors of expertise. Social workers have spent decades trying to "prove" the effectiveness of their interventions, and have fought for professional prestige through this "expert" label. However, social workers have come to recognize that the expertise on the experience of any particular relationship, oppression, or phenomenological event belongs to the person who has lived it, and social workers can honor this wisdom by sharing power within the professional relationship. In the contemporary world, many strong client voices in the realm of human service interchanges, situations, and relationships have made it known that adopting the role of expert is not necessarily helpful, particularly when it obscures their own ownership or participation in the work. Social workers recognize the disillusionment and anger in people, such as people with disabilities, women, and people of color, who have experienced service systems that rely on expertise as humiliating, insulting, or patronizing. The development of many contemporary social work perspectives reflects this reality in their deliberate attempt to reduce the centrality of expertise and to substitute an enhanced commitment to partnership, or shared power.

Minimization of Distance   Such theoretical perspectives as the strengths perspective, nearly all feminist approaches, and empowerment approaches make a conscious effort to minimize the rigid boundaries of expertise between social worker and client. The client is seen as the expert on her or his life, whereas the social worker is seen as skilled in various arenas that will help the client get to a client-defined destination. This shift is played out in practice approaches that involve visiting clients' homes or meeting in community facilities like coffee shops, houses of worship, and community centers. These approaches emphasize client comfort in familiar surroundings, in contrast to the 50-minute clinical hour in the agency office, which tends to send a message of worker authority and increased distance between social worker and client.

This trend toward more fluid relationships represents a deliberate shift in thinking and a transformation of values. For some theoreticians, the shift signals a new way to respect the experiences and views of people who previously had never been listened to closely. For others it is a critical response to a power analysis that results in the elimination of abusive or oppressive authority relationships. In broader terms, this shift to a more egalitarian relationship affirms the right of the client to enter into a relationship of shared power in which the assets of both the client and the worker are recognized, valued, and used. Accordingly, in this conception, one person does not have power over the other; rather, a collective power is facilitated through sharing.

One of many Native American views of shared power has much to offer contemporary practitioners (Lowery & Mattaini, 2001). This approach conceptualizes shared power as shared responsibility allocated according to the particular strengths of each participant. It does not mean equality of responsibility, but recognizes that skill levels differ and emphasizes the worth of all contributions. Social workers are responsible for maintaining professional ethics, and clients are responsible for making changes in their lives. For example, in a situation of client substance abuse, social workers would ask questions such as, "How does your drinking affect your children?" rather than making a statement such as, "You need to quit drinking" (p. 116). The relationship created by this approach discourages hierarchies in the client–worker relationship and encourages a sense of joint investment in both the process and the results.

Global Citizenship and the Local Community   The final tension considered here, between local and global investments of effort, is now more relevant for social work practitioners than ever. Traditionally, U.S. social workers have carried out most of their activities in local or neighborhood contexts and have regarded global developments as remote. The social and cultural history, geography, and resource wealth of the United States have all contributed to national and professional insularity. The tendency to judge U.S. culture as preferable is shaped by and reflected in such devaluing language as "third world" or "underdeveloped" that is almost exclusively based on the U.S.-defined dimensions of economy in any given culture. Traditional Western markers are primarily limited to economic assets that are important for all people but are not the totality of life for any person.

Globalization is a "set of social processes that appear to transform our present social condition of weakening nationality to one of globality" (Steger, 2009, p. 9). It consists of a complex of economic, social, and technological processes that have resulted in the formation of a single world community in which we are all citizens. The essence of globalization is the flow of information, ideas, knowledge, technology, capital, labor, artifacts, and cultural norms and values across national borders. The impact of global occurrences, such as terrorism and wars, have a direct and compelling influence on the environment and practice of social work at home and require workers to develop new skills and devise new perspectives. Social problems such as homelessness, neglect and exploitation of children, poverty, violence, health and aging issues, and epidemic infections are worldwide issues that impact all nations (Sowers & Rowe 2009; Steger, 2009).

The process of globalization facilitates the recognition of the signs of U.S. economic dominance. For example, the availability of Coca-Cola in Bangkok or the presence of Pizza Huts in Manila are signals that the U.S. economy dominates many world markets. The outsourcing of many U.S. jobs to countries in which much cheaper labor can be located has resulted in high rates of unemployment domestically as well as exploitation abroad. Yet many U.S. and other Western-based companies are thriving. These dimensions of globalization are part of the reality of expanding international trade in the Global South (Payne & Askeland, 2008).

Factors Promoting Globalization   International financial institutions, such as the World Bank, the International Monetary Fund (IMF), and the World Trade Organization (WTO) have promoted and facilitated globalization, which has been accompanied with significant challenges and questions. These organizations were originally instituted to promote economic stability and growth and to help individual countries as they experience economic crises, but the strategies they have adopted to achieve these objectives have been problematic and counterproductive. Many question the policies of these international financial institutions, as they have disproportionately benefited wealthy people around the world, failed to address concerns about the environment from worldwide development, led to abuses in human rights, and provided significant setbacks to those promoting social justice. Globalization agreements and efforts have not incorporated regard for employment creation, expanded or improved health, or educational or social services for the poor, or promoted progressive land reform. Critics charge that decision-making within these international financial institutions lacks transparency and often reflects the interests of wealthier countries (Head, 2008; World Health Organization, 2012).

As an example, the WTO has been the primary promoter of free trade policies. Although free trade policies were ostensibly designed to encourage the development of market economies, the policies often require receiving nations to accept particular conditions thought to foster trade and economic growth, including privatization of resources and restrictions on imports and exports, as well as cutbacks in many social programs such as health, education, and housing and income assistance. In many cases these conditions have had the effect of abolishing welfare entitlement programs and social safety nets, debilitating local economies, altering or eliminating substantial aspects of indigenous ways of life, and eroding the autonomy of national governments. In their place, large multinational corporations have been free to build enormous industries whose purposes relate entirely to profit, with little regard to environmental or other concerns. Globalization policies have been associated with a substantial increase in worldwide poverty, misery, violence, HIV/AIDS cases, drug addiction, and economic stratification (to name a few), as well as the exploitation of the land and other important environmental resources (Payne & Askeland, 2008).

In such a context, social work as a profession is developing models of understanding that are global in context. For example, social workers in domestic settings are working with diverse populations, including immigrants and refugees, in this country, and gaining knowledge of non-Western ideas and perceptions. Rather than focus on strictly clinical work, social workers in international settings may also be called to assist with economic and social change to change situations of economic inequality and poverty (Payne & Askeland, 2008). While globalization provides both challenges and opportunities for social work, social workers can advocate and work toward a process of globalization that has a strong social dimension and is accountable for improving such global issues as poverty, environmental degradation, unemployment, and human rights (Hong & Song, 2010).

At the individual practitioner level, a commitment to reciprocity, interdependence, human rights, and social justice in practice is one way to incorporate globalization into practice. Social workers must be familiar with global events through media sources, as well as the ways in which events that occur within the U.S. impact individuals within other countries and other nations. The implications of globalization for social work practice are likely to increase in number and in impact, and social workers must be involved in helping to strengthen local structures within communities, such as nonprofit organizations to ensure civil rights (Hong & Song, 2010). Ironically, increased localization has been one of the important responses to globalization (Payne & Askeland, 2008) as people seek local alternatives that are appropriate for their particular circumstances. Social work, therefore, must focus on both global and local perspectives and make meaningful connections between global inequities and domestic social justice issues that will discourage any artificial local/global polarity (Hong & Song, 2010), as well as initiate educational and political initiatives to ensure respect for human rights, democratic decision-making, environmental protection, and meaningful benefits to local populations.

Perspectives on the Conceptualizations of the Social Work Profession   Although each of these conceptualizations provides a useful way of thinking about social work practice, each also has limitations. To use them for the best interests of the client system, the models must be used flexibly. For example, knowledge, values, and skills can overlap, and over-focusing on one dimension, such as factual knowledge, may lead practitioners to apply knowledge without adequate consideration of the appropriate skill level and social work values. The integration of many kinds of knowledge, values, and skills as applied through competencies is the goal of social work education. Likewise, dividing the work into constituent groups of beneficiaries may suggest that the background, values, and skills necessary for each kind of practice vary more significantly than the reality of practice.

Generalist social work practice requires that the social worker be proficient at all levels of client groupings, as the competencies can be applied to all system levels. The similarities far surpass their differences; work in any of them is grounded in the practitioner's consistent set of values, theoretical perspectives, and goals. Similarly, a social worker does not build a relationship, then cease utilizing engagement skills (i.e., become "all business") because she or he has moved on to the assessment phase. Like all relationships, the professional connection between social worker and client must be attended to, nurtured, and encouraged to grow throughout the time of their work together, or it will wither. Likewise, assessment occurs throughout the course of the work, beginning with the first interaction and continuing through the last. The phases of helping are highly connected, with special emphasis paid to one phase at any given time during the work. The tensions discussed in social work, while useful, may fall short of describing the full, day-to-day realities of social work practice within specific practice settings and with specific populations. Therefore, while the structures are helpful, flexibility in the application is important.

Importance of Self-Knowledge   To practice competently, social workers must have a clear understanding of themselves, including their own biases. One type of bias that can impact practice is bias about socioeconomic class. For example, you may notice that you respond differently to people who are living in poverty than to people who are socioeconomically middle-class. To grow in cultural competence, you can explore the reasons for this difference; perhaps you learned as a child that "poor people" are somehow less worthy than affluent people, or that they are not hard-working. Your family and/or the wider culture may have supported these views. To uncover your biases, you can ponder the following questions: How does this view clash with the view that people living in poverty are resilient in the face of debili-tating exploitation? How is your view implemented in your practice? Are you less empathic, energetic, or invested when you work with people living in poverty? Are you less likely to advocate or seek out resources? Do you have an opinion about a client that is a "truth" in your mind, or do you treat it as one interpretation out of many?

Practice decisions are shaped by our experiences, motivations, values, attitudes, and other factors. Sometimes practitioners are unaware of influences on practice decision. Therefore, social workers need to continually explore the ways their values are prioritized, the patterns they develop in making decisions and exhibiting prac-tice behaviors. This exploration may include recognizing idiosyncratic approaches, quick responses, and patterns of reaction. Are you patient and likely to stand by your client, even when things do not go well? Are you quick to assume someone is judging you or wants to bring you harm? Are you likely to blow off steam or sulk when you are rebuffed? Do you assume that you are competent to deal with any crisis? Are you inclined to address an interpersonal problem privately and quietly? Are you more likely to "sound off" in a meeting? All of us have idiosyncratic ways of dealing with relationships, stresses, and social interactions that are personality variations. Social workers must recognize their behavioral patterns, understand how they impact their work, and identify those responses that, if changed, would help them to develop a higher degree of competency and skill.

Culture has a strong influence on social work practitioner biases. For example, culture may impact practitioner beliefs and attitudes about a situation in which an adolescent's choice of vocation may clash with the family's ideas about suitable profes-sions. In another example, cultural bias may influence a practitioner belief that, in social relationships, participants are equal and all should speak out honestly about tension in the relationship. These two examples are part of mainstream U.S. culture, and are cultural variants that are not necessarily shared by all people and therefore do not represent a universal truth. The emphasis on individual drive and ambition as well as equality and forthrightness are Western beliefs, and clash with other cultures. In another example, an Asian client, who is 20 years older than a practitioner, may relate in very formal terms with the practitioner and appear reluctant to share details about her family life, even though she has been referred for parent support. Rather than labeling her "resistant" or "closed," the practitioner can recognize potential bias in the

QUICK GUIDE 1

Social workers work with many different types of client populations. Examples include:

LGBTQ youth, adults, and older adults
Range of religions (such as Jehovah's Witnesses, Jews, Catholics, Mormans, and Muslims)
People with Disabilities (Physical, emotional, behavioral, developmental)
Minority Ethnicities (such as Native Americans, Hispanics, African Americans, Asian Americans, and Pacific Islanders)
Immigrants (from such countries as Mexico, India, China, Vietnam, and African nations)
Refugees (from such areas as Africa, the Middle East, and Asia)
Older Adults, Active Military and Veterans
Rural Populations
People in Poverty
Children, Adolescents, and Youth

Questions to explore your self-knowledge about diversity

- What is my race, ethnicity, religious affiliation, and socioeconomic status?
- How does my background related to diversity impact my knowledge about this client's background?
- How is "family" defined by my culture? Other cultures?
- Do I assume my culture is the worldwide norm?
- To what degree do I understand the ideas of other cultures other than my own?
- To what degree is my language respectful about other cultures?
- Do I think that I will be less empathic or energetic when working with persons from one or more particular populations?
- Do I acknowledge that cultural differences exist?
- Do I recognize the thought patterns of other groups as equally valid as mine?
- What is diverse within my culture? What diversity exists between my culture and others?

Questions to ponder about building cultural competency

- How might my biases impact my practice?
- Do I think that I will be less likely to advocate for or seek out resources for particular types of clients?
- Do I think that I will advocate for effective services that differ by culture?
- Do I stereotype client from specific groups?
- Am I aware that some clients may not relate closely to their cultural group?
- Am I aware of different types of natural support systems that differ from my own (e.g., folk healers and religious leaders)? Do I understand how to utilize natural support systems in practice?
- Do I know how to respectfully obtain personal and family background information from clients to determine their ethnic/community sense of identity, rather than making assumptions?
- Will I look for working environments that positively impact my cultural competence?

Sources: Lum, 2011; Mason, 1995; NASW, 2001

perspective applied about such relationships and style and evaluate the "fit" of the practitioner and client perspectives.

The emphasis on our own biases and assumptions is closely related to values and how these values influence our ideas of ethical practices. These constructs are critical to social work practice. The next chapter will deal more explicitly with the ideas and contexts relating to values and ethical frameworks and the emphases placed on them in social work practice.

THEORETICAL PERSPECTIVES FOR SOCIAL WORK PRACTICE

Social work has a history of adopting, adapting, formulating, and integrating various theoretical perspectives. A theory is an explanation of some event or phenomenon. Theories usually have clear principles and propositions that provide a framework for predicting events and a supporting body of empirically-based evidence. Theory assists social workers by providing the flexibility to see client systems and social problems from many points of view, anticipating outcomes of different interventions, and in applying the results to future situations (McNutt & Floersch, 2008). Theories can bring order and coherence to practice situations so that activities are based on a logical assessment of the fit between a client system situation and the assertions of the theory. Theories about the social world evolve and adjust in order to fit new ideas or to accommodate further evidence that adds or contradicts theories. Consequently, some theories are discarded when they are no longer acceptable to the professional community (for example, theories asserting that some races are superior to others) or when they are unable to be grounded in reliable research. Other theories evolve to incorporate additional refinements or methods, such as the contemporary interpretations of Freud's 19th-century psychoanalytic theory.

A perspective is a view or lens through which to observe and interpret the world. A perspective is generally less structured than theory but is similar in that it is often based on values and beliefs about the nature of the world. For example, a belief that people are generally good is a perspective that will guide practice and interpretations of client system situations. In that way the ideas behind theory and perspective are similar. In social work practice, the terms theory and perspective are often viewed as synonymous and may also be called theoretical perspectives. These three terms are used interchangeably throughout this book.

The theoretical positions of the social work profession have received important contributions from biology, ecology, sociology, anthropology, psychology, and many of the humanities and human service professions. More recently, social work scholars have expanded the development of the profession's own understandings of practice to create theories specific to the provision of social work services. Some of these theories focus on psychosocial systems, problem-solving, ecological/ecosystems, and empowerment, as they represent uniquely social work emphases on the interface between a person and her or his environment.

Theories are important to the social work profession, as they are driven by values. The ways social workers select, implement, and evaluate a theory or theoretical perspective will be highly influenced by their value orientations. For example, creating an alliance with a client that results in complete dependence on the social worker could be the result of an adherence to a theory that supports "reparenting," because the theory allows for directing the client's goals to carrying out the will of the social worker. On the other hand, using a theory in which autonomy and self-determination are prioritized will result in a different outcome. The role of theory in this scenario would be critical in directing the work of the practitioner.

This book presents five practice perspectives that guide the focus and are central to the content. They are: (1) the ecosystems perspective; (2) the social justice perspective; (3) the human rights perspective; (4) the strengths perspective; and (5) postmodern perspective approaches, including critical social construction, narrative theory, and solution-focused interventions. Other perspectives are included in subsequent chapters as they relate to specific levels of practice. Some of these perspectives have a longstanding history within social work education and practice (such as social justice), and others are more contemporary (such as postmodern approaches).

Ecosystems Perspective

Using concepts from systems theory and ecology, the ecosystems perspective is an often-used framework for generalist practice. This perspective examines the exchanges between individuals, families, groups, and communities and their environment. Systems theory utilizes constructs to organize complex activity in the social environment, while ecology seeks to explain how people adapt to and influence their environment. Taken together, these two concepts describe the functioning and adaptation of human systems in a dynamic interchange with each other. Interactions or transactions between the two help explain how people influence their environment, and vice versa. The ecosystems perspective is very useful for social work practice because it encourages increasing the degree to which people and their environments fit one another. Increasing the degree of fit between the two can lead the social worker toward direct services with individuals, families, and groups to mobilize and draw on personal and environmental resources for effective coping. Increasing the degree of fit between human needs and environment can also lead the social worker to influence the client system's social and physical environment to adequately address human needs. Influencing the environment can include working to influence organizations to develop more responsive policies and programs, as well as working to influence legislation, regulation, and implementation of laws (Gitterman & Germain, 2008a).

Social Justice Perspective   Social justice refers to the manner in which society distributes resources among its members, including material goods and social benefits, rights, and protections. Although there are various theories of social justice regarding the criteria for the distribution of wealth, the profession of social work generally accepts an egalitarian focus that is concerned with the fair distribution of both material and nonmaterial resources (Rawls, 1971) and equal access for all people (Reichert, 2011). From this perspective, developing or distributing social and natural resources based on political or social power rather than social justice or human need is unacceptable (Saleebey, 2012). The social justice perspective has important implications for social workers and their practice as it requires a response to the injustice that is status quo in our society.

As required by the National Association of Social Workers *Code of Ethics* (NASW, 2008), social workers invest a significant amount of time and effort in championing individual, group and community rights, working toward more effective institutional responses, and influencing major social policy shifts. Social work is, therefore, a political profession. Social workers identify individual, family, group, and community needs, and seek to address the areas of injustice that have a negative impact on people. For example, if a social worker is working at the community level with a group of older adults who have been wrongfully denied access to services or government support programs, it is the social worker's responsibility to confront that injustice either directly or indirectly by using community resources to address the issue. The issue can be addressed individually, or, if the problem rests in policy or procedures, at the group-level.

Human Rights Perspective

The United Nations describes human rights as rights that are inherent in our nature and without which we cannot live as full human beings. Human rights and fundamental freedoms allow us to completely develop and use our human qualities, our intelligence, our talents and our conscience, and to satisfy our spiritual and other needs. They are basic for humankind's increasing demand for a life in which the inherent dignity and worth of each human being will receive respect and protection (Lundy, 2011: United Nations, 1997).

The principle of human rights offers a powerful and comprehensive framework for social work practice; it not only recognizes needs but strives to satisfy those needs (Lundy, 2011). The human rights perspective provides a moral grounding for social work practice and reflects an ongoing commitment to the belief that all people should have basic rights and access to the broad benefits of their societies. Many social workers evaluate their practice based on its contribution to an environment in which universal human rights are honored.

Contemporary perspectives on human rights emphasize the social understandings of our commonality rather than the rigid assertions about specific human rights. Accordingly, human rights can be viewed as socially constructed, reflecting differing contexts, ideas, and cultures (Gregg, 2012). Ife (2000) likewise suggests that the human rights discourse should de-emphasize the focus on legalistic,

EXHIBIT 1.4

A summary of the Universal Declaration of Human Rights

1. Everyone is free and we should all be treated in the same way.
2. Everyone is equal despite differences in skin colour, sex, religion, language for example.
3. Everyone has the right to life and to live in freedom and safety.
4. No one has the right to treat you as a slave nor should you make anyone your slave.
5. No one has the right to hurt you or to torture you.
6. Everyone has the right to be treated equally by the law.
7. The law is the same for everyone, it should be applied in the same way to all.
8. Everyone has the right to ask for legal help when their rights are not respected.
9. No one has the right to imprison you unjustly or expel you from your own country.
10. Everyone has the right to a fair and public trial.
11. Everyone should be considered innocent until guilt is proved.
12. Everyone has the right to ask for help if someone tries to harm you, but no-one can enter your home, open your letters or bother you or your family without a good reason.
13. Everyone has the right to travel as they wish.
14. Everyone has the right to go to another country and ask for protection if they are being persecuted or are in danger of being persecuted.
15. Everyone has the right to belong to a country. No one has the right to prevent you from belonging to another country if you wish to.
16. Everyone has the right to marry and have a family.
17. Everyone has the right to own property and possessions.
18. Everyone has the right to practise and observe all aspects of their own religion and change their religion if they want to.
19. Everyone has the right to say what they think and to give and receive information.
20. Everyone has the right to take part in meetings and to join associations in a peaceful way.
21. Everyone has the right to help choose and take part in the government of their country.
22. Everyone has the right to social security and to opportunities to develop their skills.
23. Everyone has the right to work for a fair wage in a safe environment and to join a trade union.
24. Everyone has the right to rest and leisure.
25. Everyone has the right to an adequate standard of living and medical help if they are ill.
26. Everyone has the right to go to school.
27. Everyone has the right to share in their community's cultural life.
28. Everyone must respect the 'social order' that is necessary for all these rights to be available.
29. Everyone must respect the rights of others, the community and public property.
30. No one has the right to take away any of the rights in this declaration.

Source: Human Rights Education Association (n.d.).

Western views of entitlements in favor of a more reflective dialogue among the world's peoples. This dialogue would center on exploring the meaning of human existence, and how humans want to recognize our interdependence, protect our futures, and safeguard our survival in the current and future global era. This focus is consistent with social work's emphasis on community, quality of relationship, and concern for the future of the world's inhabitants. It also recognizes the dynamic nature of living one's principles, such as those inherent in human rights, as they continuously evolve to remain meaningful.

Human rights have the potential to occupy a central place in social work practice and can provide a focal point in the evolution of people's understanding of their place on the planet. Viewed as a set of dynamic principles that can provide a substantive contribution to guidelines on international relations as well as community-based practice, human rights constitutes a vital focus for work with people. Social work as a profession can take up the lead started by many of our global colleagues in making human rights a central discourse and in exploring its relevance more thoroughly to our everyday work (Healy & Link, 2011).

The Strengths Perspective

An increasingly widespread approach in social work practice, the strengths perspective is explicit in its emphasis on affirming and working with the strengths found both in people seeking help and in their environments. Like human rights practice, this perspective emphasizes basic dignity and the resilience of people in overcoming challenging obstacles. Contemporary advocates have established some significant breaks with social workers' historical and cultural tendency to focus their work and energies on problems and pathology (Kim, 2008a).

Principles of the Strengths Perspective   Saleebey (2012) identified six key principles for the strengths perspective (pp. 17–21):

1. *Every individual, group, family, and community has strengths.* Social workers must view their clients as competent and possessing skills and strengths that may not be initially visible. In addition, clients may have family and community resources that need to be explored and utilized.
2. *Trauma and abuse, illness and struggle may be injurious, but they may also be sources of challenge and opportunity.* Clients can not only overcome very difficult situations, but also learn new skills and develop positive protective factors. Individuals exposed to a variety of trauma are not always helpless victims or damaged beyond repair.
3. *Assume that you do not know the upper limits of the capacity to grow and change, and take individual, group, and community aspirations seriously.* Too often professional "experts" hinder their clients' potential for growth by viewing clients' identified goals as unrealistic. Instead, social workers need to set high expectations for

their clients so that the clients believe they can recover and that their hopes are tangible.

4. *We best serve clients by collaborating with them.* Playing the role of expert or professional with all the answers does not allow social workers to appreciate their clients' strengths and resources. The strengths perspective emphasizes collaboration between the social worker and the client.

5. *Every environment is full of resources.* Every community, regardless of how impoverished or disadvantaged, has something to offer in terms of knowledge, support, mentorship, and tangible resources. These resources go beyond the general social service agencies in the communities and can serve as a great resource for clients.

6. *Caring, caretaking, and context.* Strength perspective recognizes the importance of community and inclusion of all its members in society and working for social justice. This is built on the basic premise that caring for each other is a basic form of civic participation.

Although these key principles may evolve and be refined, the focus of the strengths perspective is on clients' personal assets along with their environmental resources rather than on their pathology and limitations is a core element. Using the strengths perspective, social work interventions are centered on helping the client system to achieve their goals, affirming and developing values and commitments, and making and finding membership in or as a community. The strengths perspective does not preclude the need to validate the suffering and pain of the client system (i.e., the physical, emotional, or existential) nor the seriousness of the situation or distress. Rather, the strengths perspective seeks to acknowledge clients' expertise regarding their own lives and to focus on their resilience and capacities to survive and to confront such seemingly overwhelming obstacles.

Postmodern Perspective and the Social Construction Approach

Any contemporary perspective that questions the way in which knowledge is attained and valued, and distinguishes belief from truth, comes under the broad heading of postmodernism. Postmodern approaches invite examination of the cultural assumptions that underlie many of the arrangements of power and politics, and it tends to be critical of assumptions or theories that claim absolute or authoritative truth. To illustrate, the contention that "all people in the United States can get ahead if they work hard enough; therefore, all poor people are lazy" is an example of an assumption or belief that is asserted as truth. Postmodernism invites people to challenge the truth of such claims (Logan, Rasheed, & Rasheed, 2008).

Social Construction   A useful postmodern perspective for understanding the social realities of the lives of the people served by social workers, social construction suggests that people construct reality based on experiences in the social world, which occur within the context of culture, society, history, and language. Practicing from this perspective means that rather than being a product of an objective external world, or the result of the individual mind, social realities are actually beliefs formed through social interchanges (usually starting with the family) that regulate learning to make sense of things and, often, to form judgments. For example, when children experience people as good-hearted and the world as a kind place throughout their youth, this social experience informs their adult belief that the world is a relatively decent place. On the other hand, when their social experiences tell them that the world is full of evil, hurtful people, their view of reality is often very different. This observation of how people decide the nature of their world is meaningful for social workers when they work with people who have lived in a different reality than one then they experienced (Lee & Greene, 2009).

Noticing your social location in this way is consistent with postmodernism and social construction. Recognizing that a person's reality is based on her or his perceptions and experiences in the social world and that there can be more than one reality of an experience because each person perceives the world differently and has different experiences is one way to implement postmodernism and social construction. Social workers, using these approaches, approach clients from their realities, rather than the reality of the practitioner. For example, if a client appears to have little ambition to financially support themselves or to participate in civic life, an understanding of the reality of the client's culture and environment may help you to understand the realities of institutionalized racism that may leave clients bitter and non-participatory, with little vision of what their life might be. Recognition of this postmodern principle of multiple realities impacts the purpose of practice and may expand one's practice to advocate for or take up the causes of people who have little power. In practice, postmodernism and social construction approaches require acknowledgment of the practitioner's boundaries and allow practitioners to enter the world of the client system.

Deconstruction   Constructionist thinking opens the way to explore particular beliefs, such as that the world is a kind place, through a process of taking those beliefs apart to examine the way they developed, or deconstruction. Through the process of deconstruction, you can explore the factors that helped these beliefs to develop and determine which ideas that had power and dominated. The following is an example of both social construction and deconstruction.

For many years in the 20th century it was a "reality" that adults with severe disabilities were neither capable of negotiating nor entitled to engage in adult relationships, particularly sexual ones. Institutions and community-based facilities worked diligently to prevent relationships and sometimes even kept couples with disabilities from being together alone in the same room. That particular social

EXHIBIT 1.5	Some of the many assumptions by those practicing eugenics were:
Eugenics Assumptions	• People with severe disabilities are like children who should not enter into adult relationships.
	• People with severe disabilities are not adequate as human beings and should not have children.
	• People with severe disabilities are not entitled to define their needs or desires.
	• People with severe disabilities do not have the capacity for love.

understanding and sense of reality contributed to the infamous practice of eugenics, which sought to eliminate the reproduction of people who were seen as genetically deficient. As advocates for the rights of persons with disabilities began to deconstruct and debunk the assumptions of the social construction that supported this callous practice, eugenics became less acceptable.

Although these assumptions are by no means extinct now, they are situated in a specific economic, historical, and cultural setting that is no longer dominant. The concept of contextual setting, or social location, refers to the time, place, and prevailing ideas or discourse that influence standards, particularly about what is considered "right." All ideas emerge from within a context that includes attitudes or beliefs, and ideas must be recognized as contextual rather than as absolute, moral truths for all situations and contexts (Kelley, 2009). Consider, for example, a new social worker who is judged and silenced if she or he counters the so-called truth regarding persons with disabilities and, based on human rights or social justice concerns, objects to such social restrictions. You might consider those beliefs or customs that are so strongly held in your own community or agency that deviance from them has significant professional or personal consequences and is interpreted morally.

Questions that expose the attitudes and beliefs that underlie ideas often have the effect of creating opportunities and visions for people. Using social construction, social workers are likely to question arrangements that bring harm to people and to be suspicious or wary of presumed expertise that denies people their own experience. For example, a social worker may advocate for an older adult whose acute medical distress is being dismissed by a physician as insignificant. The social worker, with support from a supervisor or others, may even need to challenge the adequacy of the medical care offered to the client. In this way, the social worker refutes the idea that any particular set of assumptions or positions cannot be challenged, and operates on the idea that all beliefs, ideas, and arrangements can be respectfully questioned.

Narrative Theory Consistent with postmodernism are narrative and solution-focused interventions (Kelley, 2008). Narrative theory guides practitioners to utilize "stories" to understand the lived experience of clients. A therapeutic theory founded on the notion that people's sense of their identity consists of interacting narratives, a social worker using narrative theory would use a conversational approach to surface the cultural discourses about identity and power that have shaped the client. This approach involves helping clients make sense of the meanings they give to events in their lives. A narrative is the thread that binds disparate events together. Stories involve events linked together in a time sequence and have a plot. Narrative approaches view problems as separate from people, and people possess the qualities and skills needed to change their relationship with the problem(s) they are facing in life (Madigan, 2010).

Solution-Focused In contrast, a solution-focused approach invites clients to explore and determine the concrete change desired in their lives and the resources and strengths clients possess to make the change occur. Using a solution-focused approach, clients identify a specific goal that they believe will make their lives better in some fashion, and to attach specific strengths to a plan to reach that goal. Often clients are asked to think about an exception, times in which the challenge did not exist, and to identify the condition under which the challenge did not exist, as indicators of potential ways to address the problem. For example, if a child is struggling to focus in school, a discussion of the exceptions to this situation, or conditions under which the child is ever able to focus on schoolwork, may shed light on the factors on which to focus. Is the child able to focus early in the school day? When seated with specific classmates or away from specific classmates?

Critical Social Construction The *critical* in critical social construction is derived from a contemporary analysis of power, which recognizes that any social group can shape its beliefs to its own benefit at the expense of other groups. The critical dimension then allows the social worker to question power arrangements, especially if many people's voices have been silenced or "subjugated" (Hartman, 1994). Those who were able to arrange such systems as slavery had the power to privilege themselves, while oppressing others. In this way the privileged group not only benefits from but also perpetuates the privilege by using their power to assert their beliefs, which come to be regarded as truth. Although many groups today who benefit from privilege would not consciously choose such advantage if presented with a choice, such groups as whites, heterosexuals, Christians and those who are able bodied, benefit from privilege (Diller, 2011).

Critical social construction can be utilized in social work practice through critical viewing of power relations and dominance of specific beliefs that result, which especially impacts non-privileged groups with whom social workers often work. This critical viewing has significant repercussions for social work practice, which will be addressed throughout the book.

QUICK GUIDE 2   SUMMARY OF APPROACHES

Ecosystems perspective – Examines the exchanges between individuals, families, groups, and communities and their environment.

Social justice perspective – Focuses on the manner in which society distributes resources among its members, including material goods and social benefits, rights, and protections.

Human rights perspective – Reflects an ongoing commitment to the belief that all people should have basic rights and access to the broad benefits of their societies.

Strengths perspective – Affirms and works with the strengths found both in people seeking help and in their environments.

Postmodern approaches – Examines the cultural assumptions that underlie many of the arrangements of power and politics, and is critical of assumptions or theories that claim absolute or authoritative truth.

> *Critical social construction* – As people construct reality based on their experiences in the social world, examines power relations and dominance of specific beliefs that result, which especially impacts non-privileged groups with whom social workers often work.
>
> *Narrative theory* – Through telling "stories," helps clients make sense of the meanings they give to events in their lives.
>
> *Solution-focused interventions* – Explore and determine the concrete change desired in their lives and the resources and strengths clients possess to make the change occur.

Complementary Aspects of the Theoretical Perspectives

Each of the five perspectives—ecosystems, social justice, human rights, the strengths perspective, and the postmodern perspectives—offers the social worker unique and complex challenges. In many respects they complement each other by placing a slightly different emphasis on the work that fits well with the other views. The strengths perspective, for example, is used in the service of bringing social justice to people who have experienced inequities. Human rights, in turn, may be seen as the moral grounding for undertaking justice-oriented work. The constructionist ideas regarding the influence of historical and cultural standpoints allow the social worker to question previously unexamined ideas and approaches that marginalize people and characterize them as pathological or undeserving. Taken together these perspectives provide a comprehensive and compelling approach to social work.

STRAIGHT TALK ABOUT THE TRANSLATION OF PERSPECTIVES INTO PRACTICE

The relevance of theoretical perspectives to actual practice situations has always raised questions from students and social work educators. Some of the frameworks and perspectives may seem abstract or far removed from client systems, such as a destitute older adult, or a politically active group of adults with psychiatric disabilities. On the other hand, when perspectives reflect and honor the experience of people and fit the situation, they can be a critical guide to social work practice. For example, using the strengths perspective may mean focusing on the support networks and clean-up efforts of a struggling, poverty-stricken neighborhood to focus on the capabilities and assets, rather than focusing on the dilapidated housing, trash, and abandoned cars that line the streets. While using the frameworks and perspectives and integrating theory and practice can be very challenging, the frameworks and perspectives offer guides and a foundation for social work practice.

CONCLUSION

This chapter has presented several different perspectives with which to view social work practice. Social work practice is defined in the following ways: (1) by type of practice or range of practice settings; (2) as a set of activities; (3) as a set of roles; (4) as a set of competencies and practice behaviors; (5) by types of client grouping; (6) by practice frameworks; (7) as a licensed profession; (8) by purpose; and (9) by the tensions experienced in the profession. Self-knowledge is also a critical component of effective social work practice.

Exploring social work practice through this array of methods allows for many interpretations of situations and various foci of attention, some of which will be discussed in subsequent chapters. Coverage of the tensions experienced by the profession allows you to consider the profession in its ongoing concerns and have a sense of reference and context. Also discussed were five explicit theoretical perspectives, and you are encouraged to use them flexibly to offer an empowering and liberating approach to people who have lived their lives in the margin of society. Although the challenge of translating these approaches into ethical, sustainable, effective social work competencies will always be present in your practice, it will also sustain your growth. Building on this frame, Chapter 2 will delve into the role these values and ethics play in social work practice.

MAIN POINTS

- Social work practice can be conceptualized in at least nine different ways, to include by type of practice or range of practice settings, by the activities of the profession, as a set of roles, and as a licensed profession.

- Tensions that persist in contemporary practice include those between social control and social change, acceptance and change, clinical and nonclinical perspectives, experts and shared power, and global citizenship and the local community.

- Effective social work practice requires questioning your assumptions, understanding the impact they have on your work, and framing your methods.

- Theoretical perspectives play a significant role in guiding practice activities.

  The five perspectives, or frames, presented in this book are ecosystem perspective, the social justice perspective, the human rights perspective, the strengths perspective, and postmodern approaches, such as narrative and solution-focused, and critical social constructionism.

- The ecosystems perspective provides a frame for generalist practice, and examines the fit between individuals and their environment.

- The social justice perspective addresses the manner in which resources are allocated.

- Guided by the United Nations Universal Declaration of Human Rights, the human rights perspective is a frame that encompasses the conviction that all people have civil, political, social, economic, and cultural rights simply because they are human beings.

- The strengths perspective is a practice approach that highlights and works with individual, family, group, and community resilience and assets, honoring the client system's goals and dreams, and asserts that all communities possess resources.

- Critical social constructionism is a contemporary, postmodern frame based on the assertion that social truths are agreed-upon beliefs shaped in a common history and culture and perpetuated through assumptions, language, and stories. It also notes the power arrangements responsible for the prevalence of such truths.

- The translation of perspectives into practice is challenging and necessary for effective social work to maintain its mission, integrity, and consistency.

EXERCISES

a. Case Exercises

1. Go to www.routledgesw.com/cases and explore the Sanchez family interactive case study by reviewing the introduction and the tasks of each of the four phases. Click on the "Start this Case" button (under the Engage Tab), and complete Tasks #1, 2, and 3.

2. After completing Exercise #1, imagine that you are working with the Sanchez family as a social worker employed by the U.S. Bureau of Citizenship and Immigration Service. You are assigned to Celia and Hector Sanchez as they

receive their permanent resident cards, or "green cards," which authorize them to live and work legally in the U.S. Your role is to help them understand their new rights and responsibilities. What knowledge, values, and skills do you think would be particularly important for your work with the couple? Identify at least one *specific* area of knowledge, one value, and one skill. What challenges might result from emphasizing one (i.e., knowledge, value, or skill) over the other two?

3. After completing Exercise #1 and #2, consider the social justice issues involved in the family's permanent resident status and the ways in which the issues are related to human rights. Respond to the following:

   a. From your knowledge of human rights and the Sanchez family, identify two human rights that apply particularly to Hector and Celia as immigrants.

   b. How might you expect that having a green card would impact the family?

4. Consider the consequences of different social work practice discourses, and respond to the following:

   a. Describe the ways in which a social work discourse of strengths differs from a discourse of pathology when applied to the Sanchez family.

   b. Describe the way in which the strengths-based social worker would approach the family. How would you describe the professional relationship? How would you describe the focus of the work? Be as specific as possible.

5. Go to www.routledgesw.com/cases. Click on the Sanchez family, then "Start this Case," then review the case file for Emilia Sanchez (including Client History, Client Concerns, and Goals for the Client). Click on "Mapping this Case" (under "Case Study Tools"). Explore Emilia's relationship with her relatives by reviewing the family genogram. Also review Emilia's ecomap, and interaction matrix (under "Case Study Tools"). Answer Emilia's critical thinking questions.

6. After completing Exercise #5, consider the following case, which occurs prior to the time depicted on the website:

   You are employed as a hospital social worker, and have been assigned to Emilia Sanchez. Emilia is 24 years old and has just given birth to Joey. She used crack cocaine during her pregnancy, and Joey tested positive for the drug. A child protection social worker assigned to the case believes that Joey will likely be taken into the custody of the state, and probably placed with Celia and Hector, who have already indicated their willingness to take him as a foster child. As Emilia's social worker, your specific responsibility is to work with her on a plan to address her substance abuse and to adjust to the placement of Joey in foster care with her parents. You have been told by the nursing staff that she has just had an angry outburst and is now quietly sullen. You are currently heading to her hospital room.

   a. What do you notice about your own responses to the information you have received about Emilia? How will they influence your attitude about her? How will your attitude influence your work with her?

    b. What are you thinking as you walk to her room to meet with her? What emotions, if any, are you bringing with you?

    c. From the information you have at this point, what areas of strengths can see in Emilia?

    d. Balancing the safety of the child with the rights of the parent, would you suggest that Emilia have interaction with Joey while in foster care? If so, under what circumstances?

    e. You are anticipating that Emilia will be sullen when you first meet with her. What will you say to her? If she later expresses anger at the situation, what might you say then? What might be the next step in this situation?

7. Go to www.routledgesw.com/cases. Click on the Sanchez family. Go to the "Assess" menu, then click on "My Values." Take the Values and Ethics assessment.

b. Other exercises:

8. Write a two paragraph reflection journal entry to answer the question, "Why do I want to be in the helping profession?"

9. Reflect on three concerns that you have about being an effective helper. How can those also be strengths? (For example, Concern: "I am afraid that I will become too attached to my clients." Strength: "I am capable of developing strong relationships.")

10. Imagine a scenario in which a friend or family member is expressing doubts that social work is really a profession (She says "a person without a college degree can be a social worker"). Write a one-two page paper summarizing the reasons that social work can be considered a profession, using the main points from this chapter.

CHAPTER 2

Applying Values and Ethics to Practice

© marekuliasz

Deeply earnest and thoughtful people stand on shaky footing with the public.

Johann Wolfgang von Goethe

Key Questions for Chapter 2

(1) How can I identify as a professional social worker and conduct myself accordingly (EPAS 2.1.1)?

(2) How can I apply social work ethical principles to guide my professional practice (EPAS 2.1.2)?

(3) How can I apply critical thinking to inform and communicate professional judgments (2.1.3)?

(4) How can I advance human rights and social and economic justice through my professional practice (EPAS 2.1.5)?

CONSIDER THE DEGREE TO WHICH YOU AGREE WITH the following statements related to social work practice:

- People who choose to smoke cigarettes or cigars, chew tobacco, or drink excessive amounts of alcohol are responsible for resolving and paying for their own health problems, since the risks of these behaviors are widely known.

- Under some conditions, physical discipline of children is acceptable.

- Children who are sexually abused should never be returned to the residence of the person who abused them.

- Social workers who agree to an agency's terms of employment do not have the right to criticize the agency's practices.

- Most homeless people want to live out on the streets.

- Under particular circumstances, suicide is an acceptable strategy.

- All world citizens, regardless of what they have done, deserve to be free from hunger.

- HIV-positive people should be tattooed on the heel so potential sexual partners can take precautions.

- Perpetrators and victims of crime are equally deserving of social work services.

Social work is a value-laden profession. While social workers may not agree about all values, the importance of professional values is central to the practice of social work.

Values are best defined as beliefs, while *ethics* are the rules of conduct that embody those beliefs. This chapter focuses on the way that social work values shape professional ethics; specifically, ethical codes and their application to practice; the relationship between ethics and the law; and the ways social workers manage ethical conflicts and dilemmas.

A BRIEF HISTORY OF SOCIAL WORK ETHICS

When social work first became a profession in 1917, social workers were more concerned about their clients' morals than about the conduct of individual social workers (Reamer, 2006). In the first two decades of the 20th century, the emphasis on social reform shifted to address social problems, such as those related to health, employment, and poverty (Reamer, 2008). In the late 1940s and early 1950s, social workers made several attempts to develop an official, written code that articulated the current, collective thinking of the profession. In 1947, the American Association of Social Workers (a predecessor of the National Association of Social Workers) adopted the profession's first formal code that established guidelines and standards for its ethical conduct (Reamer, 2006). The National Association of Social Workers adopted the first formal code of ethics in 1960 (Reamer, 2008).

The second wave of ethical development occurred in the late 1970s and early 1980s. Advances in biotechnology (such as organ and bone marrow transplants) changed the medical profession. These advances were especially influential in sensitizing social workers (and other human service workers) to the importance of their values and the relationship between their values and their professional decisions. New concerns about decision-making developed: Who could make certain decisions? For example, who decides whether a dying relative is taken off life support? Which decisions were acceptable? Was it all right to abort a fetus diagnosed with a severe developmental disability? Other societal changes, such as developments in computer technology, globalized ecopolitics, and growing interest in human rights, also shaped and intensified the concern for values in ethical practice (Reamer, 2008).

Recently, social work ethics have been concerned with risk management. Ethical standards guide conduct with a concern for liability risk and professional malpractice. A body of literature has developed around risk-management strategies that protect clients and prevent ethics problems (Reamer, 2008). Over time, social work has developed standards to guide workers, prioritize principles, affirm its values, and distinguish social workers' responsibilities. The *Code of Ethics* (NASW, 2008) (referred to hereafter as "*Code of Ethics*") reflects the profession's official views on appropriate conduct with clients, with one another, with organizations, and society at large.

PROFESSIONAL CODES OF ETHICS

The social work *Code of Ethics* (2008) is a comprehensive code of ethical standards and guidelines that serves to: (1) affirm social work as a legitimate profession; (2) provide guidance for practice circumstances; and (3) explicate the standards to which the public may hold the profession accountable. Social work practitioners generally appreciate the structure the *Code of Ethics* provides in today's increasingly complex practice situations.

The *Code of Ethics* is the official code of the social work profession in the United States. This chapter will review the *Code of Ethics*, discuss the *Code* as a tool of

<table>
<tr><td>EXHIBIT 2.1

Other
Subgroups'
Ethical Codes</td><td>Sub-groups of the social work profession, such as those working with specific populations, have developed their own ethical codes, including (Reamer, 2008):

• National Association for Black Social Workers
• Clinical Social Work Federation
• International Federation of Social Workers (IFSW)
• Canadian Association of Social Workers</td></tr>
</table>

practice, and highlight the most recent International Federation of Social Workers (IFSW) code.

The NASW *Code of Ethics*

The six core values in the preamble form the basis of the *Code of Ethics* (see Exhibit 2.2). The core values represent a mix of the worker's activities, skills, principles, character, and attitudes. The ethical standards cover and are organized according to the following six relationship categories or standards:

1. Social workers' ethical responsibilities to clients.
2. Social workers' ethical responsibilities to colleagues.
3. Social workers' ethical responsibilities in practice settings.
4. Social workers' ethical responsibilities as professionals.
5. Social workers' ethical responsibilities to the social work profession.
6. Social workers' ethical responsibilities to the broader society.

Within each of these categories the *Code of Ethics* contains from two (category five) to 16 standards (category one), along with varying substandards to total 155

<table>
<tr><td rowspan="7">EXHIBIT 2.2

The
Foundation of
the Social
Work
Perspective</td><td>CORE VALUES</td><td>ETHICAL PRINCIPLES</td></tr>
<tr><td>Service</td><td>Social workers' primary goal is to help people in need and to address social problems.</td></tr>
<tr><td>Social justice</td><td>Social workers challenge social injustice.</td></tr>
<tr><td>Dignity and worth of the person</td><td>Social workers respect the inherent dignity and worth of the person.</td></tr>
<tr><td>Importance of human relationships</td><td>Social workers recognize the central importance of human relationships.</td></tr>
<tr><td>Integrity</td><td>Social workers behave in a trustworthy manner.</td></tr>
<tr><td>Competence</td><td>Social workers practice within their areas of competence and develop and enhance their professional expertise.</td></tr>
</table>

---

6. SOCIAL WORKERS' ETHICAL RESPONSIBILITIES TO THE BROADER SOCIETY

6.01 Social Welfare
Social workers should promote the general welfare of society, from local to global levels, and the development of people, their communities, and their environments. Social workers should advocate for living conditions conducive to the fulfillment of basic human needs and should promote social, economic, political, and cultural values and institutions that are compatible with the realization of social justice.

6.02 Public Participation
Social workers should facilitate informed participation by the public in shaping social policies and institutions.

6.03 Public Emergencies
Social workers should provide appropriate professional services in public emergencies to the greatest extent possible.

6.04 Social and Political Action
(a) Social workers should engage in social and political action that seeks to ensure that all people have equal access to the resources, employment, services, and opportunities they require to meet their basic human needs and to develop fully. Social workers should be aware of the impact of the political arena on practice and should advocate for changes in policy and legislation to improve social conditions in order to meet basic human needs and promote social justice.

EXHIBIT 2.3

Sample of the NASW Code of Ethics

---

standards. These substandards explain the appropriate conduct for social workers (see Exhibit 2.3 for a sample of the *Code of Ethics*). Social workers may deviate from these expected norms when appropriate ethical justification is available (Barsky, 2010).

International Federation of Social Workers Ethical Statement

The International Federation of Social Workers (IFSW), a worldwide professional organization of social work organizations and individuals, documents its position on ethical practice in a statement that reflects concerns that are similar but not identical to the NASW *Code*. The IFSW document, *Ethics in Social Work: Statement of Principles ("Statement of Principles")* was approved jointly by International Federation of Social Workers and the International Association of Schools of Social Work in 2001. This document contains a more explicit and pervasive emphasis on human rights than the *Code of Ethics*. For example (IFSW, 2012, para. 3):

Definition of Social Work: The social work profession promotes social change, problem solving in human relationships, and the empowerment and liberation of people to enhance well-being. Utilising theories of human behaviour and social systems, social work intervenes at the points where people interact with their environments. Principles of human rights and social justice are fundamental to social work.

As a federation, IFSW assumes that member organizations, including the National Association of Social Workers (NASW), adhere to the standards in the *Statement of Principles*. Therefore, a comparison between the *Code of Ethics* and the *Statement of Principles* reveals some of the different emphases of the U.S. and international ethical practice. The following section of the *Statement of Principles* highlights seven international conventions that form "common standards of achievement and recognise rights that are accepted by the global community" (IFSW, 2012, para. 4). These include the Universal Declaration of Human Rights (1948), the International Covenant on Civil and Political Rights (1965), and the International Covenant on Economic, Social, and Cultural Rights (1965), as well as some conventions that the U.S. has not yet ratified.

The next two sections in the IFSW statement, on human rights and social justice, are the equivalent of the first two principles in the *Code of Ethics*. These address social work's commitment to human rights and human dignity. The full text of the Principles is shown in Exhibit 2.2. The values in these statements are similar to those in the *Code of Ethics*. However, the Universal Declaration of Human Rights, which is central to the IFSW statement, includes emphases on a global (rather than national) perspective and on the importance of economic, social, and cultural rights for all people.

Limits of Ethical Codes

Although the *Code of Ethics* offers helpful guidance to social work practitioners wrestling with difficult situations, the code is often difficult to apply because it offers general statements, can be difficult to interpret, and may offer unrealistic guidance given a social worker's situation (Strom-Gottfried, 2008). As discussed through examples in this chapter, issues of context, risk taking and creativity, and diversity also provide challenges to ethical codes.

The Role of Context   Ethical decisions are made within a practice context, which may shape the decision-making process in subtle ways. For example, if a social worker is unaware of attributing negative qualities to a client because of missed appointments, the social worker may believe that the client does not want services, is unable to take full advantage of scarce resources, or that the client does not deserve these resources. If the social worker is in a position to choose the clients who have access to a limited service, she or he may disqualify the client without further reflection. On the one hand, this could be viewed as using good judgment as the social worker is seeking to maximize the effectiveness of the scarce service. On the other hand, the decision reflects a decision-making process that is not transparent, that is influenced by one interpretation of client behavior, and is

EXHIBIT 2.4

Final Proposal for New IFSW Ethical Document 2004

Human rights and human dignity: Social work is based on respect for the inherent worth and dignity of all people, and the rights that follow from this. Social workers should uphold and defend each person's physical, psychological, emotional, and spiritual integrity and well-being. This means:

1. Respecting the right to self-determination—social workers should respect people's rights to make their own choices and decisions, irrespective of their values and life choices, providing this does not threaten the rights and interests of others.
2. Promoting the right to participation—social workers should promote the full involvement and participation of people using their services in ways that enable them to be empowered in all aspects of decisions and actions affecting their lives.
3. Treating each person as a whole—social workers should be concerned with the whole person, within the family and the community, and should seek to recognise all aspects of a person's life.
4. Identifying and developing strengths—social workers should focus on the strengths of all individuals, groups and communities and thus promote their empowerment.

Social justice: Social workers have a responsibility to promote social justice, in relation to society generally, and in relation to the people with whom they work. This means:

1. Challenging negative discrimination—social workers have a responsibility to challenge negative discrimination on the basis of irrelevant characteristics such as ability, age, culture, gender or sex, marital status, political opinions, skin color or other physical characteristics, sexual orientation, or spiritual beliefs.
2. Recognizing diversity—social workers should recognize and respect the racial and cultural diversity of societies in which they practice, taking account of individual, family, group, and community differences.
3. Distributing resources equitably—social workers should ensure that resources at their disposal are distributed fairly, according to need.
4. Challenging unjust policies and practices—social workers have a duty to bring to the attention of policy makers, politicians and the general public situations where resources are inadequate or where policies and practices are unfair or harmful.
5. Working in solidarity—social workers have an obligation to challenge social conditions that contribute to social exclusion, stigmatization or subjugation, and to work towards an inclusive society.

Source: IFSW & IASSW, 2004

perhaps inaccurate. For example, if the client did not have reliable transportation to appointments, had an unstable home life, or had a medically needy child, the decision-making process may have resulted in a different outcome. Such judgments may not only be inaccurate but also present an ethical question relating to the professional discretion and impartial judgment that the *Code of Ethics* requires. Therefore, an ethical code provides guidance that is decontextualized.

Professionals must be aware of their perspectives and judgments as important influences in making ethical decisions (Barsky, 2010). Thoughtful social workers frequently come to the conclusion that a resolution of an ethical question depends on many factors, in recognition of the importance of the surrounding circumstances, social location, and current pressures (Harrington & Dolgoff, 2008). Ethical decisions also benefit from consulting with knowledgeable colleagues (Reamer 2006). An in-depth conversation between professionals and their colleagues regarding their thinking, the implications of varying choices, and an analysis of their values can be very helpful. For example, many health care facilities have committees that review situations that involve ethical questions. In other situations, social workers engage in a discussion with a knowledgeable colleague or supervisor about a situation, and arrive at a decision through dialogue (also discussed later in this chapter).

Risk Taking and Creativity   Another concern about rigid interpretation of ethical rules is the potential for oversimplification. Social workers may rely on the *Code of Ethics* to avoid using their creativity and professional judgment. Using critical thinking to inform professional judgment is a valuable component of social work practice and sometimes requires risk taking (Strom-Gottfried, 2008). There is rarely one correct answer to an ethical dilemma. Social workers must have a tolerance for ambiguity, and work to apply ethical principles in practice (Dolgoff, Harrington, & Lowenberg, 2012).

The following case vignette illustrates this idea:

A school social worker named Cora, in strict adherence to the ethical principle of practicing only within her area of competency (Code Standard 4.01a), at first declined to see a young student who clearly was struggling with substance abuse. Cora felt competent to work with adolescents with developmental issues, but she was not educated to treat those struggling with substance abuse, so she decided it would be unethical to work with the student. The student, however, was in great need of assistance and was ready to work on the problem, and had no other obvious or realistic options for services in her rural community. The substance use issue was intertwined with other challenges related to growing up in difficult circumstances, and developmental issues were involved. When the student asked her to reconsider, Cora decided to discuss the matter with a colleague. The colleague reminded Cora that she had skills and experience that would directly benefit the student. For example, she knew about adolescent development, had strong relationship skills with adolescents, and was committed to a hopeful, resilience-based perspective. Cora's colleague also advised that she might find a resource in her former supervisor, who was a certified substance abuse counselor. Cora contacted her former supervisor, who was recommended available training and the supervision of a clinician who also was a certified substance abuse counselor.

Substance abuse in school populations is a common, yet a potentially life-changing issue for youth. Cora was able to negotiate with the school administration for time and reimbursement to obtain training that would allow her to better serve her clients, and the school population in general. Cora was creative and willing to explore options in response to important needs, and her client(s) were likely to receive competent services from an experienced worker who sought both training and supervision. There is some risk in Cora's assumption of the role of providing substance abuse treatment; however, taking on the issue is more closely aligned with social work's values than denying help to the student.

Strict and nonreflective adherence to a standard code can help preserve the current social order, which may be in direct conflict with the profession's commitment to confront social injustice and oppression (Witkin, 2000). If the student's family in the example was limited financially and belonged to an ethnic population that experienced oppression, the student would likely experience more barriers to receiving needed services than students in other families without these characteristics. The student's access to needed resources would be constricted, a situation of social injustice. Cora's commitment to responding to the student's need and providing services embodies social work's obligation to confront injustices.

While Cora made a decision using the *Code of Ethics*, colleague advice, and critical thinking, not everyone will agree that her solution was the best solution. For example, advocates for diversity empowerment might view Cora's effort as perpetuating social injustice because the situation involves non-white people receiving substandard services with a worker without the appropriate credentials. While there may be different, well-informed opinions about the best course of action in this situation, a too-narrow interpretation of the *Code of Ethics* that precludes critical thinking and creativity should be avoided. Resolving issues like Cora's is seldom easy; it is not hard to overlook worthwhile concerns, such as competency of the worker. Practitioners are encouraged to consider the *Code of Ethics*, and talk to colleagues to fully explore issues. This decision-making process contributes to the ongoing development of the profession's larger ethical stance.

Diversity A postmodern recognition of diversity and multiple realities raises another concern about the use of codes. Universal codes meant to apply to everyone in all situations can fail to address the contexts of both social workers and clients, or to recognize differences among people and cultures. For example, the ethical mandate for confidentiality, which refers to the social worker's obligation to keep information about clients from becoming public in any way, is often regarded as absolutely critical in maintaining a professional relationship. However, a universal application of confidentiality as an ethical mandate may present an obstacle in cultures less individualistic than that in the U.S., in which the community may be both the reference point and a rich source of resources that may assist individual clients. In such communities, maintaining confidentiality can be experienced as secretive, alienating, divisive, and harmful. For example, some Latino families may not be receptive to individual work with a social worker, but become engaged and respond to family systems work (Diller, 2011).

Briskman and Noble (1999) suggest that social work needs an ethical model that is more affirming of difference. One possibility is to develop multiple codes, grounded in client-based and service delivery schemes. For example, the authors cite an organization that works to prevent and treat sexual assault in Melbourne, Australia, that shapes its code based on the perspective that sexual assault is a violation of human rights (p. 64). This is an example of a specific code tailored to the organization's commitments to a particular client population. Another way to recognize difference is to develop constituency-based codes founded on specific concerns, such as tolerance for sexual orientation in schools. In this case, norms would develop with specific reference to the issues raised by gay, lesbian, bisexual, transgendered, and questioning groups. The development of specialized codes for different groups departs from the universal application of U.S. and Canadian codes.

In some cases a bicultural code of practice may work best. For example, the Aotearoa New Zealand Association of Social Work (ANZASW) currently uses a Bicultural Code that reflects an effort to deal justly with its native Maori people through recognition of the independence guaranteed to them in a treaty negotiated in 1840. The historical emphasis on the Maori people's independence has led New Zealander social workers to recognize the contemporary demands that this principle places on their work with Maori people. This code is based on the IFSW code and explicitly requires a bicultural focus for all of New Zealand's social workers. See Exhibit 2.5 for an excerpt from the ANZASW Code of Ethics.

Despite the challenges discussed above, the NASW *Code of Ethics* provides a valuable structure for the social work profession. The *Code* articulates expectations and provides overall guidance about conduct so that social workers have clear expectations and can identify points of departure from the norms. The *Code of Ethics* is value-based and is consistent with social work history. Like all ethical codes, the *Code of Ethics* has limitations and is framed in consideration of experiences outside of the mainstream. As such, the *Code of Ethics* addresses growing concerns for recognition and affirmation of diverse peoples worldwide. Codes are evolving documents that reflect the consensus of the profession at the time of adoption, a consensus that can change as contexts change. Social workers are encouraged to be involved in the professional dialogue about the *Code* through professional organizations, and to contribute to its evolution.

ETHICS AND THE LAW

International, federal, state, and local laws have a significant impact on social work practice, and its constant changes create a complex, sometimes bewildering climate in which to practice (Strom-Gottfried, 2008). Ethics and the law are related and sometimes overlap, yet there are clear distinctions between the two. The following

In New Zealand, the Treaty of Waitangi, negotiated in 1840 between the occupying British and the native Maori chiefs, recognized the native Maori people as Tangata Whanua and guaranteed their right to independence. This treaty is an integral part of contemporary ethics in New Zealand social work, and it reflects an active commitment for the promotion of indigenous identity. The following is an excerpt from Section C, the Bicultural Code of the Code of Ethics of the New Zealand Association of Social Work.

EXHIBIT 2.5

Sample Bicultural Code of Practice

Independence

1.1 Social work organisations and agencies and individual social workers should acknowledge and support the whanau [here, Maori clients] as the primary source of care and nurturing of their members.

1.2 Social workers are expected to work in ways that recognise the independence of the whanau and its members, by empowering the whanau and its members to handle their own lives and living conditions, and by enabling them to take care of themselves and to develop autonomously and collectively.

1.3 NZASW [New Zealand Association of Social Workers] recognises the right of Maori clients to have a Maori worker. Social work agencies and organisations should ensure that Maori clients have access to Maori workers at all levels, and social workers are expected to open up access to Maori workers. If no Maori worker is available, appropriate referral may be made if that is requested by the client. During their social work education Maori social workers should receive appropriate training in Maori models and methods.

Liberation through Solidarity

2.1 Social workers should work with agencies and organisations whose policies, procedures and practices are based on the Treaty of Waitangi, and actively and constructively promote changes in those agencies and organisations that operate from a monocultural base.

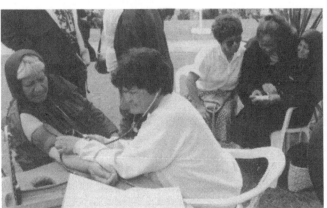
Cultural solidarity: Maori address health issues.

Non Discrimination

3.1 All social workers are expected to participate in Treaty of Waitangi education as part of their entry into social work and on an ongoing basis. This should include a knowledge and understanding of their own ethnicity and the actual history of Aotearoa New Zealand.

NZASW, 1993

EXHIBIT 2.6

Changes to the NASW Code of Ethics

The first NASW Code of Ethics, adopted in 1960, was only one page long with 14 brief declarations, such as "to give appropriate service in public emergencies." In 1979, the second NASW Code was adopted, including six sections of brief principles and a preamble. This Code was revised twice, eventually containing around 80 principles. For example, statements about solicitation of referrals, fees for referral, and dual relationships were added.

A third Code was approved by the NASW in 1996, and included four major sections: a preamble, the purpose of the NASW Code of Ethics, Ethical Principles, and Ethical Standards. There were many changes from the previous Code, including the addition of a mission statement, a section on the uses of codes, a list of social work values and principles, and almost twice the amount of practice standards from the 1979 Code. This third Code also contained standards about confidentiality limits, avoiding conflict of interest, and emphasizing social justice and social change (Brill, 1998; Reamer, 2009). The 2008 version of the Code includes new standards regarding cultural competence and social diversity, respect, discrimination, and social and political action (NASW, 2008).

discussion highlights parallels and distinctions between social work ethics and the law, as well as the potential for, and benefit of, collaboration between social workers and lawyers.

Parallels between Ethics and the Law

Many parallels exist between social work ethics and the laws that impact practitioners. For example, social workers who make derogatory comments to colleagues about clients who are living in poverty may be accused of actually violating the *Code of Ethics* (Section 1.12), as well as the spirit of the profession by expressing oppressive attitudes. In many states, clients could initiate a complaint with the state licensing board that a social worker has violated the *Code of Ethics*. If the client's culture differed from that of the social worker, a complaint could be added regarding a violation of *Code of Ethics* Section 1.05b, which requires the social worker to understand the function of culture and recognize the strengths of all cultures. State licensing boards can impose sanctions or require various forms of corrective action, such as license suspension or revocation (Reamer, 2008). NASW can also sanction social workers through a process by an ethics committee, and can impose a range of penalties such as suspension from NASW, mandated supervision or consultation, censure, or others (Reamer, 2008), which includes publishing the names of those sanctioned. While NASW's process is professional—rather than legal—sanctions can still harm a social worker's career.

Conflicts between Ethics and the Law

Social workers may experience practice situations in which there is an overt conflict between ethical practice and the law. The next section considers two legal duties: the duty to report and the duty to protect. Both of these legal mandates support ethical practice in most contexts, and create conflicts in others. These duties will be discussed more fully in Chapter 4.

Duty to Report: Child Protection   Social workers operating in the arena of child protection frequently find themselves in a contentious environment. Like many types of helping professionals, social workers have a legal duty to report their suspicions of child abuse or neglect to child protection authorities or law enforcement.

A common sequence of events leading to a report of suspected child abuse or neglect begins with a child telling a teacher, school nurse, or another school-related adult that she or he has been abused or neglected. Alternatively, a nurse or doctor may suspect abuse or neglect based on physical signs. In either situation, the professional reports these allegations to the proper authorities. The charges are then investigated by a child protection worker, who often interviews the parents/guardians as well as the child, and determines whether the charges are substantiated and further action is required. The person making the referral may be convinced that the child is being abused or neglected, but she or he may also suspect that the abuse or neglect is going to be difficult to substantiate. In such a case, reporting the abuse might place the child at much greater risk, because the parent/guardian will learn about the accusation, become angry and take out the anger on the child. Where there is no obvious way to protect the child, the referring professional may be tempted not to obey the reporting law.

© Mark Bowden

A discussion of abuse or neglect can elicit strong emotions

A similar situation may occur if the social worker is cynical about the adequacy, timeliness, or effectiveness of the response of child protection agencies. For example, a social worker who experiences an unreasonably long response time from a child protection agency due to inadequate department staffing may be reluctant to report again.

Social workers may encounter other practice situations in which compliance with the law may create more harm than good. For example, a social worker may work through a 15-year-old pregnant client's options about her pregnancy. In a discussion with the client, the social worker discovers that the client knows who the father of the baby is, and the father is terribly frightened about the possible reaction of his family to the pregnancy. The client is under 18, and legally a child; therefore, the pregnancy becomes a matter of child abuse and, by law, must be reported. However, the social worker views the reporting possibility as nonproductive: the client claims that the sexual activity was consensual; reporting will not assist the client; the father is under 18 years old; reporting will possibly alienate the baby's father, who might otherwise participate in the decision-making process; and reporting could put the baby's father at risk of a violent reaction from his family.

Duty to Report: Adult Protection   All states have reporting laws designed to protect older adults from abuse, neglect, and exploitation; however, each state has its own definition of reportable acts, those responsible for reporting, and the entity that is responsible for accepting the report and investigating (Barsky, 2010). Social workers may experience similar challenges between the law and ethical duties with situations involving suspected abuse of older adults.

Duty to Protect: Threats of Violence   Social workers also have an ethical obligation to protect people from serious, foreseeable, and imminent harm (Barsky, 2010). This ethical and legal responsibility emerged from a court case resulting from a tragic situation. In 1969, Tatiana Tarasoff, a young student at the University of California at Berkeley, had a casual dating relationship with a graduate student from India. He apparently did not understand dating customs in the U.S., and consequently was despondent that Tatiana was simultaneously dating several men. Depressed, he went to a psychologist at the University Health Services and told the psychologist that he intended to kill Tatiana with a gun. The psychologist wrote a letter to the campus police and asked that the graduate student be detained in a psychiatric hospital. The police interviewed the young man but did not feel there was evidence to prove that he was dangerous. The police required him to promise that he would not contact Tatiana. When Tatiana returned from a summer visit abroad, the man stalked her, and finally stabbed her to death. Tatiana's family sued the campus police, the University Health Services, and the Regents of the University of California for failing to warn them that their daughter's life was in danger. The trial court dismissed the case because although there was precedence for notifying the victim, there was no precedence for warning a third party (which, in this situation, would have been Tatiana's parents). The appeals court supported the dismissal, and an appeal was taken to the California Supreme Court, which overturned the dismissal, citing the therapist's responsibility to warn people who have been threatened as Tatiana had been. This case is *Tarasoff I* of 1974 and is often called the Duty to Warn decision.

This ruling opened the way for the family to sue the police and the therapist. A massive outcry from members of both police and treatment-related groups led to the decision by the California Supreme Court to hear the case again. The 1976 court decision, *Tarasoff II*, stressed that when a therapist has determined that her or his client presents serious danger of violence to another, the therapist must "use reasonable care to protect the intended victim against such danger." The court's position was that the therapist might have to take any of several steps to ensure the safety of the person threatened by the client, and the emphasis was on the Duty to Protect rather than the warning emphasized in *Tarasoff I*. Although the Tarasoff decisions were based on the work of a clinical psychologist, social workers, as well as other human service professionals, are subject in most states to the same legal precedents.

Social workers must now carefully consider any threats of violence they learn about in the course of practice. For example, if a client with a history of violence threatens to "belt" his girlfriend after discharge from a hospital, an assessment of the seriousness of the threat must be completed. Additionally, social workers must practice with consideration of protection against legal charges. A situation of this type is complex, and social workers must respond with several considerations in mind, including their employer's policies, the legal obligation of Duty to Protect, and the *Code of Ethics*. In the case of the Tarasoff decisions, the courts, not the ethics board of the profession, defined the parameters of responsibility.

Practice situations at various client system levels can involve the legal system. U.S. society has become increasingly litigious, and social workers may encounter practice situations with legal questions in policy practice, research, and community and direct practice (Barksy, 2010). For example, a social worker may encounter ethical challenges at the policy level if a new law requires adoption workers to provide the "adoption triad" (i.e., birth parents, adoptive parents, and adopted children) with access to all information about the adoption. If a birth mother had been guaranteed confidentiality at the time of the adoption, she, along with her social worker, may be distraught by this sudden breach of the document that she signed, which she considered a binding legal agreement. Another example is a social worker faced with maintaining a city zoning ordinance against a group home for those leaving prison, an establishment that she believes is critically needed in the community. Social workers working at the community level can experience ethical challenges in supporting community initiatives. For example, a community group may work toward an ordinance which would result in discouraging individuals who are not documented from residing in the community. Social workers can discover challenges in attempting to practice within the constraints of the law and the *Code of Ethics*.

Collaboration between Ethics and the Law

In the best scenarios, the law and social work professions can work together to empower people whose legal and social rights are frequently violated. There are many examples of this kind of collaboration in joint law and social work degree programs; for example, collaborations occur in legal clinics for refugee and immigrant peoples, and in family law and social work partnerships. In the U.S., the law drives much of the complex social welfare system, as well as the structures for shaping income and benefits distribution. A joint effort between the professions of social work and the law, based on common goals, can benefit a great number of people, with each profession enhancing and enriching the work of the other.

The law also places external pressure on all of the helping professions to hold one another accountable within their professions. The law protects the general public from the possibility that helping professionals can collude to cover up or obscure unethical conduct within their profession. Although the incidence of covering up unethical behavior is relatively low, all professionals may be tempted to minimize the seriousness of any offense raised. Child abuse allegations and convictions involving priests from the Roman Catholic Church, and a scandal involving the abuse of developmentally disabled in state care in New York state unveiled by the New York Times in 2012 are tragic reminders of the potential dangers to vulnerable populations of unethical behavior by people in helping professions.

DILEMMAS AND CRITICAL PROCESSES

The complex context of real client system situations often highlights the potential for competition between social work values, and/or between ethical standards in a given situation. This complexity can make it difficult to identify the appropriate ethical decision. The next section addresses the distinction between ethical conflicts and dilemmas, the idea of an ethical screen, various models for resolving dilemmas, and some examples of common dilemmas in social work practice.

The Distinction between Value Conflicts and Ethical Dilemmas

Value conflicts occur when an individual's personal values clash with those of another person or system. In contrast, ethical dilemmas occur when a social worker must choose between two or more relevant, but contradictory, ethical directives. An ethical dilemma exists when there is no clear single response that satisfies all ethical directives in a situation (Dolgoff et al., 2012).

Many practice situations involve ethical dilemmas. For example, social workers may learn that their clients are having affairs or using prostitutes, and struggle with contrasting ethical and legal obligations to maintain confidentiality and notify client partners of physical harm. In another example, clients may demonstrate racism toward their social worker, who would then struggle with weighing an ethical mandate against abandoning clients with confidentiality, client self-determination and other ethical obligations. Another situation involving an ethical dilemma is when a relative of a suicidal person asks a social worker to help their relative, but the person has not sought services.

In all of these situations, no one response is clearly the correct, ethical way to handle the situation. Social workers must struggle with their ethical and legal mandates, consult with colleagues, and engage in other activities to arrive at a decision. In the next section, structures to assist with the decision-making process are discussed.

The Ethical Principles Screen

A useful tool in considering ethical priorities is the ethical principle screen (Dolgoff et al., 2012). This screen assists in the decision-making process by highlighting the relative significance of values and the likely results of particular decisions. The elements in the screen, listed in Quick Guide 3, are reflected and prioritized in professional social work values, and are rank ordered so that Principle 1 has the

---

QUICK GUIDE 3   ELEMENTS OF THE ETHICAL PRINCIPLES SCREEN

- **Principle 1: Protection of Life** This principle refers to guarding against death, starvation, violence, neglect, and any other event or phenomenon that endangers a person's life.
- **Principle 2: Equality and Inequality** This principle reflects a commitment to equal and fair access to services and basic treatment.
- **Principle 3: Freedom and Autonomy** This principle affirms the notion of self-determination and supports people's right to make free choices regarding their lives.
- **Principle 4: Least Harm** This principle supports the idea of protecting people from harm; when harm seems likely in any event, it asserts that people have the right to experience the least amount possible.
- **Principle 5: Quality of Life** This principle confirms that people, families, and communities all have the right to define and pursue the quality of life they desire.
- **Principle 6: Privacy and Confidentiality** This principle supports the right of people to be protected from having their personal information made public. Maintaining confidentiality means the social worker must not share client circumstances, struggles, or decisions without the client's explicit (generally written and signed) permission. Information revealing any identifying characteristics (such as name and physical description) must also be kept confidential.
- **Principle 7: Truthfulness and Full Disclosure** This principle directs social workers to tell clients the full truth of any information pertaining to them, and explain whatever is needed to ensure understanding.

Source: Dolgoff, Loewenberg, & Harrington, 2009

highest priority and Principle 7 the lowest. The screen is constructed to reflect socially constructed priorities that involve values.

The following scenario demonstrates the use of the ethical principle screen. A social worker believes that a child on her caseload is at risk because his father regularly beats him. This case clearly reflects Principle 1 regarding threats of bodily harm. A competing concern is a violation of confidentiality and the continuing trust of the child that is threatened if the abuse is reported. Thus, the case also involves Principle 6. To resolve the dilemma, the ethical questions involved are identified: the protection of life versus confidentiality. These questions are prioritized by applying the ethical priorities screen, the priorities are rank ordered. In this situation, the life of the child provides a more urgent and compelling principle to guide actions than does a desire to maintain silence so that the child will continue to trust the social worker.

While not all social workers would agree with the prioritizations in the ethical screen in this situation, the screen offers a tool for decision-making. However, the principles address very broad concepts that are subject to interpretation, and their relative evaluation in the real practice world often invites critical thinking skills. The next section explores other models for resolving ethical dilemmas.

Models for Resolution of Ethical Dilemmas

Different models for resolving ethical dilemmas structure decision-making in other ways, reflecting a different ordering of social work values. Here are two examples.

Reamer (2006) suggests that social workers follow several steps to enhance the quality of their ethical decision-making:

1. Identify the ethical issues, including the social work values and duties that conflict.
2. Identify the individuals, groups, and organizations that are likely to be affected by the ethical decision.
3. Tentatively identify all possible courses of action and the participants involved in each, along with the possible benefits and risks for each.
4. Thoroughly examine the reasons in favor of and opposed to each possible course of action.
5. Consult with colleagues and appropriate experts (i.e., professional colleagues, supervisors, agency administrators, and attorneys).
6. Make the decision and document the decision-making process.
7. Monitor and evaluate the decision.

Strom-Gottfried (2008) adds several options for addressing ethical dilemmas, including researching the literature, relevant laws, and policies, consulting formally with established committees, obtaining supervision, and consulting peers. Strom-Gottfield (2003) also suggests the following strategies for solving ethical dilemmas:

1. Consider the "worst case scenario" of each option.
2. Consider the principles of least harm, justice, fairness, and the level of publicity that will result.
3. Consider clinical and ethical implications.
4. Consider the process.
5. Consider barriers to acting on the principal identified power relationships.

Both Reamer's and Strom-Gottfried's approaches help analyze the issues involved in any particular dilemma. While Reamer's model is more methodical than Strom-Gottfried's because it prescribes a specific sequence of analysis, the Strom-Gottfried model allows for a creative strategy that can accommodate specifics of the client system situation. Like codes of ethics, neither can produce an infallible result, and flexibility is needed in applying models to the specifics of the individual context.

Representative Examples of Practice Dilemmas

Of the many potential dilemmas, three types are common in social work: dual relationships, professional versus private tensions, and struggles between paternalism and client self-determination.

Dual Relationships   Relationships between social workers and clients that exist in addition to and distinct from their professional contacts create dual relationships. Such relationships are often only publically known in extreme circumstances, such as when the worker has a sexual relationship with the client. Sexual relationships, although an egregious affront to the profession, are fairly easy to resolve because there is a clear prohibition against them.

Nonsexual dual relationships pose another set of questions. For example, consider the situation of Lara, a new social worker.

> *Lara works in a local community health center with young adults whom legal authorities have identified as delinquent. Her responsibility is to support their integration into the community's work-study program that the center sponsors. Lara finds her work interesting and rewarding, and particularly enjoys working with one client, Joe, who is progressing in the program. Lara's 20-year-old sister recently called her to tell her about her new boyfriend. Lara realized that her sister's new boyfriend is her client, Joe.*

The dilemma created by this dual relationship is between personal values and professional responsibilities. There are many reasons for Lara to maintain the current situation. Lara enjoys her job, values her work with Joe, and does not want to resign. On the other hand, Lara also wants to participate in her family gatherings, which are likely to include her sister's new boyfriend. The *Code of Ethics*, however, cautions against dual relationships, which includes a professional and a personal relationship. Because of the ethical mandate about confidentiality, Lara is not free to tell her family that Joe is a client, or reveal anything about his situation.

Specifically, Section 1.06c of the *Code of Ethics* stresses that social workers should not engage in dual relationships with clients or former clients in which there is a risk of exploitation or potential harm to the client. Reamer (2003) classifies dual relationships as unethical when they

- Interfere with the social worker's exercise of professional discretion.
- Interfere with the social worker's exercise of impartial judgment.
- Exploit clients, colleagues, or third parties to further the social worker's personal interests.
- Harm clients, colleagues, or third parties (p. 129).

In this scenario, the dual relationship Lara has with Joe has the potential to affect her professional discretion and impartial judgment. For example, if Lara discovers personal information about Joe that suggests he might not be a viable candidate for marriage, Lara may experience difficulty conducting herself in a professional manner.

In some situations, the social worker may have justification for a dual relationship with a client. For example, in rural areas dual relationships may be very difficult to avoid, given the small number of service providers (Dolgoff et al., 2012). In a rural area, clients may have worked for a member of a social worker's family, attend the same faith community as a social worker, or be behind the social worker in the checkout line at the local grocery store. In a world characterized by expanding notions of community and multicultural experiences, social workers may find that they are asked to maintain relationships in several domains with others and to perform multiple roles. For example, in some cultures, social workers would risk insulting a family by refusing an invitation to a family dinner party.

Some social workers disagree that rigid boundaries are necessary or desirable. Zur and Lazarus (2002) maintain that some aspects of dual relationships—such as greater connectedness and an increase in the client's self-determination—can actually benefit rather than harm the client. They discuss a specific framework that separates the idea of exploitation into more manageable components to determine whether a certain aspect is potentially beneficial or hurtful in any particular relationship. The continuum, displayed in Exhibit 2.7, allows the social worker to explore the areas and degrees to which a specific relationship's dimension (for example, a social relationship with a client) might benefit or harm a client.

Applying this to the situation described earlier, Lara might evaluate the risk of exploiting Joe rather than empowering him if she were to continue working with him. Lara might also consider the likelihood of increasing Joe's vulnerability through her knowledge of his past struggles. Some variables, such as control of resources, might not be applicable; unless, of course, Lara favored Joe as a potential

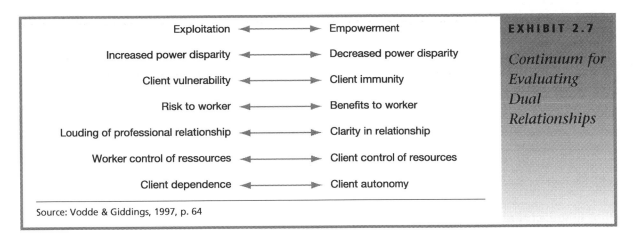

Source: Vodde & Giddings, 1997, p. 64

brother-in-law by allocating an unfair share of resources, such as work stipends. Lara may be at high risk for clouding the professional relationship by working with him while he is dating her sister. Lara will need to evaluate whether the level of risk is too high to continue working with him.

A second framework that includes the idea of "boundary crossing" as different than "boundary violation" (Zur & Lazarus, 2002, p. 6) allows the social worker to decide if the dual relationship occurs naturally (with no intention of the social worker) or is contrived (arranged by the social worker for her or his own purposes). The framework also allows the social worker to examine whether the relationship is exploitive (for the worker's benefit), essential (for survival in the community), or enhancing (beneficial to the client). Returning to Lara's situation, although her relationship with Joe is naturally occurring (she did not arrange for the relationship), the dual relationship is neither essential for the community's well-being, nor empowering for Joe. The dual relationship has the potential to be exploitive. Analysis using this framework can be helpful in situations in which dual relationships are impossible to avoid; this framework uses the specific content of the case rather than applying a one-size-fits-all model that simply prohibits dual relationships.

Responsibility to the Larger Society and Client Well-Being Relationships that involve child or elder abuse or neglect, exploitation of older adults, and intimate partner violence were once considered private family issues outside the jurisdiction of the law or of any public sector interest. Through the legal and social welfare systems, the larger society has wrestled with the question of whether certain events are private matters or public issues that impact society. While social workers often tend to value the private experience of families to recognize and acknowledge clients' perspectives on their experiences, social workers are also obligated to intervene with the client's environment to the extent that public issues impact and reflect the quality of lives of all citizens. The dilemma in this situation arises when the social worker values both the client's right to privacy and the potential benefit to the community of making the issues public.

For example, a social worker might struggle with his or her obligation to encourage a rape survivor to report the incident to the police (Dombo, 2011). Authorities often treat survivors of rape or sexual assault poorly. Survivors are reluctant to pursue legal recourse for many reasons. These may include: (1) wanting to heal and to avoid dredging up the past through legal proceedings; (2) fear of rejection if their experiences are made public; and (3) fear that actions of the authorities could re-traumatize them by implying that they are the guilty parties. Due to these reasons and more, rape is often underreported. Many social workers encourage reporting, so the crime is reported and appropriate resources can be devoted to the issue. This benefit to society of reporting the crime must be balanced with the experiences and wishes of individuals who have been raped. Attempting to force a person to report an experience of rape to the authorities may jeopardize mental and physical wellbeing. Striking a balance between promoting the public's general welfare and honoring the wishes of individuals is very challenging in practice, and must be the result of critical thinking and practice experience. When making decisions, social workers must consider the following questions: What is a public issue and clearly a private issue? Is violence against people an individual event or a violation of a larger social order as well? Does violence affect the entire community? Does a zealous social worker have the right to raise a public issue that could exploit the client's pain?

Paternalism and Client Self-Determination Long a cornerstone of social work, self-determination is the right of a client to make choices and exercise control over her or his life. In the *Code of Ethics*, self-determination is listed in Section 1.02, second only to a general commitment to clients. In contrast, paternalism refers to a process of interfering with clients' self-determination because the social worker believes she or he has a better understanding of what is in the clients' best interests. Reamer (2006) identifies three different forms of paternalism: (1) withholding information from clients for their own good (for example, not communicating a medical prognosis or the extent of injury in an accident); (2) deliberately lying to clients, which is more overt than withholding information (for example, assuring a distraught family member that a dying client will improve); and (3) intervening to prevent clients' behavior by controlling their physical placement (for example, facilitating involuntary hospitalizations). A paternalistic approach is also when social workers provide too much assistance to clients who avoid developing their problem-solving skills, give too many suggestions, or interact in a controlling manner. For example, a social worker who contacts a resource for a client, when the client is capable of making the contact, is acting in a paternalistic manner.

Questions about what rights and responsibilities social workers have to interfere in client situations have been debated since the profession began. In some situations, social worker rights and responsibilities are dictated by the law. However, social workers are sometimes asked to make an ethical decision concerning client self-determination without such a clear-cut guide, such as intervening with a client who is disorderly or does not adhere to social norms in public. Another example is whether a client should be forced to conform to mainstream standards of cleanliness in a residential setting, or whether a social worker must facilitate the involuntary hospitalization of a person with a mental illness who is not adequate in her or his self-care. In such a case, someone must make a judgment about the standards of "adequate." In another example, some social workers may struggle with the question of whether to intervene with clients who are eating in an unhealthy manner, and who have a health condition that is exacerbated by the eating habits. Situations in which social workers question whether to intervene are questions involving social control and social change, and raise the broader issues about the degree to which U.S. society tolerates eccentricities or even mere differences. Social workers are obliged to "promote the well-being of clients" (Section 1.01 of the *Code of Ethics*), and must decide about the balance between responsible caretaking and excessive interference with client self-determination.

A subtler variation of this dilemma occurs when the social worker urges clients to engage in an activity that the worker values. Reamer (2006) has called this "pseudopaternalism" because the paternalistic interference does not arise out of professional values but from personal self-interest, even if the social worker has good intentions. For example, most social workers would agree that it is consistent with the profession's values to encourage a young person to achieve the highest level of education possible. At the same time, most social workers also value education personally. If a social worker encourages an adolescent with great academic potential to achieve scholastically beyond the goals of her or his family, the social worker could be acting paternalistically if the client is not interested in such achievement. The social worker could be imposing her or his dreams and goals on the client. In contrast, accepting the adolescent's limited view of life may not be upholding the social worker's commitment to promote an individual's well-being. In another example, a social worker could encourage a community to enact a community-wide recycling program, when the community is not interested in recycling currently. The social worker could again be imposing his or her values about environmentalism on the client – in this case, the community. The issue of client system self-determination versus paternalism occurs frequently in social work practice; social workers must be constantly aware of this dilemma and make decisions based on critical thinking.

STRAIGHT TALK ABOUT EXPECTATIONS AND STANDARDS IN A LITIGIOUS WORLD

Many issues concerning ethics and values are complicated by context and are difficult to resolve in real practice situations. Social workers must consider values and ethics thoughtfully throughout their practice career. The following sections consider ethical social work practice in terms of postmodernism and in the context of a litigious environment for professionals. While different, social workers must navigate social work practice with consideration of both perspectives.

Thoughtful Practice in a Postmodern World

The current substance of professional ethical inquiry tends to be focused on ethics enforcement and risk management (Reamer, 2008). Rather than serious client issues, such as poverty, or the collective responsibilities of the social work profession, the *Code of Ethics* and most ethical inquiries focus on the individual conduct of the practitioner. Social workers, however, can explore other dimensions to enrich the professional discourse about ethical practice. For example, Walz and Ritchie (2000) suggest that Gandhian principles (based on the work of Mahatma Gandhi), could make a valuable contribution to Western thinking about ethical issues in social work practice. These principles include service, social justice, nonviolence, priority to the disadvantaged, and the notion of the heart as unifier of all things. Social workers interested in expanding the notion of ethical practice might examine the degree to which their practice is consistent with nonviolence, maintaining material simplicity, or prioritizing the needs of the disadvantaged.

Risk Management in a Litigious World

Some situations require social workers to be concrete and decisive about their actions. U.S. society is increasingly litigious, particularly in the helping professions. Licensed social workers are obligated by a state regulating body to uphold the *Code of Ethics*, even if practitioners question certain aspects and/or challenge the *Code of Ethics* to encourage change. While an uncommon occurrence, social workers face risks of censure and lawsuits charging malpractice or negligence. Social workers are held accountable for upholding the *Code of Ethics* by the profession through the NASW ethics complaints process, and by the courts through lawsuits and state licensing boards. In a very small number of cases, social workers are also indicted on criminal charges about ethical misconduct (such as an allegation of fraudulent billing) (Reamer, 2008). Boland-Prom (2009), in a study of 27 state regulatory boards about their actions against 874 certified and licensed social workers found that the most frequent violations were dual relationships, license-related problems, problems with basic practice, crimes, and practice below specific standards of care. State regulatory boards typically sanctioned social workers with letters of reprimand, revoked certificates or licenses, imposed probation or instituted the supervision of practice, and accepted the social workers' surrender of their licenses. Strom-Gottfried (2000), in an examination of 781 ethics violations, discovered that boundary violations (with sexual relationships as the most numerous) composed the greatest number of ethics violations (254). "Poor practice" (160) was the second, and "Competence" (86) was the third most common violation, with "Record keeping," "Honesty," "Breach of confidentiality," "Informed consent," "Collegial violations," "Billing," and "Conflicts of interest" accounting for the remainder violations.

Social workers are encouraged to take advantage of the knowledge and experiences of other competent and concerned social workers through supervision when wrestling with decisions that have an ethical component. Strategic thinking and planning with another professional, particularly one with more experience in recognizing various perspectives and aspects of the situation, can be an invaluable decision-making aid. Further, receiving input on practice through regular social work supervision can assist in the prevention of inadvertent ethical violations.

Social work practice occurs within a context of reflective, contextual practice, in which values are explored and rules about them are challenged, and the litigious structure of absolute rights and wrongs of ethical violations. The ability of social workers to transition from a postmodern perspective to one in which the legal and professional obligations are clear requires that social workers be "bilingual" in a sense. Social workers must be aware of the expectations of the law and the profession, as well as maintaining a personal obligation to engage in critical thinking.

CONCLUSION

This chapter has examined many governing morals and values rules. Consistent with the postmodern ideas of challenging taken-for-granted precepts, social workers are encouraged to explore the implications of the profession's values and ethics.

As social workers gain skills in ethical decision-making, they encounter increasingly complex ethical dilemmas. Although social workers strengthen the profession through active participation in questioning the values and ethics principles, social workers must also understand contemporary concerns about ethical violations, and the need for critical thinking about societal expectations.

MAIN POINTS

- Ethical codes are useful in providing standards of professional conduct, but may also restrict creativity when dealing with different contexts and multiple realities.

- Social workers have an ethical responsibility to contemplate, challenge, and work to change the profession's ethical awareness.

- Social work ethics and the law are different, although they have parallels and offer the potential for interdisciplinary collaboration. Ethics and law can conflict when legal duties, such as the duty to report, come into question in certain contexts.

- Ethical dilemmas arise out of competing values—for example, dilemmas involving dual relationships, privacy versus public issues, and paternalism versus self-determination.

• Social workers need to develop sophistication in risk management practices that protect them from ethical violations by learning from others' experiences and familiarizing themselves with basic legal mandates.

EXERCISES

a. Case Exercises

1. Go to www.routledgesw.com/cases and select the Sanchez family case. Review the client history of Celia and Hector Sanchez. Imagine that you are a social worker employed by Our Lady of Guadelupe Church, where the Sanchezes are parishioners. You have noticed that when Celia comes to pick up commodities, she talks about the desperation of her family. When asked about food stamps, she replies that her husband will not allow her to enroll in the program, even though she thinks it would be a good idea, and is considering enrolling despite her husband's wishes, if she could think of a way to do this. You have helped other families access the program, even when one member does not wish to enroll. Should you encourage Celia to use your assistance to enroll? What social work values are involved in this situation? Is it an ethical dilemma? How might you resolve the situation? What might be the implications?

2. Go to www.routledgesw.com/cases and select and become familiar with the Riverton case. Under the "Engage" tab, answer the Critical Thinking Question #1.

3. Go to www.routledgesw.com/cases and select and become familiar with the Riverton case. Under the "Assess" tab, take the Values and Ethics Assessment (under the "My Values" option).

4. Go to www.routledgesw.com/cases and select and become familiar with the RAINN case. Under the "Engage" tab, answer Critical Thinking Question #2. Also consider the following question: The RAINN website combines services and fundraising. Using the NASW Code of Ethics as a guide, do you believe that it is ethical to combine these activities on the same website? Explain.

5. Go to www.routledgesw.com/cases and select and become familiar with the Hudson City case. Under the Engage tab, answer the Critical Thinking Questions.

6. Go to the www.routledgesw.com/cases and select and become familiar with the Brickville case. Under the "Assess" tab, "My Values" option, take the Values and Ethics Assessment.

b. Other Exercises

7. You are a social worker in a psychiatric hospital setting. While most patients are discharged from the hospital setting after only a few days, some patients are able to stay for longer periods of time. Due to the mandate from insurance companies

to discharge patients as soon as possible, discharge planning begins as patients are admitted.

Your client, Bea, was admitted for psychiatric reasons and is terrified to leave the hospital after a stay of over three weeks. Your responsibility is to locate long-term housing and outpatient care for her in the community. As you meet with her one morning, you find her tearfully pleading with you to be allowed to stay longer, as she does not feel able to live independently. You are not sure that she is ready either, although the medical staff state that she is ready for discharge. You feel caught between the demands of your organization and the wishes of Bea.

    a. What three primary values are represented in your thinking about Bea's situation?

    b. Is this an ethical dilemma? Justify your answer.

    c. In what section of the *Code of Ethics* would you look for guidance?

    d. Apply one strategy for resolving ethical dilemmas discussed in the chapter to Bea's situation.

    e. How will you go about resolving this situation?

    f. Compare your response with other students. Do using different strategies lead to different resolutions?

8. After reading the following scenarios, consider these questions:

    a. What is the ethical issue?

    b. What are the values of the client system?

    c. Are they in conflict with your values? Society's?

    d. What strategy would you use for resolving the situation?

    e. What would you do?

After answering these questions, discuss findings with classmates.

Scenario A In a public setting over lunch, one of your co-workers begins speaking disparagingly about a challenging client. The co-worker does not use the client's full name.

Scenario B You are new to the area in which you have taken a job. Your new client, a hairdresser, offers to cut your hair.

Scenario C You are seeing a couple for marriage counseling. Both parties report that they are committed to remaining in the marriage. While you are out socially, you see the wife holding hands with someone other than her husband.

Scenario D While out with friends at a restaurant, you notice one of your clients is sitting at the next table.

Scenario E You are working with a teenage girl in a youth shelter. She reveals to you that she was raped by her neighbor, who is involved with gangs and drugs. She reports that she told her father and they agreed not to press charges due to fear of retaliation.

Scenario F Your client of six months reveals that he/she has begun to experience romantic feelings for you.

Scenario G One of your clients reports symptoms of a mental illness to you. She reports that she wanders around her neighborhood at night in the winter without appropriate clothing, is hearing voices, and refuses to take her medication or go to the hospital. You are concerned about her safety.

Scenario H You are working with a family in a family preservation program. In the home, the teenage daughter is violent towards her mother. During an altercation, the mother hit the daughter, presumably in self-defense, leaving a mark.

Scenario I You are leaving the agency for another position. On your last day, a client brings you a goodbye gift and asks if you will still call her.

Scenario J You receive an emergency call from your client. He reports to you that he just got fired from his job. He states, "I am not going to let him get away with this. He is going to be sorry. He has not seen the last of me." You know that this client has access to firearms and a history of assault.

Scenario K You receive a call from the spouse of your client. He wants to discuss his wife's case with you. The wife's chart does not contain a release of information.

Scenario L One of your adolescent clients invites you to his graduation and family party.

Scenario M A client's insurance benefits have ended but you feel she needs continued care.

Scenario N You are working with a teenage client. She reports to you that she is having sex with her boyfriend when her parents are not at home. She asks you not to tell her parents, and if you do, that she will feel as though you have betrayed her trust.

9. In a journal entry, reflect on a time when you faced a situation that challenged your personal values. What did you do? Why? How did you decide what to do? What were the results of your decision? If you could re-live the situation, would you handle it differently?

10. In a journal entry, reflect on a time where you observed a person doing something that was unethical, according to the NASW Code of Ethics. Describe the situation, and the part(s) of the Code that was violated. What did you do? Why? How did you decide what to do? What were the results of your decision? If you could re-live the situation, would you handle it differently?

CHAPTER 3

Individual Engagement: Relationship Skills for Practice at All Levels

What do I need out of a relationship? What can I give in a relationship? I am no different than any of you out there today. I have the same heart, I have the same feelings, I have the same aches and pains and the same hope and dreams that you do. I have suffered disappointment in relationships, as have you. I have been hurt too, but through all of this I have grown . . .

Resa Hayes, disability activist, in Mackelprang & Salsgiver, 2009, p. 284

Key Questions for Chapter 3

(1) How can I prepare for action with all client systems? (EPAS 2.1.10(a))
(2) What are the key interpersonal skills and techniques that I need to know to work with any client system? (EPAS 2.1.10(a))
(3) What communication mistakes can I avoid? (EPAS 2.1.10(b))
(4) How do I utilize a social justice and a human rights perspective when working with client systems? (EPAS 2.1.5)

THE ABILITY TO CREATE A RELATIONSHIP is at the very heart of social work practice. The idea of building relationships with people was probably a big part of what attracted you to social work in the first place and will likely be an abiding component that sustains your commitment. Social work scholars and practitioners have long recognized relationship as a crucial element of the profession's work with all types of client systems. Relationship building, or engagement, is the first step in the professional helping process and leads to the other processes of assessment, intervention, and evaluation.

When you think of relationships, you may first think of one-to-one partnerships that characterize much of direct practice. These relationships, however, rarely stand alone and are usually enhanced by other connections in the client system, which can include teachers, landlords, therapists, case managers, clergy, friends, and family, and anyone helping the client to reach her or his goals. Relationships are also as critical in group, family, organization, and community development work as it is in interventions with individuals. At the community and global level, relationship skills are necessary for work with decision-makers, community members, coalition partners and funders.

This chapter explores aspects of engagement, the process of building relationships across direct and indirect practice settings, and its importance to the overall success of social work practice. The first section examines the importance of listening to the situation and perspective of a client system (i.e., individual, family, group, community or organization). You will learn about interviewing skills and approaches that will both help your client share with you and help you enlist the assistance of others. This section also includes skill combinations that will help you establish productive connections with your clients. Later sections in the chapter show the engagement process through the lenses of strengths-based, social justice, and human rights practice as the perspectives that ground the work.

HEARING THE CLIENT'S SITUATION AND PERSPECTIVE

The most important moments of a social work relationship often occur at the beginning. Regardless of your agency's focus or your theoretical perspective, you first want to hear the client's situation and perspective. The length and the type of listening that you do will vary from practice setting to practice setting and agency to agency. For example, some agencies will expect you to complete a comprehensive psychosocial history that details a client's whole life, which may seem much more focused on the past than on the present. Despite this focus, you will want to prepare for your intervention work with the client system by carefully listening to them first. Quick Guide 4 provides questions to consider as you first listen to the client system.

QUICK GUIDE 4   LISTENING TO A CLIENT SYSTEM

As you listen to a client system, ask yourself the following questions:
- What brings the client system here today?
- How does the client system describe the situations, and what meaning do these situations have for them?
- What will life look like when the situation is better?
- What strengths and talents do they have?
- What are their expectations of me?
- What do they want to happen in their work with me?
- What can we accomplish together?

Although these may seem like general considerations, they are helpful in establishing the respect and affiliation that lead to successful work together. Rather than asking questions in a rote manner, you will want to demonstrate genuine curiosity in questioning, as well as your overall approach to the relationship with the client. Seeking understanding of the client's situation and perspective helps keep the conversation fresh and honest, as you are less likely to ask questions designed to confirm what you already think. You personally may have experienced frustration with people who assume they understand your thoughts or situation because they knew a lot about others' thoughts and situations in a similar situation, or have pre-formed opinions about people in your situation. It is challenging to remain open to and hear the story of your client if you have made conclusions mid-conversation based on information in a textbook or from your experience.

The ability to carefully listen is a critical social work skill because you want to hear the client's situation as she or he has experienced it, which may be different from your expectations. Relationship and listening skills are essential to a genuine approach to working with people, and are just as meaningful for work with client family members, community members, and staff as they are with clients.

For example, you may have made a connection with a client because you have heard their situation and perspective, conveyed professional and warm acceptance, demonstrated confidence and hope, and worked with your client to set tasks and goals. If one of those tasks is to help the client locate and secure housing that better suits her needs, you may want or need to work with a reluctant landlord. In such a situation you might perceive the scenario only in the client's terms—that is, you might be inclined to criticize a landlord who resists renting to your client. For this reason, in your beginning contacts with the landlord, you want to use the same relationship approach and listening skills that you used with your client. You will work to avoid stereotyping the landlord by entering the conversation with an open mind and avoid jumping to conclusions about "slumlords." You will avoid assuming that you know about the rental business based on your experience working with other landlords, and try to understand the difficulties the landlord has in maintaining rental properties in a profitable manner.

Social workers are often so invested in clients' rights that they are sometimes tempted to leap to unfounded conclusions about perceived adversaries. In some cases, that person may be the only person who can help the client; therefore, you need to put effort into developing an effective relationship. The social worker's role is to be fair, articulate, and open with all parties concerned. You will need to recognize your own biases and understand the client population.

Core Relationship Qualities

In a seminal work that still applies to social work practice, Carl Rogers (1957) defined unconditional positive regard and empathy as necessary to forming and maintaining effective professional relationships. Later, genuineness and warmth were added as essential items. These core attitudes or conditions are the basis for caring, and assist in forming a positive, non-possessive relationship with a client system.

Warmth   The demonstration of **warmth** is "non-possessive caring" (Shebib, 2003, p. 72). The expression of warmth entails the expression of caring and concern without expectations for the individual or relationship, as well as goodwill by the social worker. Examples of warmth are caring facial expressions, soothing tone of voice, and appropriate pacing of verbal interaction. Social workers also show their concern and warmth through extending courtesies, such as making sure that the client system is physically comfortable, offering them something to drink, making eye contact, and offering well-timed, appropriate humor. However, social workers will want to consider ethnic and cultural considerations about the expression of warmth, and offer flexibility in the amount of warmth demonstrated to coincide with the comfort level of the client system.

Empathy   is the "act of perceiving, understanding, experiencing, and responding to the emotional state and ideas of another person" (Barker, 2003, p. 141). **Empathy** is a core condition for all helping relationships, and social workers must develop a capacity for working with feelings, even intense feelings, without changing the subject, offering quick solutions, or moving the dialogue to the intellectual level. An empathic attitude is characterized by a willingness to learn about the emotional world of another. Suspending judgment by controlling personal biases, assumptions and reactions is a first step toward adopting empathic attitude. An essential element to empathy is the ability to enter into the emotional state of another person without actually feeling the same feelings as the person. For example, if the client has experienced a death in the family, rather than making assumptions about how the client feels and the intensity of the feeling, an empathetic stance toward the client is to attempt to understand the nature and intensity of the client's feelings about the situation. The client may have a number of reactions to the death, such as being grief-stricken, relieved, or satisfied at some sense of justice being served. A social worker can also express empathy to a colleague who experienced significant challenges in his/her work through attempting to understand the colleague's feelings about the situation. Also important is the ability to make an empathetic response by putting into words the feelings as expressed by the client. You will learn more about this later in the chapter.

Genuineness   Sincerity and honesty are essential elements of **genuineness**. Developing an effective relationship requires these elements, as well as being unpretentious, acknowledging your limitations, and providing only sincere reassurances (Barker, 2003). Being authentic and reliable provides the basis for a trusting relationship, which is the core of a helping relationship. Social workers who are genuine provide information that is timely, helpful, and accurate, and avoid hidden agendas and games with client systems (Shebib, 2003).

Unconditional Positive Regard   Each human being is considered to be "of worth whose rights and dignity are to be respected without reservation" (Barker, 2003, p. 445). **Unconditional positive regard** requires acceptance and nonjudgmentalness of client systems, regardless of whether a social worker approves or accepts individual or collective client system actions.

While these qualities must exist for a positive relationship, these qualities do not guarantee that client systems will always respond positively. That is, social workers should expect to be rejected by clients at times, despite their best efforts. For example, clients may interpret empathy as manipulation, and view genuineness as fake. A mature acceptance of the limits of possibilities will help social workers avoid burnout.

Specific Skills for the Dialogue

Communication is complex, and demonstrating oral communication skills is important to professional practice. In this section, you will learn about approaches and skills that are useful in purposeful communication with your clients. This includes both nonverbal and verbal communication skills that facilitate professional relationships with clients. Consistent with the strengths approach, the discussion below will emphasize skills used in facilitation of a dialogue with clients (Saleebey, 2013) rather than interviewing skills.

Preparing to Listen   **Attending** is a term used to describe a method of conveying interest in communicating with clients to promote exploration of ideas and challenges. Attending involves verbal and nonverbal components such as verbal following, eye contact, posture, and disciplined attention (Barker, 2003).

Prior to meeting with clients, you need to ensure that you have a physical and psychological commitment to a professional conversation in the best interest of the client. This can mean ensuring that any psychological needs you may have do not interfere with your listening to the client system. Additionally, attending means that you fully disengage from prior interactions with others before engaging with a new client system. You also need to mentally prepare yourself to avoid reacting to client system communication in a verbal or nonverbal way that conveys judgment or impatience. Third, ensure that you practice nonverbal behaviors that facilitate dialogue. See Exhibit 3.1 for nonverbal behavior guidelines.

While these are general guidelines, you will need to alter your nonverbal behaviors to match the comfort level of your client. In most cases you will want to sit relatively close to the client. When the client is not upright (i.e., the client is in a hospital bed), the social worker will usually want to minimize the height differential by sitting on a low chair at an angle to empower the client to control the amount of eye contact and full face contact.

EXHIBIT 3.1

Nonverbal behavior guidelines

Culturally competent social workers will utilize verbal and nonverbal behavior consistent with the cultural expectations of their clients (NASW, 2007b). Some modifications of these general guidelines may be needed, depending on the culture of the client.

- Turn your shoulders and legs toward the client.
- Sit with an open body position, with uncrossed arms and legs.
- Slightly lean toward the client.
- Smile and use head nodding to provide positive reinforcement.
- Maintain eye contact, within the comfort and cultural norms of the client.
- Use responsive facial expressions.
- Speak in a warm, pleasant tone.
- Give brief, encouraging comments
- Avoid the presence of large objects and heavy furniture between you and the client.

Professional working with a client at eye level

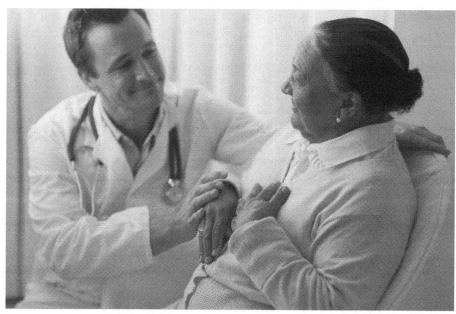

© Yuri Arcurs

A facet of the interchange that can be both highly variable and powerful is the degree of eye contact that you maintain with the client. In most cases you will make strong intermittent contact to indicate interest and connection. For clients of almost any culture, eye contact that is too constant and intense may be acutely uncomfortable; to some clients, even minimal amounts may seem intrusive. Individualizing your eye contact to maintain client comfort is an important practice skill.

Other important aspects of client interaction include dress and time of day of meeting. In general, your dress will be guided by the policies of your agency. Some agencies ask their staff to dress formally or "business casual," while others ask that staff match the style of dress of the clients, which may be informal. Clothes that are relatively modest in nature facilitate the interaction focus to be on the client, rather than on the social worker. Social workers work different sets of hours (i.e., day, evening, overnight, and weekend), depending on the nature of the work. If meeting clients in a community setting outside of the office, social workers must exercise some flexibility because there is less control over the environment of the interaction than in an office setting. In settings outside of offices, such as in client homes, community centers, and public locations, social workers must maintain professional verbal and nonverbal communication, or work to minimize distractions.

Diversity Considerations   Your preparation will include consideration of elements of culture or style—both yours and your client's—to which you need to attend. Supervision, peer consultation, and a general assessment of the particular culture in which you work will be helpful. Weigh whether your dress, posture, and language are appropriate to the person you are working with and the setting in which that work takes place. In general, the goal is to communicate respect for your client and for the nature of the event in which you are accompanying her or him. For example, in most practice contexts do not wear jeans to a courtroom appearance or use "dorm room" slang to the judge. Conversely, you might dress more informally in making an outreach home visit to a client and family, or while engaging in community organizing or community outreach. Continually be aware of the customs and contexts of the client system with whom you are working, and work to match the nonverbal behavior of the client.

Specific Interviewing Skills

Discovery-Oriented Questions   In general, your role is to hear the client's situation and perspective in the way in which she or he wants to tell it, without making assumptions and filling in information gaps yourself. **Discovery-oriented questions** are designed to invite your client to communicate her or his purposes in communicating with you and to express goals for the relationship. Discovery-oriented questions assist you in getting to know the client, and they can put the client at ease. Such a process may mean waiting a minute after you make welcoming introductions to allow your client an opportunity to begin. If she or he does not take that cue, you can invite her or him with such phrases as, "Where would you like to begin?" or, "Can you tell me what brings you here?" Some clients may be prepared for a much more directive stance on your part, so you may need gently to encourage their ownership of the dialogue.

Silence As you dialogue with a client, moments of silence should be permitted. There are many interpretations of silence, and some people feel more comfortable with silence than others. While some practitioners and clients struggle with their ability to allow silence in a dialogue, you will want to develop a comfort with silence. Exhibit 3.2 describes the possible meanings of silence.

During silence, you can attend to the silence through the use of eye contact, maintaining your psychological focus on the client, maintaining strong nonverbal focus (i.e., avoid shifting your body, checking your watch, or other means of communicating discomfort), and using self-discipline to minimize external and internal distraction. Conversely, some silences should be broken by the social worker to discern the meaning of the silence or to shift to another topic. Silence during your dialogue can be encouraged to allow for self-reflection and for slowing the pace of the conversation. Use of silence can also lead clients to answer their own

---

EXHIBIT 3.2

Meanings of silence

There are at least six meanings to silence:

1. *The client is thinking.* Some clients need more time than others to gather and organize their thoughts. Allowing time for this helps clients feel empowered, and worthy of the patience of the social worker.

2. *The client is confused.* Your questions may be unclear, or clients are unsure what you expect from them. If you suspect this, you might try asking whether the client is confused and needs the explanation or question again. Clients may demonstrate confusion or uncertainty during the beginning stages of the relationship or time together.

3. *The client is experiencing uncomfortable thoughts and/or feelings.* Providing silence gives the client time to process pain or anxiety, and consider proceeding further in the discussion. If you think the client is silent because of powerful emotions, you may wish to provide empathy. A statement such as, "I sense that the topic of your daughter's life provokes some strong feelings," provides support for the client and conveys understanding.

4. *The client is working to develop trust with you.* Silence provides clients with a sense of dignity and control over their lives, a way to avoid rejection, and a way to maintain control over the conversation. To move the relationship toward more openness, you could proceed slowly with a discussion of less personal matters, or you might choose to raise the issue of trust directly with clients.

5. *The client is a quiet person.* You might utilize open-ended questions to draw out the client, or, in a supportive manner, discuss her or his silence directly in terms of your working relationship. Consider using other methods for the discussion, as some clients are more expressive when combining an activity with a discussion, journaling or when using the arts.

6. *The client has achieved closure.* If you think this may explain the client's silence, you can ask the client if there is anything else to discuss at the moment.

questions and discover their own next steps toward a resolution (Shebib, 2003). The appropriate length of silence between the social worker and the client must be determined by the situation and context. If, for example, a client is giving extensive thought to a question asked by the social worker, a long silence may be appropriate. Long silences may not be appropriate in settings where the social worker may only have a few minutes with each client, such as some hospital or school settings, or when a client is clearly suicidal or homicidal. Ultimately, professional judgment must be the guide to interpreting and responding to silence.

> Example:
> Social Worker: *You mentioned that your daughter was really angry at you the other morning, and you had a fight before she left for school. What was the argument about?*
>
> Client: *Well, I think she and I. . . . .(lapses into silence for a minute). (The client begins to cry).*
>
> Social Worker: (After a few minutes). *This argument really upset you.*

Following Responses   This type of response gives clients immediate feedback that their message has been heard and understood. This immediate feedback can be conveyed through paraphrasing, summarizing, conveying empathy, and showing attentiveness through short verbal statements and questions such as "Go on," "I see," and "Can you tell me more?" (Barker, 2003). You will use **following responses** to provide only enough response to inspire the client to continue speaking. Providing a more complete response at the time may interrupt the client's direction and distract the client from fully sharing her or his situation and/or perspective.

Paraphrasing   Expressing an idea of the relevant points of the immediate past statement of the client in your own words is **paraphrasing**. When the social worker states the essence of the client statement, clients are assured that the social worker heard and understood them accurately. Paraphrasing can also assist clients to clarify their own thoughts to them (Barker, 2003). After paraphrasing, you can invite the client to correct you if you are mistaken. An invitation to correct can communicate both that you care enough about them to want to accurately understand their situation and perspective, and that you recognize that you may have the wrong understanding of the specific issues at hand. Paraphrasing is presented without judgment and without an attempt to solve any issues. There is no attempt to add meaning to or change the meaning of the client's statement (Shebib, 2003).

> Example:
> Client: *I don't really get what is going on. I mean, I put food on the table, keep a roof over our heads, and try to keep her going to school. I don't understand what her problem is!*

Social Worker: *You are working really hard to provide for her, and cannot understand what your daughter is talking about.*

Client: *Yeah! She is skipping school, running off, and now I am in trouble!*

Social Worker: *In other words, she is not grateful for everything you are doing for her. In fact, she is getting you in hot water with the law.*

To avoid monotony, social workers can use a variety of lead-ins for paraphrasing, to include the following:

As I understand it . . .

It sounds a little like . . . As I hear it . . .

The picture I am seeing is . . .

Clarifying   **Clarifying** is closely related to paraphrasing, but the social worker is directly asking for client feedback to clarify a point. Clarifying could be used at the end of a paraphrase. For example, the worker might say, "What I understand you to be saying is. . . . Is that right?" Clarification contributes to understanding the uniqueness of the client's message rather than generalizing or framing it in a way that matches your own perceptions. Clarifying increases the accuracy of your assessment while communicating a respect for the complexity of the client. Clarification may be needed at any point in the planned change process, especially to clarify each other's intent, interpretations, and meanings. For example, a client may demonstrate tearfulness or giddiness that you might assume is joy, but the tears may also be a sign of regret, loss, or confusion. Clarifying the meaning of a client's communication is an important component of the work.

Example:

Client: *The last time she ran off, she told me that she was never coming back, and that she hated me. Can you believe that?*

Social Worker: *Did I get this right? You are saying that she refuses to ever come back home?*

Client: *Yeah. What does she think, that she can live at a friend's house forever?*

Summarizing   A summary can provide closure and consensus either after a segment of the session is complete or at the conclusion of the whole session. A summary is a way of confirming your understanding of the client's message thus far and checking the validity of your assumptions. **Summarizing** can also help to establish organization for the entirety of your work together as well as frame a particular interaction or series of interactions. In situations in which the issues evoke significant client emotion, or in which the work is directed at some complex task, summing up the interaction can demonstrate manageability and hopefulness. Summing up can provide both you and your client with a snapshot of the topics discussed thus far, which can help to clarify the future direction of the dialogue. Summarizing can also assist to focus a conversation that wanders off topic.

Examples:

Social Worker to Client: *So far we have talked about your job responsibilities and my responsibilities, and how they overlap. We have discussed this vaguely in prior staff meetings, but no one has really picked up on the idea that this is a problem because we are wasting some time. Do you agree with this?*

Social Worker to Client: *Let me try to sum up our discussion so far. Your daughter has run away many times, and refuses now to come home. As far as you can tell, you have not done anything to cause her to run away and skip school. No one at the school believes that you are trying to fix the problem. Is that a fair summary so far?*

Social Worker to Client: *You have been working to get this street fair organized, but you are really in need of some people from the business community to get involved in the organizing and recruiting phase. Do you agree?*

Direct, Closed Questions   Direct [closed questions] encourage the client to provide factual information in a concise manner, and can be helpful in gaining specifics about behaviors, such as frequency, duration, and intensity. This type of question can be answered "yes" or "no," or with a numerical answer. While such questions do not encourage clients to open up and share, they are useful in situations in which precise information is needed. These situations include times when you will need a numerical number to get a sense of the situation, or involve dangerous conduct or some imminent threat of harm that must be dealt with directly in order to ensure safety. An example of a situation in which you would seek concrete information is a situation in which a client is expressing an intention of hurting her or himself or someone else and you need to know if the client has a real plan to engage in the dangerous behavior (for example, suicide) and the means to carry out the plan (Shebib, 2003).

Examples:

Client: *I have not eaten in quite a while.*

Social Worker: *How many days has it been since you had a meal?*

Community Member: *We had a community organizer working on health care in this neighborhood a while back.*

Social Worker: *For how many years did the organizer work in this neighborhood?*

Open-Ended Questions   Open-ended questions are designed to elicit extensive answers (Barker, 2003). While both open- and closed-ended questions are useful for social workers, open-ended questions are especially helpful at the beginning of a dialogue to encourage client systems to share their experiences and perceptions in the manner which is most comfortable to them, and in the way that makes the most sense to them.

Examples:

Client: *My brother and I always fight after school.*

Social Worker: *What are your fights like?*

Community Member: *We are trying to push the drug dealers out of our community.*

Social Worker: *How do the drug dealers affect the neighborhood?*

Indirect Questions   **Indirect questions** are questions phrased as sentences, rather than questions. Indirect questions allow clients the freedom to choose to respond or not, as well as provide flexibility in the type of response. In a conversation, mixing direct and indirect questions can help clients to feel less pressured and avoid monotony in the types of questions asked (Barker, 2003).

Examples:

Client: *I would like to save some money to buy a house some day.*

Social Worker: *I wonder how you would be able to save money, given all of your responsibilities.*

Client: *My teacher is picking on me! She punishes me for things that other kids get away with!*

Social Worker: *Wow! That sounds really upsetting. It must be so hard to try to get your work done in class while you are worrying about your teacher looking over your shoulder all of the time.*

Empathic Communication   **Empathic communications** assist client systems to identify and label their feelings, as well as provide the support to process feelings that previously were overwhelming. Social workers must, therefore, gain familiarity with the range of feelings, and become skilled at accepting feelings at face value, rather than placing judgments on feelings or demonstrating disapproval about client feelings (Shebib, 2003). To demonstrate empathy, social workers identify the feelings expressed as well as communicate their understanding of the feelings to the client. Social workers also must be aware of their own emotional state at the start of a relationship, so that the social worker is aware of the potential impact of her or his emotional state on the interaction with the client (Shulman, 2009a).

Three types of empathy are: (a) basic empathy, in which the social worker mirrors clients' statements; (b) inferred empathy, in which the social worker makes guesses about feelings based on clues in clients' statements; and (c) invitational empathy, in which the social worker encourages clients to talk about their feelings. Invitational empathy provides an opening for the clients to talk about feelings, without making a demand to do so (Shebib, 2003).

Example of basic empathy:

Client: *I am ready to kill my daughter. This whole thing is very embarrassing. I have not done anything wrong!*

Social Worker: *So you are angry with your daughter for putting you in this situation where you have to defend yourself.*

Example of inferred empathy:
Client: *My husband tells me that I am no good, and beats me all the time. But last week, he started beating my daughter too!*

Social Worker: *This is a really tough situation. I suspect that you might have been scared at what he might do to her.*

Example of invitational empathy: Client: *My son died in a fire last week.*
Social Worker: *Oh, I am so sorry for your loss! A lot of people in your situation might feel in shock about their sudden loss.*

A common challenging scenario with beginning social workers and the use of empathic communication is the challenge of working with clients who are older than they are who have had different types of life experiences. For example, a young social worker who does not have children may be working with a mother of several children. If the social worker is asked whether she has any children, the social worker may be tempted to respond defensively by talking about her training, or suggest that they are not here to talk about her background, but instead about the situation with the mother. Instead, this scenario provides the opportunity to utilize inferred empathy by thinking about the likelihood that the client might be wondering about her ability to understand the client's situation. A response of, "No, I do not have any children. Are you wondering whether I am going to be able to understand what it is like for you to raise your children? I am wondering that too, and I am wondering if you will explain it to me so I can understand?" might help to uncover the true concern that underlies the question (Shulman, 2009a).

As mentioned earlier, when clients experience or verbalize emotions, sometimes the use of silence is the best response. When you sit with a person who is experiencing powerful emotions and simply wait until that person is ready to speak, you are respecting her or his pace of processing feelings. In a profession characterized by verbal purposefulness, respecting your client's silence can be a challenge. Clients generally experience this process as non-pressuring, accepting, and highly respectful. This silence can be needed at any point in the planned change process (i.e., engagement, assessment, intervention, or termination and evaluation).

Avoiding Communication Pitfalls

Communication is an art and a skill. When dialoguing with client systems, social workers need to keep in mind the possibility of creating communication errors. Some of the more common errors are described below:

Jargon   Social workers, like most professionals, have their own **jargon**, or verbal shortcuts, to describe their activities that can be confusing to clients. Jargon can include abbreviations, distinctive words, and routines. You are encouraged to keep your communication with clients as free from jargon as possible (Shebib, 2003).

Examples of jargon:

Social Worker: *After the intake process, you will be on level one for a week. Then your DJO will assess you, and decide whether you can go to level two or you need to go back to court.*

Social Worker: *We have been discussing my ecomap and genogram for some time, so I am now redirecting the discussion.*

Leading Questions   The way in which a question is asked can shape the response. **Leading questions** manipulate client systems to choose the preferred answer. Leading questions can mask the preferred approach of the social worker. Clients who have a need to be liked and/or those who are compliant are especially vulnerable to leading questions (Shebib, 2003).

Example of leading questions:

Social Worker: *Given all that you have tried in the past, don't you think it is time to call the police to get the drug dealers out of the neighborhood?*

Client: *Well, I guess so.*

Social Worker: *Do you want to call now to report what you just saw?*

Excessive Questioning   While questions are an essential part of the helping dialogue, the asking of questions also puts the social worker in control of the dialogue. Client systems can feel resentful of answering questions, which can interfere with the relationship. Employing a diversity of responses can avoid a situation where clients feel interrogated. When dialoging with client systems, mixing different types of questions with empathic communication, summarizing, silence, and other types of approaches may help to avoid defensiveness, frustration, avoidance, and other responses that interfere with building a relationship. In those situations in which the social worker must ask many questions, it may be helpful to have a periodic break to check in with the client to convey respect and recognition that the questioning may be taxing (Shebib, 2003).

Example of check-in:

Social Worker: *I have asked a lot of questions of you today. How are you doing so far?*

Multiple Questions   Asking two or more questions at the same time, or **multiple questions**, can be problematic, because the client can become confused. If the questions are complementary, or the second question clarifies or adds to the first question, the multiple questions are not a problem.

Example of complementary multiple questions:

Social Worker: *What happened after the police arrived? What did the police do?*

Irrelevant Questions   Social workers need to have a clear idea of the purpose of the dialogue, and avoid asking **irrelevant questions**, which are questions that do not relate to the topic at hand. While you might be curious about details of situations or perceptions, you must consider the relevancy of the information before asking a question.

> Examples of irrelevant question:
>
> Client: *My boyfriend broke up with me for the second time this week. He already asked my friend out!*
>
> Social Worker: *How did he tell you that he wanted to break up?*
>
> Client: *My husband and I have been swingers for several years, and now he tells me that he wants to stop. I don't want to stop – I really enjoy swinging.*
>
> Social Worker: *How often do you two swing?*
>
> Client: *My mother wants me to wear clothes that are so out of style and boring. I don't want to wear her stupid clothes!*
>
> Social Worker: *Exactly what type of clothes does she want you to wear?*

Using Children or Neighbors as Interpreters or Translators   In an effort to communicate respectfully and effectively with clients from all ethnic, cultural and linguistic backgrounds, social workers have a responsibility to provide services in the language chosen by the client, which may include the use of qualified language interpreters (for example, certified or registered sign language interpreters). Interpreters generally need proficiency in both English and the other language, as well as orientation and training. At a minimum, social workers also have a responsibility to ensure that interpreters maintain confidentiality, are properly trained to the ethics of interpreting in a helping situation, and understand key terms and concepts specific to agency programs. Therefore, the use of children or neighbors as translators is problematic, and should be avoided. Social workers may need to prepare themselves for work with various populations by learning at least the basics of another language and customs of another culture, and completing training about how to work with professional interpreters who are linguistically and culturally competent (NASW, 2007b).

Integrating the Core Qualities and Skills in Dialogue and Interviewing

Social work practice activates the core qualities and skills through dialogue and interviews. Interviews are "purposeful conversations between social workers and clients . . ." (De Jong, 2008, p. 539). While many skills are important to social work, meta-analysis of controlled research studies indicate that relationship factors may be more important to positive client change than specific practice models used. Social workers spend more time in dialogue and interviewing client systems than any other professional activity, including individuals, couples, families, small groups, in supervision, organizational task groups, and with colleagues in other organizations (De Jong, 2008).

ARTICULATING PURPOSE: SOCIAL WORKER ROLE AND AGENDA

In addition to demonstrating the core qualities and basic skills, social workers must be able to articulate the purpose of their involvement with the client. Transparency refers to social workers' communication about the intention of their involvement, and displays a lack of an agenda for the client (Miller, 2011). More specifically, transparency can be demonstrated through a social worker's openness in discussing the reason why a particular question or line of questioning was asked. This kind of openness can help the client understand the social worker's thought process about the client's situation and address possible obstacles in the working relationship. Transparency helps to remove the mystery from the work and lessen the distance and power difference that may exist between the worker and client. Social workers who are transparent with their clients can, if asked, acknowledge that they are a student who will leave at the end of the year, or are young and have no children, or are not a person of color.

The beginning of the professional relationship is often the most appropriate time to articulate your purpose through a clear, concrete description of roles or job responsibilities, as well as describe the types of activities that are outside of your roles and responsibilities. For example, if you are a court advocate for a family violence shelter, you will want to be clear with your clients about the type of work that you do as an advocate, and that you cannot provide counseling, medication, or other types of services for them or their children.

On a more subtle level, the **overall purpose** of the work should also be transparent and explicit. It is sometimes tempting to engage the client in work on one issue and attempt, somewhat surreptitiously, to work on other issues you may think are more worthy. For example, if you and your client agree to work on the client's troubling relationship with school authorities regarding her child, you will focus on this task without trying to covertly intervene in her parenting style, or her negative relationships with men or any other issue. To maintain a genuine and honest relationship with clients requires that you are open about the jointly agreed-upon direction of the work, the methods you are employing, and the goal of your intervention. If you believe that another construction of the issue is more relevant or that the goals should be different, you should raise this issue early as a tentative point of negotiation about the nature and scope of the work. You would then take your cue from the client's response. Avoiding secret agendas is important to enhance the overall genuineness of the initial engagement, which will make a great deal of difference to the client's sense of being valued and accepted in the relationship.

This respectful stance regarding the client is one way to demonstrate a strengths-based approach. The following section will elaborate on the strengths approach in the engagement process.

MOVING FROM SPOTTING DEFICIENCY TO RECOGNIZING STRENGTHS

A central hallmark of the profession is the **strengths perspective**, or an orientation that emphasizes client systems' resources, capabilities, support systems, and motivations to meet challenges and overcome adversity to achieve well-being (Barker, 2003). The strengths approach to engagement can be challenging due to the need to identify and intervene in practice-based challenges. Social work as a profession developed during the late 1800s, in which moral conversion was a focal point. Poverty was often seen as a reflection of moral deficiency or laziness rather than as a structural failing that places and maintains people in adversity. As professional fields of helping, including psychology, developed more sophisticated and complicated assessment schemes (including the *Diagnostic and Statistical Manual of Mental Disorders DSM IV-TR Fourth Edition,* the first edition of which was published in 1958), the focus on pathology became more pervasive. Although social workers have a long history of recognizing their clients' strengths, this emphasis on pathology, problems, or dysfunction is deeply embedded in the culture and the traditions of the helping professions. An emphasis on pathology also persists in the current social service delivery system. Agencies often offer services around specific types of problems (for example, substance abuse or major mental illnesses), and these problems define the focus of the agency. Given the agency's purpose and milieu, developing the ability to focus on strengths when meeting a client may seem like an enormous challenge.

Nevertheless, when there is support for, and commitment to, a departure from illness-based models, social workers can implement strategies based on client strengths rather than deficiencies. For example, an increasingly visible body of literature reflects a different perspective from the various models of deficit, damage, or blame. Much of this literature focuses on **resiliency**, or "the human capacity to deal with crises, stressors, and normal experiences in an emotionally and physically healthy way" (Barker, 2003). Resiliency is an important area of potential strength in client systems, and social workers consider resiliency in assessing the coping skills of their client systems. Most people raised or living in the most daunting of circumstances survive, and sometimes even thrive. Communities can demonstrate resiliency after a traumatic event or natural disaster by processing grief and shock through utilizing the resources of community systems and assets to hold events and rituals, help victims, and demonstrate the ability to cope and function as a community. Resilient client systems can generate strengths and coping capacities even in some of the most oppressive and compromising scenarios.

In the first encounter with a social worker, client systems are often wondering about several aspects of the social worker, such as whether the social worker will be trustworthy, whether the social worker will be understanding or judgmental, and whether the social worker will be able to help. Clients are also often unsure whether they really need or want any assistance (Shulman, 2009a). Strong engagement skills and methods, discussed below, will assist to build an effective professional relationship and overcome these complex challenges to working together. For example, sending a clear nonverbal and verbal message that you will not make negative judgments or try to change the client as a person, but rather will affirm their aspirations and work to make them a reality can help build client trust. Engaging in enjoyable activities together, when possible, and seeking to incorporate humor, joy, and laughter into the helping process is another strengths-based approach. Lastly, being sensitive to cultural factors, honoring diversity, and seeking to assist people in activities and involvements that hold meaning for them will serve you well in a great variety of practice settings (Kisthardt, 2013).

Skills and Methods

Rapp and Goscha (2006) identify a series of methods for building a strengths-based relationship. Beyond exhibiting the core conditions described earlier, they suggest mirroring, contextualizing, self-disclosure, accompaniment, reinforcement, and celebration.

Mirroring   Reflecting the client's talents and capabilities so that the client can see himself or herself from a strengths perspective is called **mirroring**. The worker focuses on these strengths, emphasizing their presence in the client's daily life and stressing their importance. This metaphor counters the negative reflections many clients have of their own value and worth. While all mirrors distort images to some degree, the social worker uses a mirror that is more positively focused while building the relationship and throughout work with the client. For example, if family members identify specific issues that they experience with one another, the social worker might reflect on the care that they are demonstrating for one another by wanting to improve their family functioning, rather than split up the family.

Contextualizing   Social workers use **contextualizing** to put clients' issues into the wider community, society, and/or global context in order to discourage them from blaming themselves for their problems, and encourage them to take environmental and structural roots of their challenges. For example, when landlords discriminate against clients of the mental health system simply because they do not want "those people" to live in their building, some clients blame themselves for somehow being undesirable. In such a situation, the social worker can explore with the clients the effects of stigma in an intolerant society or provide evidence that they are acceptable to other, more positive landlords. In another example, a social worker might point out to a community group experiencing drug dealing and violence that their community is struggling with funding cuts to city social and community services. Rather than blaming themselves completely for their problems, community residents must also consider that their security challenges are at least partly rooted in lack of sufficient city-sponsored security resources and a lack of sufficient job opportunities. The goal of contextualizing is to point out the environmental roots, causes, and contributions of client system challenges, while working with clients to maximize their strengths and take advantage of their opportunities and resources toward addressing their specific challenges.

Self-Disclosure   The extent to which social workers should reveal their feelings, values, and personal information to their clients remains a controversial issue in social work practice and is called **self-disclosure** (Barker, 2003). Rapp and Goscha (2006) point to the inequities of expecting clients to disclose the most personal and profound aspects of their hearts and souls, while professionals sit back and reveal little about themselves. A strengths-based approach, based on normalizing the relationship in genuine ways, calls for some degree of worker self-disclosure to establish trust, validate the quality of the relationship, and model effective ways of managing emotions for the benefit of the client. Although there are no accepted standards governing the amount and nature of the disclosure, social workers should only reveal themselves to clients to assist to achieve a client goal. Social workers should carefully consider the purpose of self-disclosure, and avoid self-disclosure that strictly meets their personal needs and goals, such as for expressing strong feelings about something. For example, the social worker might share a story from her or his life to demonstrate appropriate behavior in a specific situation to meet a client goal of learning this skill, rather than for bragging about how well they handled a certain situation.

While caution should be exercised about self-disclosure, social workers can strive to foster open communication and a caring relationship, which may mean sharing relatively unimportant aspects of one's life routinely with clients, while maintaining a focus on the client. You can, for example, converse with clients by talking about your hobbies and interests, or share information about your familial or parenting status if asked. Social workers are warm and genuine with client systems and avoid strict formality by sharing information about themselves to further the relationship, yet avoid self-disclosure of significant information unless it serves a therapeutic purpose or is designed to achieve a client goal.

Accompaniment   Rapp and Goscha (2006) advocate for accompanying a client in the performance of a task. This can be **accompaniment** in the literal sense of going with a client to court or to a landlord's office, and/or in a metaphorical sense, as in joining the client in her or his journey of change. Rapp and Goscha (2006) acknowledge the concern for dependence on the social worker that can emerge when accompanying clients. Social workers may wonder whether helping a client through a task is serving the best interests of the client in the long-term, or whether clients should be encouraged to complete tasks on their own most of the time. The need or desirability for accompaniment should be assessed critically according to the specifics of the situation. Social workers should take into account such aspects of the situation as the emotional, physical, and intellectual capabilities of the client, as well as the working relationship with the client, when making decisions about accompaniment. Social workers should be careful to take cultural considerations into account as well, and avoid making decisions about clients based on *their* cultural or personal sense of the optimal level of independence, especially when the client either does not share or cannot attain their ideal. When literal accompaniment is not appropriate, the metaphor of accompaniment can be vital to the working relationship and resonate with the client long after the work is over.

Reinforcement and Celebration   To carry out a strengths-based approach successfully, the social worker needs to understand the meanings the client attaches to such behaviors as praise or recognition of particular events or accomplishments. **Reinforcement**, a procedure that strengthens the tendency of a response to recur, can assist in building a strong working relationship, but the social worker cannot make assumptions about the client system's view of reinforcement (Barker, 2003). Some expressions of support, for example, can affirm one person and embarrass another. By the same token, indiscriminate praise can be insulting, whereas purposeful, immediate, and specific positive feedback may be highly valued and strengthening. Although nearly everyone wants to feel appreciated, the worker needs to anticipate client responses in a particular context.

Logistics and Activities

To establish a collaborative relationship with the client, the strengths-based social worker will give the client the choice of the location, day, and time to meet, and, if possible, offer the option of meeting in a community location, rather than an office setting. Some clients may request that the meeting be held at their home, either because they have small children, they lack transportation, or they are simply more comfortable in this setting. As is true of much of social work, the client's choice of location depends on the nature of the contact, the context of the interaction, concerns regarding confidentiality, and the agency purpose.

Incorporating recreational or fun activities can also help the client to engage and feel comfortable. Some clients, such as children, adolescents, immigrants, and others, may struggle with sharing aspects of themselves in an office, face-to-face setting, and may be more comfortable in another setting, or talking while engaged in another activity. For this reason, playing basketball with an adolescent, playing a board game with a group of children, or walking in the woods with a client struggling with mental health issues can create an environment that is much less intense (and therefore more tolerable for some clients) than an interview-type meeting in an office.

An interesting variation on this theme is a strategy of using metaphors (Sims, 2003). This approach assumes that some clients will use metaphors as a way to communicate without exposing the specific details of their own lives until they are ready to do so. For example:

Client: (Her small business has recently failed) *Have you ever thought about how trying new activities is kind of like wearing a new pair of shoes?*

Social Worker: *With new shoes, you are never really sure whether they are going to fit or be comfortable.*

Client: (looking at the social worker while biting her lip and nodding).

Social Worker: *Wearing new shoes can make you feel maybe a little vulnerable.*

Client: *I just want to wear my old slippers! I know they fit me.*

Social Worker: (Nodding in agreement) (Gives her a chance to be more direct if she is ready to be)

Client: *It does not always feel good to feel vulnerable.*

(Sims, 2003).

As you can see, the social worker responds by using the same metaphors that the client suggests, and is sensitive to the client's readiness to be more direct. The purpose of using and encouraging metaphors is to respect the client's communications rather than seeing them as a form of resistance. The social worker can use the information conveyed by the client to support the client's efforts to communicate the depth of her or his experience. This strategy honors the client's meaning and validates her or his approach to engagement in the work.

RECOGNIZING AND ARTICULATING POWER

Social work practice almost always involves issues related to power. The relationship of power, or the possession of resources that enable a client system to accomplish a task or to exercise influence and control over others (Barker, 2003), may not be readily apparent to you. Because you are eager to engage with clients in their struggles and help them achieve their goals, you may not immediately see why a discussion about power is relevant. In most instances, the obstacles clients face involve power, and the effective management of power can benefit the client. Power is a component in many types of social work relationships, from political advocacy to one-to-one counseling.

Sources of Power

There are four sources of power with which social workers engage: agency resources, expert knowledge, interpersonal power, and legitimate power (Hartman, 1994).

Agency Resources   Social service agencies have access to and control over a number of resources. These include **tangible resources**, such as clothing or money for emergency housing, and **intangible resources**, such as individual counseling and education groups. Traditionally, social workers and administrators have allocated resources to client groups or individuals depending upon their evaluation of the fit with agency purposes and funding source(s) guidelines and constraints. However, some contemporary organizations, and even some federal programs, have experimented with arrangements that empower clients to determine and control the resources they receive. For example, an agency that serves children with disabilities might encourage the child's family members to identify the services they need, both within and outside the agency, rather than undergoing an agency-driven assessment in which a professional tells them what they need. By the same token, some Medicaid provisions include a waiver that permits a family member (usually a parent) to act as case manager, thereby coordinating services for the child and eliminating the costs of professional case management. Services that the family chooses and obtains with assistance are called **client-directed resources**. The concepts of power sharing and client-consumer advocacy threaten the notion that social workers and administrators should continue to control the disposition of agency resources. Most agencies struggle with the idea of giving up their authority in this arena.

Expert Knowledge   Many social workers find that some clients see them as possessing **expert knowledge** because they have credentials and experience. Social workers can counter that assumption in part by using the strengths perspective to clearly articulate that clients are the experts on their own lives, and work to develop a partnership with clients that focuses on reaching client-defined goals.

Interpersonal Power   The personal attribute characterized by the ability to build strong relationships, develop rapport, and persuade people is known as **interpersonal power**. Both social workers and clients may have this type of personal power, which is closely related to charisma. Although the social worker's interpersonal power can benefit clients when used to attain resources or access to services, the social worker's interpersonal power can also perpetuate the power imbalance between the client and social worker by diminishing the client's efforts to reclaim power. Social workers must strive to reduce their interpersonal power by establishing more egalitarian relations and genuine collaboration with clients (Hartman, 1994).

Legitimate Power   The term **legitimate power** refers to legal power to perform actions to control the behavior of others. Social workers need to be very careful about this power. The responsibility to protect clients and others (as in cases of child or elder abuse) must be acknowledged, but social workers must also recognize that legitimate power must be exercised very carefully, and have limits. Social workers must be careful not to replicate the power abuses of the past by exercising legitimate power beyond what is absolutely necessary to protect clients. Although social workers may choose to retain their various powers to use for individual, family, group, or community client benefit, they may alternatively try to work to change structures that impact client systems at the policy and organizational levels to empower clients.

Power in Client Lives: Jasmine Johnson

There are many other ways to think about power with clients. One approach focuses on the power relationships with which the client system is struggling that are external to the social worker relationship, either in interpersonal, community, policy, or larger cultural terms. For example, a community may be struggling to convince their local government to allocate and spend city funds on trash pickup in their area. The second involves power issues experienced directly between the social worker and the client. This vignette about Jasmine Johnson provides examples of the types of power.

Jasmine's Situation   You work in a family support agency.

> *Your client, Jasmine Johnson, is an African American mother who comes to you with a concern about parenting. Her teenage son is difficult to manage behaviorally both at home and, increasingly, at school, and Jasmine is unsure how to deal with him. He often does not seem to respect her authority, and ignores her attempts to discipline him. He is "sassy," talks back, and is occasionally quite rude to her. He ignores the limits she sets, and he does not obey school-night curfews or help with any household chores. Jasmine struggles financially to support him with only sporadic help from his father. Her job pays poorly, carries little status in the work world, and does not provide extra money for either her or her son to enjoy any recreational activities. Overall, Jasmine struggles with low self-esteem.*

Jasmine and Power Relationships   At first glance, Jasmine's challenges appear to be strictly personal. Jasmine knows that she does not feel good about herself or her situation, and she assumes she needs to improve at something. You might assume that she needs to address her self-esteem issues, or you might even conclude that she is depressed and needs medical attention. Yet, from another vantage point, it is likely that power, or lack of it, plays an important role in her experience. She is a woman who performs many interpersonal roles (i.e., ex-wife, single mother, daughter, worker, friend, and neighbor) and a member of a cultural group that has experienced pervasive and persistent oppression in our culture for more than 300 years.

These external conditions and stressors can be deconstructed and examined toward the goal of empowerment (Broussard, Joseph, & Thompson, 2012). Although

Jasmine may have interpreted her experiences as signs of her own deficiencies, she has also exhibited remarkable resilience in dealing with disadvantage and oppression. She has managed to survive in trying circumstances that have had a far-reaching political repercussion in her life. If she is to be empowered, she will need to recognize those political events in which her life is embedded. You can bring these events to her consciousness through purposeful (and skillful) articulation.

This does not mean that Jasmine's own sense of her problem is erroneous and that the real problems are racism and sexism. Rather, as a social worker, you want to validate her experience and recognize the meaning she makes of it. However, Jasmine has genuine feelings of inadequacy, and she has experienced the negative side of the power differential. That realization can open many doors. For example, Jasmine may begin to separate her feelings of inadequacy from her sense of identity and start to look at her experiences as a function of her social location. In turn, this new perspective might inspire her to engage in some action, such as forming an informal support group for single, African American mothers. This group can share stories and experiences, provide day care arrangements for one another, or coordinate grocery shopping. Adopting a different outlook also might encourage Jasmine to become a spokesperson for more stringent requirements regarding child support payments or increased benefits for working women with children. The possibilities for Jasmine's roles and activities are endless, and these types of activities may affirm Jasmine's experience even as they are instrumental in changing the quality of life for her and others.

Jasmine and the Social Worker   There are nearly always noticeable differences between social workers and clients. As long as the profession sustains the concept of the social worker as the expert, there will be a felt power differential based solely on distinct roles in the social worker-client relationship. There are also likely to be additional differences related to gender, age, race, socio-economic status, and other dimensions of diversity. These differences, if perceived as problems, can complicate the engagement process.

For example, Jasmine may find it challenging to think of social work students or younger workers in general as a genuine source of help to her. Students may come from a different cultural or ethnic background, and they may not have partners or children. They may also seem to her to be so privileged by their race and education that she thinks that they cannot relate to any experiences in her life. Yet, social workers are supposed to be experts, and as such they have some level of power that she may resent, admire, or barely recognize. In most cases, these differences related to worker and client roles or attributes are, at the core, about power and power differences. It is necessary, then, to discuss these differences openly when they get in the way of the work. Simply raising them can open up the whole relationship. For example, in the early stages of the relationship, a social worker working with Jasmine might ask whether she has any hesitation about working together.

"Look, Mr. Cook, I know you think I'm a nice kid and that you like me," I said.

"I do like you," he confirmed.

"But you know I'm too young to have experienced what you go through every day!" I said.

Silence.

"And you may even figure that a kid like me can't help you."

Silence.

"Right?" I continued.

A nod.

"But you want to get out of this depression real bad, don't you." I said.

"I sure do," he said. Then he sighed and added, "I'm probably being foolish. You

youngsters are right out of school with the latest techniques. I guess I'd just be more comfortable talking to someone closer to my own age. But here I am already talking to you, aren't I? So I guess I already decided to try it out with you."

I nodded. "How about if anytime you feel uncomfortable, you say so and every time I think you may be a little uncomfortable I'll say so?" I suggested.

"It's a deal," he said.

Source: Middleman & Wood, p. 163

The social worker might acknowledge that the two of them have experienced different life circumstances in the past and present, and ask that Jasmine be willing to tell her about her background and current life situation so that the social worker can try to help her. Asking the client to share this type of information, and giving the client the opportunity to talk about any hesitations that they may have in working together, provides the client with the opportunity to talk about differences, and the impact they may have on the relationship.

Jasmine Johnson: Conclusions Situations that involve power differentials are sometimes awkward or even embarrassing to confront, particularly for you as a student or an otherwise humble person. You may understand that the client sees you as having power simply because of your role as a helping person, an employee, or a student. At the same time, you may secretly wonder yourself about the extent to which you can help an exasperated parent who might be of a different race and remote social class, when you might not even be a parent yourself. You may feel some hesitation at working with an oppressed client when you have enjoyed so much privilege. Thus, the concept of power in the relationship can become quite complex and problematic.

When these situations arise, you might hope that the concern will pass or that clients will just trust that you know what you're doing in spite of these differences. Unfortunately, once these issues are perceived as problematic, they will not simply go away. You and your client should confront them directly through open acknowledgment and exploration. Exhibit 3.3 provides an additional case example. In this excerpt, consider the ways in which the worker addresses the issue at hand. What impact do you think this approach will have on the future of the relationship?

This vignette in Exhibit 3.3 conveys an extended message. First, the social worker addresses the immediate obstacle, namely, that the social worker is both much younger than the client and comparatively inexperienced. The social worker also paves the way for ongoing honesty in the relationship by addressing the obstacle openly and directly. The scenario is a relatively complex one in which both participants seem to feel an initial lack of power that is subsequently alleviated by open acknowledgment. Ultimately, the case suggests that there is a broader context for the work than the one-on-one issues that emerge between the social worker and client, leading to recognition of social justice and human rights issues in client stories.

VIEWING THE CLIENT SYSTEM SITUATION AND PERSPECTIVE FROM SOCIAL JUSTICE AND HUMAN RIGHTS PERSPECTIVES

When clients describe a situation that involves hardship or oppression, or is emotionally challenging, social workers use their professional skills and methods to respond to the client. However, social workers are human, and are often personally moved by the situation as well. There is something very powerful about talking directly to an individual who is sharing a powerful story with you. This experience typically creates a bond. In such circumstances, the larger perspectives of social justice and human rights may seem like remote, intellectual concepts in the face of the personal pain that your client is experiencing. Nevertheless, these perspectives provide the rationale for your practice, and they help to connect client system experiences together. The perspectives thus provide an organizing frame for your work so that your practice transcends the parameters of individual emotional responses or simple sympathy. Instead, your practice consists of carefully planned and cohesive activities in response to clear, principled commitments.

Although you will always want to recognize personal pain and respond on a human level, you will also want to go beyond the intensity of the direct relationship and see your client's experiences as social justice and human rights issues as well.

Full Participation in Culture

As discussed previously, work toward social justice is a cornerstone of social work practice and is mandated in the NASW *Code of Ethics* (2008). A number of "isms"—racism, elitism, sexism, heterosexism, ageism, and others—that you have probably already studied reflect the degree of social injustice that is prevalent in our culture. These prejudicial attitudes toward the "other" are generated by our society and suggest a discriminatory standpoint that most social workers find profoundly distasteful. The attitudes also carry concrete repercussions for individuals, families, groups, and communities that take many forms of exclusion, including exclusion from resources and limited class mobility within society. The term "exclusion from society" refers to the process through which people are unable to participate in the benefits of public and cultural resources. This inability can be felt quite literally (for example, when there is no access ramp to the public library for people with disabilities) or more indirectly (for example, when policies and poverty limit public investment in public schools). Clearly, these exclusions are injustices because they represent arbitrary allocation of access to public benefits, and are violations of the principles of human rights. Such exclusion tends to be pervasive and on all levels—individual, organizational, and structural. Quick Guide 5 examines these three levels of social and cultural exclusion. Considering client situations relative to all three levels is necessary for the practitioner's work to have an impact on larger systems, such as policies, communities, organizations, and societies. Although many practitioners see their labors in terms of individual, one-by-one achievements, the collective achievements make a difference in the larger environment. Social workers respond to clients on a personal, one-to-one level while acknowledging and addressing social injustices and human rights violations that exist on a variety of levels.

Strategies and Skills for Promoting Social Justice and Human Rights

Social justice and human rights issues call for well-organized strategies and skill sets that usually fall within the realm of policy practice. The following strategies are particularly applicable in such situations.

- Understanding the repercussions of social injustices on clients, and helping clients understand.

- Helping clients gain access to their legal entitlements through social advocacy.

- Convincing legislative bodies to adopt, amend, or repeal laws when such changes would benefit clients.

QUICK GUIDE 5

Individual Exclusion

During your dialogue, clients may make an explicit reference to the way exclusion has influenced their situation. Individual exclusion refers to the perception of being left out of or barred from participation in interpersonal situations. This situation could include, for example, that your client was a victim of racist harassment by peers in school or was dismissed by teachers as having no future because of her or his ethnicity or ability status. Other clients may not give voice to any strong or concrete sense of their own exclusion or the violation of their human rights but rather describe it as "fate" or the "way things are."

Organizational Exclusion

Like individual exclusion, organizational exclusion, the phenomenon of being prevented from participation in activities by an organization, can be obvious to clients (and others), or obscure. For example, even hiring practices that appear to be fair may actually favor some groups over others through their written or unwritten rules, regulations regarding promotions, or subtle differences in work assignments. In fact, many battles over such efforts as affirmative action arise from a concern for the organizational structures in which unjust practices and advantages have taken hold. For example, given that there is no equality in access to this country's top-rated educational institutions, the organization that automatically hires the candidate with the most prestigious degree—even when the requirements of the job do not mandate it—is engaging in preferential practices that are rooted in injustice.

Just as Jasmine Johnson interpreted her difficulties as individual failings, your clients may not be sensitized to the role that discrimination plays in their place of work. It is important in connecting and engaging with clients to recognize manifestations of organizational exclusion and work with clients regarding the meanings they attach to these life events. In some situations, this alone can be a liberating activity.

Structural Exclusion

Institutionalized arrangements such as poverty tend to maintain and perpetuate themselves. Structural exclusion refers to the interconnecting role of institutions and societal forces in preventing participation or limiting access, such as the connections among poverty, poor schools, limited achievement, limited employment options, restricted housing, poor health care, and shortened life expectancy. Recognizing poverty as a structural problem directly conflicts with presumptions about equal opportunity and the myth that anyone can get ahead. Social workers may need to work with client systems over a period of time before clients and others can begin to see the roots of their personal challenges as stemming from structural exclusion.

- Educating the community regarding certain populations, for example, giving a talk on the needs of refugee children or families with disabilities.

- Developing resources (this involves identifying and procuring resources that are needed but do not exist currently).

- Facilitating the redistribution of resources.

- Testifying in court hearings regarding issues that affect clients.

This list is by no means exhaustive, but it suggests the flavor of work focused on systems larger than the individual. These strategies do not negate the importance of individual connections with human beings on a personal level, but enrich the engagement and the worker's understanding in a way that is consistent with the complexities of people in the contemporary world.

STRAIGHT TALK ABOUT THE RELATIONSHIP: INTERPERSONAL PERSPECTIVES

Being transparent, or explicit about your work, means being clear about the restrictions you encounter as a practitioner. A goal of the work is to establish a meaningful and trusting relationship with the client system, and the tasks of discussing constraints or inviting critical evaluation may not seem very appealing. In fact, the social worker may tend to postpone or avoid them altogether. Nevertheless, attending to the boundaries of your role early in the development of the professional relationship, along with confidentiality and maintaining client privacy as discussed below, is important to developing a trusting relationship.

Confidentiality

There are at least three areas in which social workers may be required to break confidentiality, including cases of child abuse/neglect/exploitation, older adult abuse/neglect/exploitation, and when a client threatens harm to another person. First, as mandated reporters, social workers, among many human service professionals, are legally required to report cases of child abuse, neglect, or exploitation. Second, most states have mandated reporting laws for adult protective services for older adults and persons with disabilities. Third, as mentioned in Chapter 2, social workers in many states have a duty to protect as a result of the *Tarasoff II* court ruling. Legal ramifications regarding these reporting requirements may vary somewhat by state or locality; however, social workers must report incidents of abuse, neglect, or exploitation that are witnessed or described to them, and must consider taking steps to ensure the safety of a person threatened by a client.

Depending on the setting of the work, the mandate to break confidentiality can appear to be a significant obstacle to establishing a relationship of trust. For example, you may be concerned that clients will not share any information with you if told that you must call the authorities. Although this concern is legitimate, to represent yourself fairly, you must communicate your responsibility in that area early in the work. Informing clients of the limits of your confidentiality may affect the relationship in unexpected ways. Clients may understand that they need help from authorities. For example, consider the social worker who worked with a woman who struggled with substance abuse and had a difficult time keeping track of her five children, for whom she was the single caretaking parent. During the sixth meeting, she told the social worker that she felt sure she was grossly neglecting the younger children and that she often struck the oldest child "hard" when he "mouthed off" to her. After discussing this situation, she informed the social worker that she had finally confided the information *because* the social worker was a mandated reporter and she knew she could get help.

Social workers must be honest with clients about their roles and not assume that an adversarial relationship will evolve simply because of the mandated reporter obligations. Most clients who struggle with caring for their children want to be good parents. Even if the client is guarded about what she or he tells you (which you might expect), you still have significant opportunities to build a relationship. You can address the obstacles to the client's successful parenting (or caretaking for an older adult), model a genuine relationship in which you support the client's parenting competence rather than search to discover deficits, and build the foundation for further work. Adopting this approach does not suggest that the imminent safety of a child or adult should be compromised. Rather, it is a strategy of beginning to work in those countless scenarios in which there is concern but no clear mandate for legal intervention.

Privacy

Client privacy can be challenging for many reasons; for one, actions that some people consider invasive can seem caring to others. When negotiating with clients about the nature and parameters of the work during engagement, you and your client should address the issues of privacy and invasiveness. For example, a client who perceives your care and enthusiasm for the work may get the impression that you want to know everything about her or his life and become an active participant in it. If this is not your intention, you should make the client aware of this. Similarly, in a group setting, you will want to discuss plans for allowing individuals to maintain some information about themselves as private, while sharing pertinent information of their choice with the group. Being straightforward about your involvement and the parameters of privacy is important to avoid the client system feeling disappointed and perhaps even betrayed. For example, if your role involves monitoring client activities and you have established a mutual, client-driven relationship, you will likely be the initiator in situations that the client may not expect. For instance, if you have suggested that the client call you when she wants to meet again, but you then drop in unannounced "just to see how she's doing" (or to see how the children seem or if they have eaten that day), she may rightly perceive your visit as an invasion of privacy.

On a deeper level, as a practitioner in your particular role and setting, you might consider the extent of privacy to which clients are entitled. For example, in a residential or correctional setting, what is important to know? If your work involves a social control function, do your clients deserve less privacy? How does your agency's purpose affect the degree of privacy granted? These questions all relate to the importance of knowing yourself and the values, skills, and roles that you bring to the setting, and noticing how you (and/or your agency) may or may not be influenced by predominant social norms about the rights (or lack thereof) of clients.

Ongoing Evaluation

All through the engagement process, you have invested your energy and skills in establishing a solid initial connection with your client that will grow as your work together progresses. Another task you must undertake in this early stage, and throughout the work, is evaluation of the effectiveness of your efforts. Relationships can be easily misunderstood. You may be concerned that your client is hesitant to be as open as you would like, or that she or he seems uneasy in some way. Conversely, you may feel wonderful about the productive beginning of a professional relationship. In either case, it is important to find out how your client feels about your relationship and to make any needed changes in your approach that are indicated by your client's feedback. Although you are not likely to use formalized tools at this point, you will want to ask your client about how the work is going. For example, you might ask an individual client whether he or she is comfortable talking about the issues discussed thus far. How does it work for him that you are a Hispanic woman and he is a black man? Is the process of meeting with you similar to, or different from, what he thought it would be? If it's different, how does he feel about it? What can you do to be more supportive or clearer or helpful? For group work, you might ask the group for feedback about your role as facilitator or teacher. At the community level, seeking feedback from individuals and colleagues in the community as well as from committees can elicit important suggestions.

At the individual level, the process of continuously monitoring your work also applies to your engagement with the people who are significant in your client's life. You will want to validate your understanding of their role in helping to achieve your client's goals and to check for any misunderstandings. This strategy will be particularly important in situations that involve contentious feelings, such as the reluctant landlord considered earlier in this chapter. While landlords might assume that you will take sides with your client, you will want to take care not to alienate them from your client's goals.

CONCLUSION

Now that you have explored the importance of various aspects of engagement and building relationships in social work practice, you are well prepared to support your connection with your client system as you enter into the assessment arena of the work. Although you are not likely to have addressed all the issues presented here in the first interactions with client system, you have the framework to go beyond the initial connections and respond to an ongoing, dynamic association. You have taken care to anticipate the meaning for the client of your particular work together, and you will find that this dimension grows and is shaped by the nature of your shared activities and experiences.

One further caveat about skills: At first you may feel that thinking about and trying to use them interferes with your spontaneity and/or responsiveness. However, as you use the skills and develop your practice style, you will become much more relaxed in using and personalizing the skills, and develop your own style. While the engagement stage is often thought of as the beginning stage, engagement is a process that will occur throughout your work with client systems. You will continue to notice engagement dynamics, your own growth as a worker in establishing engagement, and how engagement assists your client as the work continues into the assessment, planning, and implementation of the work. In particular, the next step, assessment, will build on the relationship foundation that you build during the engagement phase.

MAIN POINTS

- The social worker's first and probably most important activity in the engagement process is careful listening to the client's situation and perspective. This activity requires the worker to initiate a skilled, purposeful dialogue that nurtures the relationship.

- Negotiating the purpose and direction of the work enhances the trust between the practitioner and client. For this process to be successful, the agenda must be open and must not contain any hidden aspects.

- Using the core relationship qualities of warmth, empathy, genuineness, and unconditional positive regard, along with strong interviewing skills, will assist the social worker to establish a strong, professional working relationship with the client system.

- Respecting the strengths and resilience of the client system has an enormous impact on the work. Respect for strength is critical in establishing the relationship, and must be pervasive throughout the relationship.

- Power and its relationship to social work practice present both obstacles and potential. The various sources of worker power, the power in client lives, and the power between the worker and the client are all part of the relationship. These sources should be recognized and articulated as explicitly as possible.

- Although client stories may appear to be private and are certainly unique, they can always be seen from the perspective of social justice and human rights. This perspective gives the entirety of the work meaning and frames your commitments.

- The worker and client should discuss the issues of confidentiality and privacy openly and directly, even if these topics make them uncomfortable. The client may not welcome some of these constraints on practice, but the worker owes it to her or him to be respectful and clear about them from the beginning.

- Evaluation of the work, including the relationship, should occur at all stages, beginning with engagement.

EXERCISES

a. Case Exercises

1. Go to www.routledgesw.com/cases. Select the Sanchez case and review the Engagement phase on the interactive case study of the Sanchez family and the tasks of that phase. With classmates, discuss the needs of the Sanchez family.

2. After completing Exercise #1, review the Client History, Client Concerns, and Goals for the Client for Alejandro Sanchez. Next, click on the "Explore the Town" (under "Case Study Tools") to review the neighborhood, review Alejandro's Critical Thinking Questions, and explore his Interactional Matrix. Consider the following scenario:

   Alejandro is one of your clients. He presents as melancholy although pleasant and respectful. He says he is "unhappy" and seems to carry with him an existential sort of sadness that relates to his family. He notes on the first interview that his father Hector was also 19 when he came to this country as an undocumented worker.

   With a classmate assuming the role of Alejandro, role-play for 10–15 minutes the scenario that you are meeting Alejandro for the first time. In the session, you attempt to engage Alejandro and begin an assessment. (The class-mate who is playing the role of Alejandro should prepare by reviewing his concerns and goals, as well as his strengths.) Other classmates should observe, and consult during the role-play as needed. Afterwards, the entire class can debrief by considering the following questions:

   a. Which attending skills were used? Which were not used?

   b. Were empathic responses used? What were they? What was most challenging to you about the use of empathy?

   c. Was the social worker in the role-play able to validate his feelings of unhappiness, and identify strengths and verbally share those strengths with Alejandro? Were there other strengths that were not mentioned? How can a social worker emphasize a client's strengths when the client is not receptive to hearing them?

   d. What was the experience of the student who played the role of Alejandro? Did the student, in character, feel that the social worker demonstrated specific listening skills? Which skills?

   e. What elements of Alejandro's experiences reflect social justice and human rights concerns?

3. Go to www.routledgesw.com/cases and, under the Sanchez family case, watch the videotaped interview with Emilia and the social worker. While watching the interview, note where in the interview each of the following skills were demonstrated:

   a. Open-ended question

   b. Closed-ended question

   c. Reframing

   d. Paraphrasing

   e. Attending

   f. Nonverbal communication

   g. Clarifying

   h. Summarizing

   i. Empathic communication

   What were the strengths of the interview and what could the social worker have done differently?

4. Go to www.routledgesw.com/cases and become familiar with the Riverton case file (under the "Assess" tab). Using the questions provided by Quick Guide 4, consider the scenario of a client coming to her first appointment with you at the Alvadora Community Mental Health Center when she is clearly drunk. In a one-two page paper, describe which of the questions might be more significant than others. Would you use some of the questions to guide your interaction? Why or why not?

b. Other Exercises

5. You are a social worker in a neighborhood community mental health center. You are awaiting the arrival of a new client, Jasmine Johnson (discussed in this chapter), who lives near the center. After you introduce yourself and she relaxes somewhat, she states, "My life is a mess; nothing I ever do is right; sometimes I think I can't go on."

   Using relationship building and interviewing skills, indicate how you would respond to her statement by giving a very brief verbal (one sentence or less if possible) or a behavioral example (if appropriate) if you were:

   a. Attending

   b. Responding nonverbally

   c. Responding with minimal verbalization

    d. Paraphrasing

    e. Clarifying

    f. Summarizing (make any needed assumptions about information provided prior to the statement above)

      What other skills do you think are important for this situation?

6. For these role-playing exercises, create groups of three students so that one is the client, one is the social worker, and one is an observer.

    a. Set a stopwatch for three minutes. The client tells the social worker a peculiar story. The social worker may not speak for three minutes. However, the social worker conveys nonverbally that the client is being heard. At the end of the three minutes, the observer provides feedback to the social worker, and the social worker and client share their perspective on the process. Each student should have the opportunity to play each role.

    b. Set a stopwatch for three minutes. The client will tell the social worker about a serious concern in their life. The social worker reacts in each of the following ways:

      1. Disinterested

      2. Inappropriate affect (forced smile, blank stare)

      3. Distracting behaviors (e.g., foot tapping, excessive gesturing, fidgeting, head nodding)

After three minutes, the client provides feedback regarding the process. Each student should have the opportunity to play each role.

7. Review the following case to prepare for a role-play: Gina is your 16-year-old female client at a local teen drop in center. Gina is usually talkative and outgoing with staff and other participants. Today you notice that Gina is sitting in the corner alone and she looks as though she has been crying. When you approach Gina and inquire about her day she wipes her eyes and says in a quiet voice, "I can't do this anymore. My parents are always fighting and I just can't take it. I am not going back there."

      For the role-playing, create groups of three students so that one is the client, one is the social worker, and one is an observer. Using the above case, the client begins the interview. The observer will buzz the social worker whenever he/she becomes aware of the social worker using the following:

    a. Excessive questions

    b. Closed-ended questions

    c. Jargon

    d. Leading questions

    e. Multiple questions

    f. Irrelevant questions

Each student should have the opportunity to play each role.

8. Using the case about Gina (provided in Exercise #7), the class instructor will take the role as Gina. Various students fulfill the role as a social worker. Each social worker will demonstrate appropriate skills for the interview. When the

helper has a point when he/she feels stuck, s/he may return to the class and the next student begins where the previous student left.

9. The instructor creates flashcards with various interviewing skills. For the roleplaying, create groups of three students so that one is the client, one is the social worker, and one is an observer. Using the case of Gina (provided in Exercise #5), during the interaction between Gina and the social worker, the observer randomly presents a card with a skill listed on it to the social worker, who must demonstrate the skill in the interaction. Each student has the opportunity to play each role. After a 10–15 minute role-play, discuss the degree to which the skills were utilized appropriately.

10. The instructor creates flashcards with various interviewing skills. For the roleplaying, create groups of three students so that one is the client, one is the social worker, and one is an observer. Using the case of Gina (provided in Exercise #7), during the interaction between Gina and the social worker, the observer selects the card on which the skill presently being used is written and creates a pile. At the end of the role-play (10–15 minutes), a pile of cards that state the skills that were used is compared to the pile of cards that state the skills that were not used. The client, social worker, and observer engage in discussion about how the skills were used, and whether any skills not used may have been appropriately used during the interaction. Each students has the opportunity to play each role.

CHAPTER 4

Social Work Practice with Individuals: Assessment and Planning

Barbara, a sixteen-year-old mother of a baby, is on public assistance, and lives alone in one room. She dropped out of school when she became pregnant, her family and the father of the baby have abandoned her, and her only social contact is a neighbor who works during the day. One afternoon the young mother, lonesome and depressed, went out for an hour and left the baby alone. The baby fell off the bed and cut his head on an object, seriously injuring himself. When Barbara returned home she took him to the hospital, where the doctor in the emergency room, suspecting child abuse (maybe neglect?), referred her to the child welfare agency.

Carol Meyer, 1993, p. 22.

Key Questions for Chapter 4

(1) How can I prepare for assessment with individual client systems? (EPAS 2.1.10(b))

(2) What are the evidence-based theoretical perspectives that I will use to guide my assessment and planning with individual client systems? (EPAS 2.1.6)

(3) What are the skills that I need to have to work with an individual? (EPAS 2.1.10(b))

(4) How can I ensure that I engage in appropriate professional and personal self-care activities? (EPAS 2.1.1)

IN HER CLASSIC 1993 WORK *ASSESSMENT IN SOCIAL WORK PRACTICE*, Meyer notes that the way social workers think about such situations as Barbara's, or the conceptual boundaries they apply to them, includes many dimensions. Consider your conceptual boundaries; as you read about Barbara, what first comes to mind? Do you see her as an unfit mother? As a lonely young woman? What are the major issues you see? What do you want to know more about? Where would an assessment begin?

This chapter considers assessment, the meaning-making process that helps prioritize relevant factors and leads to appropriate action. Assessment is a key social work practice skill that involves collecting information about the client system to determine strengths as well as problems (Jordan, 2008, p. 178). We will also discuss the history of assessment, the importance of the client's goals, and approaches and skills through the lenses of theory, diversity, and diagrams. We will assess resources and explore approaches for when resources are inadequate. The chapter will conclude with a review of the planning process and two areas of challenges.

Consider several questions about Barbara's situation that Meyer (1993) poses: Is hers a case of a mother and child, a teenager without any family, a teenager subject to the rules of social institutions? Would one focus on child abuse? Child neglect? Adolescent acting-out? A teen in need of guidance about romantic relationships? A single parent in need of family planning? Loss of family and social supports? Poverty? (p. 22)

The risk in this situation is in taking too narrow a focus and overlooking key issues. The task then becomes how to acknowledge the complexity of the situation and at the same time focus with enough specificity to intervene in a helpful way. The assessment process, therefore, attempts to identify and gain insight into the situation within the client's context. At the same time, the social worker will want to discover Barbara's goals and keep them central to the process.

When you, as the social worker, begin a relationship with Barbara (or someone like her) through the engagement process, you will also consider many of the preceding questions to shape your focus in the case. Highly interconnected with engaging the client system, the assessment and planning process has already begun as you have taken the time to listen to Barbara's story and consider its meaning. Putting her story into the context of the values of social justice and human rights frames your understanding. As part of your careful listening, you will gain insight into the way in which she views her life, her goals, and ways in which you might help her achieve those goals. You may have reservations about the obstacles that seem to be in the way of the goals and priorities held by Barbara, your agency, or even the legal system. Thinking through these issues, including Barbara's strengths, is the assessment process.

A BRIEF HISTORY OF ASSESSMENT

Since Mary Richmond published her pioneer work, *Social Diagnosis*, in 1917, assessment has been a critical component of social work practice. Many contemporary social workers still consider assessment to be absolutely central to practice because it guides the worker's focus and directs the intervention. However, some criticize the language used in assessment and the implications it carries about the social worker's expertise. The idea of assessment seems to suggest that the social worker has the power to define the client's situation and to impose that definition. In actuality, the assessment of the client's current situation is framed by the client's perception of events, attitudes, and potential outcomes. The social worker's role is to gather and organize the information so she or he, in collaboration with the client, can interpret the meaning and implications of the data. A look at how the profession arrived at the word *assessment* may provide some perspective on this debate.

Since Mary Richmond introduced the term, *diagnosis*, into social work practice, the concept has been a part of the profession's history. However, diagnosis now firmly connotes a medicalized understanding of disease, dysfunction, symptoms, and the authority associated with the declaration of an illness. As the concept of diagnosis is not rooted in a strengths-based perspective, social workers may find it challenging to use the diagnostic term.

Contemporary social workers have adopted the term assessment to represent a more complete understanding of the client's context, one that focuses on strengths and resources as well as challenging areas. Assessment today is also considered a process by which clients can partner with a professional to make informed decisions about the work that can be done together. Client resources include the environment in which the client lives, and the history, culture, and traditions embedded in the life experience of the client system. Treating assessment as an act of client-focused discovery shifts the focus from professional expertise and analysis toward client definitions of the parameters for work. This process is a source of empowerment for clients.

A strengths-based approach is consistent with social work's systemic emphasis on the interface between the client and the environment. A strengths-based systemic perspective invites social workers to examine the whole person, whose many dimensions are never recognized in psychiatric diagnosis. Social workers then stress the importance of dialogue as a way of looking fully at client situations, considering their significance, hearing what goals clients have, and understanding the ways in which clients believe they can achieve these goals. This larger view of assessment assumes that the client, rather than the social worker, directs the decisions about the substance of the work. In essence, assessment has evolved into an integrative collaboration between the social worker and the client system from which the intervention flows (Jordan, 2008). In a scenario such as Barbara's, consider the complexities involved with assessment. As a result of her involvement with child protection services, she may be mandated to participate in services. Working with a client who may be reluctant or resistant requires sensitivity and possibly a longer period of time to establish rapport and build trust.

WHERE DOES THE CLIENT WANT TO GO?

Having established the importance of dialogue between the social worker and client system in the process of gaining a holistic picture of the client's situation and the client's central role in decision-making, initiating assessment then becomes the next step. Many assessment processes begin with long, detailed social histories; these histories have advantages and disadvantages. On the positive side, asking clients to talk about their life events can reveal important issues, such as their great resilience in the face of childhood abuse, that might not emerge right away but that are important to fully understand the client's situation. On the negative side, long accounts can seem intrusive, irrelevant, or even judgmental to clients as they cover personal aspects of a client's life when, for example, a single parent came only to talk about a child care allowance so she can attend a class. Such histories may also seem to emphasize previous difficulties or situations that the client would prefer to leave in the past. Finally, they can appear to be driven by the social worker or agency or even the profession itself, because they seem disconnected and remote from whatever sense of urgency the client brings to the first interaction.

For example, Barbara may find an extensive history-taking process invasive and beside the point, when she is being investigated by child protection services and may be interested only in getting her baby back. At the same time, it is possible to imagine that such a process can reveal aspects of Barbara's life that could assist the social worker in helping her to achieve her goals in her situation through more or better-placed supports. The major requirement in this situation is that you, as the social worker, have effective communication and relationship skills in order to make the client feel comfortable and respected. Providing information to Barbara regarding the reasons for the questions you are asking and maintaining a flexible and patient approach may enable her to comply with your requests. Whether the history is lengthy or brief, it is critical that the work center on the client's goals. The most detailed, painstaking social history will be of little use if the history—and not the goals of the client system—becomes the driving force of the work.

IMPLICATIONS OF THEORETICAL PERSPECTIVES

Despite efforts to minimize prejudices and personal biases about the client's situation, assessments are not neutral gatherings of the facts. The nature of the facts collected and questions asked suggests, at the least, your theoretical biases. For every area on which you focus, there are others on which you do not. For example, if you focus on Barbara's relationship with her baby because you see her as "case of mother and child," and you do not consider her experience as a child herself, you are choosing not to explore an area that could influence the work you do with her. Likewise, if you stress her history of delinquency but not of sexual abuse that you learn about, you are adopting an approach that will affect the nature of your work with her.

In both scenarios, you take a specific approach that is a result of your judgment about what is both relevant and important. This judgment is usually influenced by many personal attributes (for example, who you are, what you believe about the nature of people, and where you work) and by the type of information you believe helps make a story understandable. The theoretical perspective and its assumptions will also influence what you consider useful here, as will the context and function of the agency. All of these components shape the kinds of questions you ask and, therefore, the information you receive. Framed as it is by your perspectives, the client's assumptions, and your own interpretation of a situation, the assessment process is never unbiased.

Classic Theories

The theoretical perspective that you adopt strongly influences the assessment, the client–social worker relationship, and the subsequent work. Three classic theories that have throughout the history of the social work profession influenced assessments are: psychoanalytic theory, attachment theory, and cognitive theory.

Psychoanalytic Theory   Based primarily on the writings of the Austrian physician Sigmund Freud, **psychoanalytic theory** maintains that the unconscious is at the root of human behavior. Freud identified three structures that interact to determine human behavior: the *id, ego, and superego*. Each structure has a distinct function. The id is the repository of unconscious drives such as sex and aggression. In contrast, the ego is the managerial, rational part of the personality, which mediates between drives and perceived obligations. Finally, the superego serves as judge and conscience.

A social worker who believes that inner, unconscious motives and explanations determine the client's choices would orient an assessment toward interpreting the client's unconscious wishes. For example, if Barbara repeatedly describes herself as a "loser," the worker's assessment would likely be directed toward discovering the rewards and gratification she receives from perpetual failure. These rewards might include more concern from previously disinterested parents or protection from the high expectations of others. The first models of assessment were rooted in psychoanalytic theory, but as psychoanalytic theory gave way during the 20th century to more contemporary models based on evidence, approaches began to shift away from psychoanalytic origins (Jordan, 2008).

Attachment Theory   Looking through another lens, the social worker might first want to address Barbara as both a person who was parented and is now parenting. Your assessment might focus on attachment theory, which is currently becoming increasingly pertinent, especially in child welfare work. **Attachment theory**, originally proposed by U.S. psychologist John Bowlby (1969; 1982), holds that very early bonding occurs between a mother and infant and subsequently plays a critical role in the child's future capacity to provide and sustain opportunities for her or his own children to attach. Most of this bonding activity occurs within the first two years of life, and it creates the foundation for the health of all future relationships.

A social worker who uses attachment theory focuses on Barbara's relationships with her early caregivers and the way in which these relationships may have contributed to her current struggles. An assessment of the attachment between Barbara and her child might include experimental observation to identify behavior patterns both child and mother demonstrate when a stranger is introduced into the scene. In stressing these relationships of parental bonding, the social worker would de-emphasize Barbara's other relationships in the environment.

Cognitive Theory   In contrast to psychoanalytic and attachment theories, if the theory guiding the work emphasizes the importance of **cognitions**, or thoughts, then the social worker will see things differently. This approach is consistent with **cognitive theory**, which asserts that thoughts largely shape moods and behaviors. In this approach the focus in assessment is more likely to be on what Barbara thinks about herself, what she subsequently thinks about the way in which she would like to think and act differently to achieve her goals, and the ways in which her thoughts and feelings influence her behavior (Beck, 2011). The cognitive approach assumes that people are thinking beings and if they change their thinking, their emotions will also change. Further, one's feelings influence both specific behaviors and general approaches to life. For example, because Barbara felt depressed, lonely, abandoned, or hopeless, she used poor judgment in leaving her baby. The social work intervention with her might involve helping her appreciate her assets more fully, which in turn would help her feel better about herself and lead her to make caretaking choices that would be safer for her baby.

You may find the assumptions in cognitive theory more similar to your own than those of the psychoanalytic perspective or more hopeful than those concerning attachment. Nevertheless, they are still predicated on assumptions that make a significant difference in the way that the social worker perceives, relates to, and works with the client. The point is not to reject every theory: We are all guided by theories, formal or not. Rather, the point is to recognize that none of them alone is "truth" and that you may use a combination of theories to frame the approach you will select for working with a client.

Contemporary Theoretical Perspectives

The three classic theories illustrate the ways theory can influence social work practice. Next we will look at the implications for assessment approaches and skills of two contemporary approaches, the strengths-based perspective and narrative theory. Like the classic theories, contemporary theoretical approaches influence assessment practice behaviors as the focus is shaped by the worldview inherent in the perspective.

The Strengths Perspective One of the four major perspectives of this book, the strengths perspective is widely discussed in the social work literature on assessment. Strengths-based practitioners uncover client assets, resources, goals, and dreams. A strengths approach also examines the potential of the environment to nurture and support the strengths of individuals. The assessment does not focus on a history of failures but on successes, resources, and goals for the future. The strengths perspective was developed for working with clients receiving mental health services but has since been used with many populations, including families and children, youth at risk, older adults, residents of economically distressed communities, and persons experiencing addictions. Saleebey (2009, pp. 109–111) identifies two elements for strengths-based assessments that are useful with all populations:

The social worker meets the client in the struggle: People have real struggles. There can be a fine line between supporting the positive dimension of a client's personality, skills, or accomplishments and framing the client experiences in terms of negative dimensions (e.g., grief, terror, sorrow, or discouragement). For this reason, it is critical to the relationship that you validate the pain that clients feel. In starting where the client is and listening to the client relate her or his concerns, painful as they may be, it is possible to uncover evidence of potential strengths on which you can build to develop a strengths-based intervention.

The social worker stimulates the discourse and narratives of resilience and strength: A narrative approach can be helpful in eliciting the client system's strengths. This "reframing" is dependent on the social worker supplying the words to help articulate the client's strengths, being affirming, and emphasizing possibilities. With the use of supportive questioning (i.e., questions that emphasize a positive aspect of the situation or a recollection of past successes), you can return the focus to client strengths. Even in the face of repeated, entrenched stories of trouble and pain, you can help clients to recognize their inner capacities for survival and learn the language of strengths in order to uncover a small seed of hope. Quick Guide 6 provides examples of supportive questions in various situations.

Another approach for assessing strengths incorporates guidelines that focus on understanding the client's perceptions of the situation (Anderson, 2013). See Exhibit 4.1 for these guidelines. This set of guidelines are intended to empower the client to share their life stories, acknowledge their strengths and wisdom, and develop greater insight into the complexities of their lives and current situation (Anderson, 2013, p. 187).

In this approach, the assessment includes a two component model in which the social worker first explores a series of questions with the client to define the problem situation (Component 1). The questions will not only help to identify the client's strengths, but also their life experiences (both positive and challenging), and strategies for coping with adversity. Gathering this comprehensive perspective on the client's identity can aid both the social worker and the client in formulating a perception of the client in terms of strengths, versus deficits (Anderson, 2013, p. 191–192). Following this exchange, the social worker and client then together

QUICK GUIDE 6   EXAMPLES OF STRENGTHS-BASED SUPPORTIVE QUESTIONS

Client Situation	Possible Supportive Questions
Middle-aged man who was recently laid off from his job and fears he will not be able to find new employment	"How did you approach the search that resulted in you being offered the position that you just left?" "Who might be a good person or group with whom to network about employment opportunities?"
Single mother with four young children who is feeling overwhelmed with her life	"How have you managed with the demands of your kids?"
Fifteen-year-old teenager who is stressed by the daily arguments with her mother	"Can you recall times when you and your mother had fun together?" "What are those times?"
75-year-old male who is experiencing depression	"How have you handled other challenges in your life?"
33-year-old combat veteran who is experiencing post-traumatic stress disorder and is having difficulty adjusting to civilian life	"What were your best assets as a soldier?" "How might those transfer to this new chapter of your life?"
Middle-aged woman whose mother has dementia is struggling with the decision to seek long term residential care for her mother. She feels that placing her mother would be a betrayal of the deathbed promise she made to her father to always keep her mother with her.	"How has your family traditionally handled difficult decisions?" "If your father were here to help you make this decision, what might his response be?"

EXHIBIT 4.1

The Strengths-Based Perspective in Assessment

Guidelines for a strengths-based assessment include:

- Document the client's story.
- Support and validate the story.
- Honor the client's self-determination.
- Give pre-eminence to the client's understanding of the facts.
- Discover what the client needs.
- Discover uniqueness.
- Reach a mutual agreement on the assessment.
- Avoid blame and blaming.
- Assess; but do not get caught up in labels (i.e., diagnoses).

Source: Anderson et al., 2009, pp. 186–188

---

identify the relevant strengths and obstacles that the client brings to bear. These strengths and obstacles can be charted on the grid in Component 2 to provide visual representation of the assessment results. See Exhibit 4.2 for this model.

In Anderson, Cowger, and Snively's two-component model for assessing client strengths, Component 1 is a process by which the worker and client define the problem situation and clarify how the client wants the worker to help. Component 2 is a graphic representation of the analysis of the problem defined in Component 1 (i.e., the assessment). It invites the worker and client together to chart strengths and obstacles in each of the four quadrants. Quadrant 2 is highlighted and may contain subcategories relating to cognition, emotion, motivation, coping, and interpersonal.

EXHIBIT 4.2

Two-Component Model for Assessing Client Strengths

COMPONENT 1

Defining the Problem Situation: Getting at Why the Client Seeks Assistance

- *Brief summary of the identified problem situation.* This should be in simple language, straightforward, and mutually agreed upon between worker and client.
- *Who* (persons, groups, organization) is involved, including the client(s) seeking assistance?
- *How* or in what way are participants involved?
- *What* happens between the participants before, during, and immediately following activity related to the problem situation?
- *What* meaning does the client ascribe to the problem situation?
- *What* does the client want with regard to the problem situation?
- *What* does the client want/ expect by seeking assistance?
- *What* would the client's life be like if problem was resolved?

COMPONENT 2

Strengths	
Quadrant 3	**Quadrant 4**
Social and Political Strengths	Personal and Interpersonal Strengths
• Civic engagement	• Cognition
• Social networks	• Emotion
	• Motivation
	• Coping
	• Interpersonal
	• Physical and physiological
Quadrant 1	**Quadrant 2**
Social and Political Obstacles	Personal and Interpersonal Obstacles
• Systemic oppression	• Powerlessness
• Legitimization of domination and violence	• Violation
	• Intense suffering
	• Isolation
	• Secrecy/silence
Obstacles	

Environmental Factors

Help Seeker Factors

Adapted from Anderson et al., 2009, p. 192

The assessment process helps clients identify their own strengths, use the resources in their environment, and tell their story about the problem. The approach is also explicitly political in that it recognizes and articulates the power relationships that clients experience and/or in which they participate. In addition, the strengths approach encourages a complex view of the environment as a source of both resources and obstacles. Although this approach seeks to identify obstacles, they are not the primary focus of the assessment.

Narrative Theory   Recall from Chapter 1 that narrative theory, influenced by postmodern thought, focuses on the client's story as the central component in the work. The primary interest for social workers using this approach is in discovering the stories of the people who consult them and in helping them to "re-author" those stories if they wish. A story is defined as events linked in sequence across time according to a plot (Morgan, 2000, p. 5). Narrative theory is consistent with the mission and values of the social work profession as it "focuses on empowerment, collaboration, and viewing problems in social context" (Kelley, 2008, p. 291). Narrative practitioners also subscribe to a person-centered approach to working with clients that embraces the essentials of social justice, a cultural context, and collaboration with and respect for the client system (Kelley, 2008).

Narrative practitioners are interested in helping the client to broaden or "thicken" her or his story. For example, Georgia is a woman who experienced intimate partner violence. Georgia developed a negative self-story response to a pattern of abuse that was perpetrated by her partner. Whereas her story originally was that of a strong and competent young woman, it slowly began to erode, reflecting increasing doubt and finally wholesale dejection as she adopted the persona of an unworthy human being. It became a "thin" story in that it lacked complexity, reflecting only her self-rejection. Georgia's perception of herself as unworthy depicts only one dimension of her person (i.e., self-rejection) while ignoring her strengths, life lessons, and capabilities.

In addition to seeking a more in-depth story, the social worker recognizes the importance of the broader social context of storytellers' lives. In one sense this is a story just about Georgia and her partner. In a broader sense, however, it is also about a pervasive social phenomenon that results in the deaths of thousands of women every year. The intervention with Georgia is to help her rewrite the thin story of worthlessness to one that more accurately reflects her talents, attractiveness as a human being, and competence. The goal of assessment is to discover the alternative story the person seeking help wishes to author and guide the development of a person-centered intervention in which specific tasks and activities will be identified that will enable Georgia to recapture her talents, attractiveness, and competencies.

In Barbara's situation, the social worker using a narrative approach would first want to hear her story. How does she fill the day? What is it like to be mother to this baby? When are the best times? When did her loneliness first interfere with her life? When is she able to conquer it? Who would say she is a good mother?

These are a few of the most basic ideas of narrative theory and ways this approach is used to assess what needs to be done. As you can see, narrative theory differs from many of the more classic theoretical perspectives, but is similar in that the story is the emphasis of the assessment process. Narrative approaches are largely driven by the person seeking assistance, although you, as the social worker, contribute your ideas as well. A narrative approach fits well with the strengths perspective as it assumes that people are the experts on their lives and that they have multiple talents, values, beliefs, and skills for improving their lives. We will look more closely at the applicability of narrative approaches in interventions with families and groups in Chapters 6–9.

Solution-Focused Approach   As introduced in Chapter 1, the solution-focused approach is another example of postmodern-influenced practice in which the client is recognized as the expert on her or his life as well as ways to affect change. Similar to a narrative approach, solution-focused interventions build on a strengths perspective and use solution-related language to empower client systems toward self-initiated change (Lee, 2009). Three primary assumptions and principles guide practitioners using this approach: (1) language is the mechanism for clients and social workers to understand the meanings of the client's life and actions; (2) because clients are the experts on their own lives, they have both the resources and the answers that will guide the solutions; (3) clients, not the professionals, function as the "knowers" within the intervention process and are therefore in the best position to create the solutions.

Solutions to the issues that clients bring to the social worker are developed in stages and created using a series of questions. While posed by the social worker, the content of the questions themselves are rooted in the story that the client has shared during the assessment phase of work. Following on the information presented by the client, the social worker asks questions to fill in historical gaps, identify previous successes and solutions, and explore the client's perceptions for outcomes. The stages of solution-building include: (1) description of the problem; (2) development of well-formed goals; (3) exploration of exceptions; (4) provision of end-of-session feedback; and (5) evaluation of client progress (De Jong & Berg, 2013, pp. 17–18). The questions posed to the client are aimed at eliciting the client's self-evaluation of the meaning of her or his life events and a futuristic exploration of possibilities.

While similar in their strengths-based orientation, client-centered empowerment approach, and commitment to collaboration, narrative and solution-focused approaches are distinct from each other. Proponents of the narrative approach emphasize the importance of "not knowing" and listening for unique outcomes to the client's presenting concerns, while solution-focused adherents delve into the possibility of "exception" questions (i.e., questions aimed at identifying instances in which the problem did not exist) (Kelley, 2008). In pursuing information regarding "exceptions," the social worker can ask the client to describe her or his life before the crisis that brought her or him in contact with the social worker. Gaining information regarding the client's previous coping skills can provide insight into the potential intervention to be developed. With their similarities and differences, both approaches can be successfully integrated with each other as well as coupled with other theoretically driven models (e.g., cognitive behavioral interventions) (Kelley, 2008). Combining approaches requires a skillful and knowledgeable practitioner who is competent in the areas being combined.

Theory and Evidence Matters

The contrast between the *traditional or classic theories* and more contemporary models provides just one example of the ways in which theoretical orientations help to shape and define the assessment processes and skills as well as the roles of the participants. As mentioned, the idea is not that one theory is better than another, but rather that theory matters in assessment and should be consistent with the social worker's basic assumptions and beliefs concerning people.

While theory guides and informs the development of social work practice knowledge, values, and competence, social work practitioners can benefit from using evidence to determine the appropriate skills, competencies, and behaviors to be applied to the social work intervention. In fact, the Council on Social Work Education (CSWE) Educational Policy and Accreditation Standards (2012) calls for social workers to engage in research-informed practice and practice-informed research. **Evidence-based practice** (EBP) is a process that aids practitioners to "systematically integrate evidence about the efficacy of interventions in clinical decision-making" (Jenson & Howard, 2008, p. 158). Within evidence-based practice context, social workers systematically determine, use, and assess interventions based on the consideration and integration of research findings, clinical expertise, client system preferences, values, and presenting issues, and the values and circumstances that will best serve the client system (Thyer, 2009). The five-step process of an evidence-based approach includes: (1) converting practice information needs into answerable questions; (2) locating evidence to answer the questions; (3) appraising the evidence; (4) applying evidence to practice and policy decisions; and (5) evaluating the process of using evidence to guide the practice intervention (Jenson & Howard, 2008).

Using the five-step process evidence-based approach to work with Barbara, where would you begin? Consider the following (Carter & Matthieu, 2010; Jensen & Howard, 2008):

1. Convert practice information into questions about background information, effectiveness of intervention or policy

    * *Information*—Barbara jeopardized her child's safety by leaving him alone.

    * *Question*—What are the risk factors most associated with teenage parent neglect?

- *Information*—Barbara wants to be a good parent.

- *Question*—How can social workers provide assistance to teenage parents at risk of child neglect?

- *Information*—Barbara appears to be experiencing depressive symptoms.

- *Questions*—What assessment tools are most appropriate to assess for depression among teenage parents? What are the most effective intervention strategies to lower depressive symptoms in teenage parents?

2. Locate evidence to answer the questions

- Conduct a search of research and literature on child development, parenting, and assessment of and intervention with depression.

- Consult with social work practitioner(s) who possess(es) knowledge and expertise in the areas of interest and ask for their options of best practices in the areas.

3. Appraise evidence

- Assess the quality of the best available evidence, including the appropriateness of the research design for the question, sponsorship of the research, the similarity of the research subjects to your client system, and other factors.

- Considering the evidence compiled from literature review, your judgment and practitioner discussion and experience, and the client's goals and social context, develop a plan for assessment and intervention.

4. Apply evidence to practice decisions

- Implement assessment and intervention plan.

- Determine the appropriateness of the research methods for the client system situation being explored while maintaining a client-centered approach, cultural humility, and applicability to the practice approach, setting, and client system.

5. Evaluate the process of using evidence to guide practice intervention

- At each step in the change process, conduct an evaluation of the efficacy of the assessment and intervention process. Questions to consider:

- Was the assessment accurate?

- Was the intervention effective?

- Was the information compiled from the literature and practitioner helpful?

- Did ethical questions emerge in the process? How were those resolved?

- What would you have done differently? Why?

Still an evolving area for practitioners, the adoption of an evidence-based practice approach can direct the practitioner to information regarding assessment, intervention, and evaluative strategies and other key resources needed to effectively work with a client system. An increasing number of social workers are accessing evidence to inform and guide their practice, however, students must continue to focus on gaining knowledge and skills to competently evaluate the myriad of evidence that is available from a variety of sources (Pope, Rollins, Chaumba, & Risler, 2011). Effective interventions also include the use of clinical judgment and skills that are required for competent and ethical social work practice. When asked in a recent survey, most social workers were reported to use evidence in a "conceptual" way, that is, to help them better understand information as opposed to using it "instrumentally" (applying directly to practice) (Wharton & Bolland, 2012, p. 162). Each client circumstance is unique and each situation must be assessed individually. Such an approach allows the practitioner to incorporate the complexities of client situations into any practice interventions implemented.

IMPLICATIONS OF DIVERSITY AND CULTURE IN ASSESSMENT

Although theoretical perspectives play a major role in assessment, other lenses are also crucial as well. For example, the impact of diversity and culture on assessment is central. Keeping an emphasis on diversity and cultural influences impacts the approaches and skills this focus implies are critical for competent social work practice. To be a culturally competent practitioner, you must develop a set of practice behaviors—knowledge, skills, values, and actions—that prepare you to address both affective and cognitive domains, and are measurable so you may evaluate your practice (Simmons, Diaz, Jackson, & Takahashi, 2008). Social work as a profession is an institution of culture, affected by the same pressures and forces that influence other aspects of our society's culture. Because of the shifting patterns of diverse populations in U.S. society, social workers must develop appropriate competencies and practice behaviors for different cultures. Consider these diversity-related questions as they may impact your ability to competently work with Barbara: (1) What is Barbara's background, including race, ethnicity, family background, religious/spiritual beliefs, and education?; (2) How might Barbara's experiences related to aspects of diversity impact her knowledge of parenting?; (3) How does your background impact your knowledge of Barbara's ethnic, racial, and cultural background?; (4) What information do you need to work with Barbara in a culturally competent way?; and (5) What culturally competent practice behaviors will be appropriate for working with Barbara?

The social worker who chooses to embrace a strengths-based and or narrative approach must be committed to developing an approach that affirms individuals' strengths and respects their culture. The social worker must also acknowledge the implications of such a commitment on the assessment process, particularly in the areas of: (1) cultural competence; (2) connecting with the spiritual dimensions of culture; and (3) global connections. We will explore each of these areas of cultural competence.

Cultural Humility

Embracing a systemic, strengths-based perspective means including meaningful components of the client's culture in the assessment process. (NASW, 2007b). Known as **cultural competence** (NASW) or "cultural humility" (Clowes, 2005), this practice behavior is defined as "the process by which individuals and systems respond respectfully and effectively to people of all cultures, languages, classes, races, ethnic backgrounds, religions, and other diversity factors in a manner that recognizes, affirms, and values the worth of individuals, families, and communities and practices and preserves the dignity of each" (NASW, 2007b, p. 10). Cultural competence requires that social workers are aware of their limitations and have respect for the unique culturally defined needs of others. As you have learned, assessment involves discovering what is important to the client system, those cultural influences that shape her or his values, how they have affected the client's experience, and how the client's perspectives differ from the worker's.

Clearly, the social worker's ability to approach another culture in this way is predicated upon her or his self-knowledge and knowledge of her or his culture (Schultz, 2004). That is, workers must understand their cultural influences, examine the influences of their own culture on their work, and understand the role culture plays in all of our lives.

An effective assessment process depends on cultural competence; you cannot simply assess the degree to which your client differs from you. Your understanding of the issues that the client brings and the way they are shaped by culture impacts the way you will perceive the presenting "problem." For example, when your client describes communicating with the spirit of his deceased father, you might wonder if he is demonstrating psychosis or hearing voices, unless you have a clear understanding that this kind of spiritual communication is part of his Latino cultural celebration of the Day of the Dead which encourages reaching out and encouraging the spirit of the deceased to come for a moment to hear the prayers and love of the family for them.

Culturally Competent Practice Behaviors

Developing the skills necessary to conduct a culturally competent assessment takes time and effort. Social workers need skills for culturally competent assessment, including the ability to ascertain, through talking to the client, the meaning of their culture, language, cultural norms, and behaviors. You must also be able to see these attributes as strengths on which to build a culturally meaningful intervention (NASW, 2007b).

Social workers who adopt a strengths-based culturally competent model of assessment and planning begin the relationship by initiating friendly yet purposeful conversations. Rather than focusing on long social histories, the worker may consider some aspect of the client's cultural frame that is puzzling or particularly interesting and ask about it through a "global question." For example, consider the situation you encounter as you are working with your client, Emilia Sanchez (refer to www.routledgesw.com/cases for information), a 24-year-old Mexican-American woman with a history of substance abuse and the mother of a four-year-old, Joey, who lives with her parents, Hector and Celia Sanchez. Joey has lived with his grandparents for most of his four years and, while he knows that Emilia is his mother, he considers Hector and Celia as his parents. Emilia has just learned that Hector and Celia are going to take legal action to officially adopt Joey. She is furious and has come to you for your help in preventing the adoption from going forward. As a culturally competent practitioner, you inquire about Emilia's cultural background and work to understand the impact of her background on her beliefs and current situation. As the client/cultural guide answers your questions, you remain highly attuned to the language she uses and inquire about **cover terms**, which are expressions and phrases that seem to carry more meaning than the literal meaning the words would suggest. For instance, your client might respond to the question about what her family thinks with a comment like, "I don't care what those hypocrites think about me; I will never be the person they want me to be!" In this case the cover terms are "hypocrites" and "person they want me to be," because they seem to have a cultural relevance to how she understands her social location.

Although extensive cultural interviewing may seem removed from the main issue, understanding the impact of the client's culture influences the overall success of the work. Of particular importance is the need to identify early on those culturally-related areas that have the potential to become barriers later in the working relationship. For example, given what you have discovered about Emilia, you would make a big mistake if you began by suggesting a family conference with her parents.

Connecting with the Spiritual Aspects of the Client System

In recent years the social work profession has placed greater emphasis on the value of helping people define what gives their life purpose and meaning (see, for example, Hodge, 2005a). Insofar as this effort might be seen as a spiritual quest, it can encourage people to find the most sustaining areas of their lives. Such areas of meaning may be found in religious practices, outdoor activities, and in social connections like volunteering in a hospice, to name only a few. They can also be incorporated into both the assessment and action stages of the work in meaningful ways. For example, your client's volunteer work may be identified as evidence of her or his value to the community, which in turn might reduce her or his sense of isolation. Facilitating a client's connection with her or his spiritual side is likely to be a lasting and significant contribution of the social work intervention.

Global Connections

Social workers have the opportunity to work with a diverse and often international client population. Having competencies in internationally focused social work practice is essential. With the number of foreign-born international migrants currently over 13 percent of the U.S. population and growing (Grieco et al., 2012), social workers may work with individuals who have come to live in the U.S. as non-immigrants (e.g., visitors, students, and temporary workers), legal or undocumented immigrants, or refugees who fled political or religious persecution. At the individual practice level, social workers can work with international migrants in a number of different areas, but the following areas generally encompass many of the interventions social workers will have with international clients: health, mental health, family conflict, language and education, economic well-being, and interethnic relations (Potocky, 2008, p. 445). Social workers can also work with individuals in foreign countries through casework, during times of disaster, and international adoptions (Healy & Hokenstad, 2008).

Whether you are working with clients in your own or another country, you can build on your cultural competence and humility to individualize your skills to the cultural characteristics of the client . . . A strengths-based approach is particularly appropriate when working with a client who has immigrated to the U.S. Given the courage and coping skills required for permanently leaving one's home country, there will be considerable strengths to draw on when aiding the client in their social and cultural adaptation to their new life.

Working in an international area can be a learning experience for both you and your client. To begin your journey toward becoming a culturally competent practitioner, you must gain cultural awareness first about your own cultural heritage, identity, and belief systems (Lum, 2008). Quick Guide 7 offers suggestions for beginning your own cultural heritage journey. After exploring your own cultural make-up you can then begin to gain knowledge and develop skills to work with an international client system. Learn as much as possible about your client's heritage, but exercise caution about making any assumptions about the client's culture, particularly in the areas of beliefs, knowledge of you and your professional value system, language proficiency, familiarity with their new home, or openness to working with a helping professional.

While not presuming to understand the individual client's life experience or goals is important in working with any client system, it is critical with a client who has relocated to a new country. Some clients may have experienced trauma that resulted in their need to flee their country, trauma during relocation, or challenges adjusting to the new culture and environment, or they may have concerns regarding their legal status in their new country. Building rapport and trust is an important element of the social work relationship that may be challenging for a client who is unfamiliar with customs, language, legal issues, or the role of helping professionals. Strategies for developing a trusting working relationship may include: outlining

QUICK GUIDE 7   MY CULTURAL HERITAGE JOURNEY

The cultural heritage journey of both clients and social workers impact their perspectives of life and work. Consider the following:

- Where do my family's roots begin?
- If my family emigrated to the United States, how long has each parent been living here?
- If I am in an adoptive family, what do I know about my biological family's journey to the U.S.?
- What are my family's traditions that have been passed through the generations?
- Do my family's traditions relate to religion or spirituality, holidays, rituals, and or significant events for the family?
- What values have I learned from my family—may be related to ethics, wealth, religion and spirituality, race, ethnicity, health, education, sexual orientation, and image?
- Have I challenged any of my family traditions or values? If so, what challenges have presented themselves and how have I responded?
- What significant life experiences have shaped my view of the world, specifically people who are different from me?
- How will my cultural heritage impact my social work practice?
- Are there changes or additions that I would like to make related to my traditions? How might I start?

your role, boundaries, and limitations as a social worker; asking questions about the client's culture, country of origin, traditions, and beliefs (being mindful not to be overly intrusive); sharing information about the culture of the agency and community, particularly as it relates to the provision of social and or health services; and creating an environment in which the client feels comfortable asking for clarifications regarding the receipt of services.

Approaching the assessment and planning process with a repertoire of strategies will serve both the client and you well. Through research and consultation, arm yourself with culturally appropriate knowledge and skills, be open to learning from the client by asking for the client's help and guidance in understanding her or his culture and experience, and tap into the strengths the client brings.

SKILLS FOR ASSESSMENT AND PLANNING

At this point in the assessment, the social worker will begin to form a view of the future that includes the client's vision and the social worker's own opinions about how they might achieve it. This creates a shared vision between client and social worker. In developing this vision, social workers will help clients articulate the kinds of changes they want to make. The questions that social workers ask will affirm that change is possible and that it can be shaped in a way that makes the client's life better. Building on the foundation of strengths, narrative, and solution-focused perspectives, skills for assessing and planning with the individual client

© fatihhoca

emphasize a collaborative, client-focused approach. Skills are shared here that high-light each of these approaches, but first, consider the unique aspects of each approach.

Strengths Perspective

Commitment to being a strengths-based social work practitioner involves begin-ning the social work intervention from that perspective. Establishing a climate during the engagement, assessment, and planning phases in which client strengths are fully explored and acknowledged creates an environment in which both the social worker and the client can identify the strengths the client brings to the inter-vention, those resources that can be mobilized, and those that can evolve.

Earlier in this chapter, the elements of a strengths-based assessment were high-lighted. Consider now a grouping of specific strengths-based questions that you can ask to engage the client in the assessment process. Saleebey (2013) provides a grouping of questions that are designed to elicit information from the client so that

QUICK GUIDE 8 TYPES OF QUESTIONS FOR DISCOVERING STRENGTHS

Survival Questions:
How have you managed to survive this far, given all the challenges you have had to contend with?
What have you learned about yourself and your world during your struggles?

Support Questions:
Who are the special people on whom you can depend?
What did they respond to in you?

Exception Questions:
When things were going well in life, what was different?
What parts of your world and your being would you like to recapture?

Possibility Questions:
What are your hopes, visions, and aspirations?
How can I help you achieve those goals?

Esteem Questions:
When people say good things about you, what are they likely to say? When was it that you began to believe that you might achieve some of the things you wanted in life?

Perspective Questions:
What is your perspective on your current situation?
How would you describe your current situation to others?

Change Questions:
What thoughts do you have about ways your situation could change?
What things have worked well for you in the past? How can I help?

Meaning Questions:
What beliefs do you hold above all others? What gives you a sense of purpose? What are the origins of your beliefs?

Adapted from Saleebey, 2009, pp. 102–103.

both the client and you can work together to identify strengths in order that they may be transformed into planning for an intervention. Quick Guide 8 contains eight questions that can provide the basis for an assessment on which you and your client can collaborate on the development of an intervention plan.

Narrative Theory

Identifying and building on client strengths provides the basis of a narrative approach to assessment. Another of the empowerment-based practice approaches, narrative approaches are distinguished from other similar and overlapping approaches (most notably, solution-focused and strengths-based) in the way in which the perceptions of clients are elicited and interpreted. Narrative practitioners first help clients to share their stories, or "deconstruct," and then through expanding and externalizing perceptions and meanings of the client's words, the stories are "reconstructed" to provide a broader, more effective approach to functioning (Kelley, 2008).

Narrative social work practice may be viewed as a three-stage approach, encompassing: client engagement; exploring and deconstructing pre-conceived client stories and assumptions, and re-authoring new, more empowering stories (Roscoe, Carson, & Madoc-Jones, 2011; Roscoe & Madoc-Jones, 2009). Within a narrative framework, the social worker completing the assessment and planning stages of work will engage the client in reconstructing her or his reality into a new reality. Using a series of strategies and questions, the social worker and client together cast new meaning on the client's life in such a way as to empower the client to interact differently within her or his world. Key to this approach is the social worker's commitment to maintaining a collaborative relationship with the client in which inclusiveness and anti-oppressive views are at the forefront. In the sharing and re-authoring of client stories, both the client and social worker views are of equal importance which serves to empower each of the partners (Roscoe et al., 2011).

Upon gleaning from the client her or his perceptions of current realities, the social worker helps the client to: (1) "externalize" the problem by re-focusing on the outcome rather than the root cause and de-emphasizing *problem-saturated stories* (self-perception in which problems dominate the client's life); (2) discover "exceptions" (i.e., those instances in which the problem or concern did not exist for the client); (3) "re-author" or reconstruct a new reality through mapping of the domain of the issue or problem; and (4) "reinforce" the change by involving others in the client's life and/or sharing the client's change experience to identify unique outcomes (Kelley, 2009; Nichols, 2011). To this end, the social worker poses questions throughout the assessment process that aid the client in reaching her or his desired new reality. Questions are categorized as:

- *Deconstruction questions*—re-focus the issue on an outcome.

- *Opening space questions*—create the possibility for unique outcomes the client may not have considered.

- *Preference questions*—translate the unique outcomes identified through the use of opening space questions into preferred experiences.

- *Story development questions*—move the preference questions to the next stage of change by creating a new reality (or story).

- *Meaning questions*—provide an opportunity for the client to replace negative perceptions with positive interpretations based on strengths identified in the story development.

- *Extending the story into the future*—the client is empowered to see her or himself in future situations. This phase may involve bringing others into the process to support the client as she or he embarks on a new reality.

Solution-Focused Approach

Using overall strategies similar to those described in a narrative-focused assessment and planning process, solution-focused assessment is grounded in the commitment to a focusing on possible solutions as opposed to problems (De Jong & Berg, 2011). Compatible with a strengths-based approach, clients are viewed as not only the experts on their lives but on their challenges and their needs, further accentuating the collaborative nature of the client-social worker relationship (Nichols, 2011).

De Jong & Cronkright (2011) provide this overview of the approach:

- Clients who are able to define what they want to be different in their lives make more progress than those who cannot;

- Clients make greater progress (and do so more quickly) when practitioners spend more time questioning them about what they want to be different than about the details of the problems;

- Clients make more progress, build solutions, and show less resistance when practitioners ask them about exceptions (i.e., their past successes) related to what they want to be different than when practitioners offer advice or confront resistance; and

- Clients make more progress when held accountable for solutions instead of problems (pp. 22–23).

A solution-focused approach incorporates a series of questions to elicit client perception, strengths, resources, and, ultimately, the solution. To facilitate the development of a trusting, collaborative relationship with the client, the first question posed may be "How are you hoping I could help you?" (Nichols, 2011, p. 252). This question invites the client to begin to share her or his story and provide insight into the reasons for seeking your help. Utilizing the ever-important social work skill of starting where the client is, the social worker continues the assessment process by asking the client to share her or his concerns, which is then followed by goal-setting. Solution-focused questions are (Lee, 2009, p. 595; Nichols, 2011):

- *Evaluative questions*—the client engages in an evaluation of the "doing, thinking, and feeling" as it relates to the issue that brought her or him to you.

- *Miracle questions*—the client describes a vision of the future in which the problem or concern no longer exists. Miracle questions promote creativity and hopefulness while, at the same time, promote client self-determination and development of planning for concrete and achievable change.

- *Exception questions*—the client considers a time when the problem or concern was not present, thus allowing the client and social worker to identify existing assets and resources which can be clues of strategies that can be used in the present situation.

- *Scaling questions*—the client is asked to consider her or his situation on a continuum from worst (1) to best (10). Quantifying the issues enables the client and social worker to frame goals, provide feedback, monitor progress, change course, and evaluate outcomes.

Drawing on the similarities of the previously described strengths, narrative, and solution-focused approaches, many practice skills are common to all three philosophical frameworks.

Developing a Shared Vision of Assessment

Preferred reality refers to the client's goal for a changed reality. Preferred reality reflects a postmodern (i.e., client systems are the experts about their lives and situations) and narrative assumption that there are different realities. Long before postmodernism became a common theme in social work theory, however, the profession engaged in efforts to make things different—that is, to work toward a different or preferred reality.

Our discussion of the assessment process has emphasized that the social worker's role is to engage in a dialogue with the client about her or his history, goals, and dreams. This process requires the practitioner to work with the client to develop a picture of what could be. Sometimes this process is difficult for clients, who may feel overwhelmed by the obstacles they face and thus find it quite impossible to be hopeful. Even more challenging, clients may be accustomed to seeing their dreams defeated through their long experience living at the margins of society. In these cases, the social worker has an important job in helping the client see that things can be different, that there are other realities, and that the client can work toward one that she or he prefers.

Being genuinely hopeful and translating that hopefulness to the client can be challenging to social workers as well. When you hear very painful stories, you often feel disheartened. For example, you might feel discouraged after hearing about generations of violence or oppression. However, the idea of preferred realities benefits both the social worker and the client because the idea that the client's reality can be different helps to protect both from a hopeless view of the client's situation.

Remember that working toward a preferred reality does not necessarily require a grand sweeping vision of riches where there was poverty or complete harmony where there was vicious violence. As much as anyone might wish for these long-range dreams, they will appear elusive and unrealistic in many contexts if they are not shared by the client. Social workers need to start with the client's vision of how things should be different, which may only involve small changes in the client's situation. Clients will often frame this vision in clear, small-scale terms when workers are drawn toward a more grandiose transformation.

© Lisa F. Young

Support for the Client's Goals and Dreams

As an illustration of the process of moving toward a preferred reality, consider once again the case of Jasmine Johnson, the single mother whom you met in Chapter 3. Recall that Jasmine is concerned because she sometimes hits her son when he speaks disrespectfully to her. Jasmine may verbalize a request to you, as a family support social worker, for help in finding another way to respond to him. Although this request might seem like a relatively concrete goal for behavior change, it can also be thought of as developing a preferred reality because realization of this goal suggests creating a different mother–son relationship, which could have many positive ramifications. Right now Jasmine is asking for help with a behavioral response. Her goal might be simply to avoid child protection charges or to reduce the likelihood that her son will respond violently back to her. Therefore, your work with her might involve exploring her vision of a long-term goal. For example, Jasmine might want to establish a more satisfying emotional connection with the most important person in her life. She may not have allowed that kind of emphasis on feelings to enter into her interpersonal experiences, or she may never have had the time to think that way. It is even possible that she has not known anyone who articulated such a goal or she may simply be disinclined to consider relationships that way. In this scenario, it is important to start with Jasmine's meaning of the situation and then explore it further without imposing your own meanings.

The shared vision in this case will be that Jasmine learns other ways to respond to her son because she has articulated that vision and because you agree that is an appropriate area for work to which you can contribute. As you work together toward achieving this vision, other, more encompassing dreams about the possibilities for her relationship with her son might evolve. Your role in this kind of strengths-based assessment strategy is to support her dream, always affirming its potential for fulfillment. In order to help Jasmine's dream become reality, you will need to engage in a process of specifying both her goals and a plan for the two of you to work together.

Setting Goals   The reason for **setting goals** is to emphasize the usefulness of clarity of purpose and the utility for clients of recognizing the difference between central, current, behavioral concerns, and the longer view of the dream. As emphasized throughout this chapter, a thorough assessment is key to effective and ultimately successful goal-setting. Specifically, prior to establishing short and/or long-term goals *with* the client system, the process includes assessing the level of possible attainment, incorporating resources as well as limitations, and changes needed in order to experience change (Garvin, 2009, p. 309). Depending on the client's goals and needs, work takes on multiple forms. Consider the following goals in the context of your work with Jasmine:

1. Goals that are discrete (single outcome) or continuous (part of an ongoing plan)—a discrete goal is to respond differently when her son speaks to her in a disrespectful way, while a continuous goal may be to improve her relationship with her son.
2. Goals that are framed within different aspects of the client system (individual, family, group, or community)—be viewed by others (e.g., family and co-workers) with greater respect.
3. Goals that are related to various behaviors and behavior changes—get a new job or get her son to speak to her more respectfully.
4. Goals that are dependent on the individual client or need to involve others (e.g., couple or family)—invite her son to participate in family meetings with the social worker. (Garvin, 2009, p. 310)

Jasmine wants to respond differently to her son when he is disrespectful to her. This goal can be measured in a variety of ways. For example, does Jasmine want to "feel better" about their conversations? Does she want to reduce by half the number of times she is tempted to respond physically to him? Does she want to eliminate those episodes altogether? Does she want him to report that things are better? Clearly, the selection of an appropriate measure should reflect Jasmine's priorities. In starting where the client system is, setting priorities will help Jasmine and the social worker sort through expectations to determine the outcomes that are more critical within a particular time frame. For example, what if Jasmine's son is well behaved two times and then rude once, which provokes her inclination to hit him? Does his positive behavior matter in terms of the goal? Is the goal to change Jasmine's behavior or her son's behavior? As you can see, establishing goals and measuring progress toward achieving them can become complicated. Nevertheless, it is worth the effort to clarify how each party defines the goal of the work and how each one will know when the goal is achieved. Unless you are in clear agreement on an end point, Jasmine may believe you will be there to work on her particular situation until she feels it is "fixed." To avoid misunderstandings, you will want to make sure you have a mutual agreement regarding this issue. Setting goals helps both the client and the social worker evaluate the degree to which they are communicating clearly and have similar expectations, and are making progress. It is also true that agencies and organizations need to understand the purpose and goals of the work, often in concrete, measurable terms. This topic is discussed more thoroughly in Chapter 5 as a function of formal evaluation.

Contracting   The process of **contracting** is one in which the client and social worker reach an agreement about what is to happen, who will be responsible, a timeframe, and the priorities. Developing a clear, measurable and achievable contract with the client emphasizes the client's role in the intervention, and recognizes her or his right to self-determination (Rothman, 2009b). Contracts vary greatly depending on practice setting, and they can be formal or informal. A formal contract generally takes the shape of a written document signed by both parties. In contrast, an informal contract can consist simply of a verbal agreement. If a formal contract is used, details should be clarified, including frequency of meetings, goals, individual roles, ways to change the plan, monitoring of progress, and the degree to which each goal needs to be met. Whether formally or informally executed, a contract should include: (1) goals; (2) objectives that emphasize action, time frame, and strategies for determining success (or failure); and (3) proposed intervention specifying "who" is responsible for "what" aspects of the plan (Rothman, 2009a). Of critical importance to both the social worker and the client is clarity regarding the desired outcomes for the intervention. While the assessment and evaluative components of the social work intervention are dynamic, ongoing, and can change, the social worker and client both will be frustrated and unsuccessful if the goals and the path to achieve those goals are not clearly stated (and re-stated). Quick Guide 9 provides an example of a contract that you might find helpful in working with Jasmine Johnson.

QUICK GUIDE 9   SAMPLE CONTRACT WITH JASMINE JOHNSON

Client Name: _____Jasmine Johnson_____

Client Description of Issues to be Addressed:

__Relationship with my son _____

__Deal with my anger toward my son when he is disrespectful to me _____

Goals and Tasks:

Goal	Client Tasks/Timeline	Social Worker Tasks/ Timeline	Three-Month Follow-Up
1. *Improve my relationship with my son*	*Begin attending weekly family therapy with my son as soon as an appointment can be made.*	*Refer Jasmine and her son to a family therapist and communicate regularly with the therapist regarding progress (with Jasmine's informed consent).*	*Social worker made referral and has had three contacts with the therapist. Jasmine and her son are attending therapy regularly.*
2. *Learn better strategies for disciplining my son, particularly when I am angry*	*Participate in weekly parents of teens class and support group (next group begins the first of next month).*	*Refer Jasmine to parenting class and support group and communicate regularly with the group facilitator regarding progress (with Jasmine's informed consent).*	*Social worker made referral and has had one contact with the therapist. Jasmine is attending class/support group regularly and finding it very helpful.*
3. *Get Devon's father to pay child support more consistently*	*As soon as possible, contact Legal Services Child Support Enforcement office to inquire if they can help.*	*Provide Jasmine with Legal Services contact information and eligibility requirements.*	*With information provided by social worker, Jasmine has made an appointment at Legal Services.*

Date Contract will be reviewed: __*We will review the contract on a monthly basis for the next three months*

I agree with the above stated goals, to complete the contracted tasks, and participate in a review and evlauation of the contract on the specified date.

___Jasmine_____   ___Julie_____
Client   Social Worker

_____   _____
Date   Date

Adapted from Berg-Weger, 2013

Honest Responding

Up to this point you have learned how social workers and clients can identify and work toward a preferred reality. This discussion has emphasized the importance of establishing a shared vision that reflects the client's goals and dreams. However, as a social worker you might find yourself in a situation in which you are inclined to challenge your client's priorities after hearing the story. For example, if your client's overall goal is to avoid being arrested again for selling illegal drugs, you might want to contest that goal as a purpose for the social work intervention. To you, a more healthful goal might be for the client to abstain from using and selling drugs, or to separate from a drug-using peer group. Conflicts over goals raise many difficult questions for social workers. For example, do you have the right to disagree with the client's priorities? Are you inappropriately pushing your values? At the same time, how can you engage enthusiastically to achieve a goal you really do not approve of?

Questions that might serve as guidelines here include:

- Are you able to maintain objectivity about the client's goals?
- Do client goals reflect your priorities, values, or cultural practices?
- What are your legal and professional commitments in this situation?
- Will you violate an ethical principle by not addressing the issue, even if the client would rather not deal with it?
- Do you think the client's goals are either unrealistic or not extensive enough?

Although there are no easy solutions in this kind of scenario, you can respond honestly if you "own" your own biases, which might include more belief in the client's potential than the client seems to possess.

When Confrontation Is Necessary You have probably concluded from the preceding discussion that in some situations in which the goal conflict is particularly acute, the worker has to confront the client system directly. To examine how this process can be conducted, return to the Jasmine Johnson case for a moment. Jasmine has requested your help to change the way she responds when her son, Devon, talks back. However, you have discovered through reliable sources that she may actually be physically abusing him. How will this revelation ethically and/or legally affect the assessment? You may be convinced that you must call the authorities to investigate the serious allegations, with the possible outcome of a temporary separation between Jasmine and her son. This may occur before you can begin to work with her on a different way of relating to him. In this case you may decide that you must establish a new, short-term goal to carry out your professional and legal commitments, and postpone work toward Jasmine's long-term goal. The assessment then moves to an intermediate plateau in which there is a dangerous, or potentially dangerous, situation that requires a more immediate (and ethical) response. You will need to be honest with Jasmine about your understanding of the situation and what you see as your ethical obligations. This process will likely involve some confrontation. In this situation, Jasmine needs to hear that there is an immediate need that must temporarily derail her long-range goals. Exhibit 4.3 provides an example of practice skills that may be helpful when you must face the dilemma of confronting Jasmine and responding with honesty.

You may have noticed in this scenario that any decision to call child welfare authorities, which might lead to a separation between Jasmine and her son, does not necessarily negate her preferred reality of getting along better with him. In fact, a temporary separation may help her achieve this reality. It will be critical in your work with Jasmine to find the common ground and support her dream, even though you must initiate another intervention at the moment. Ideally this respite will improve her chances of working on her original goal after the crisis is past. Your honest dialogue with her will then reassure her that you share her vision, that you want to help her work toward it, and that you can be trusted to tell the truth.

When Alternatives Are Necessary In most cases, people who come to social workers for assistance are realistic about their goals and dreams. Many clients are keenly aware of the level of change that would be required to realize their vision of a more satisfying life. In other cases, social workers must help people expand their vision of what is possible because they have been discouraged, oppressed, or otherwise have had their vision restricted. In such cases the task is to affirm the client's potential and power in changing her or his life.

Occasionally, however, clients will need assistance in clarifying and articulating their goals and their understanding of the situation. Their goals might seem unrealistic or even grandiose to you. These situations are complex, particularly for strengths-based social workers. One strategy for responding in this type of situation is to help clients revise their aspirations, that is, to become more realistic. The argument for establishing realistic goals is that to endorse goals that clients cannot achieve is to set them up for failure, which is a major disservice to them.

At the same time, you are not in the position to determine goals that are realistic or the upper limits of the individual's capacity for growth and change. If social workers are to "hold high our expectations of clients and make allegiance with their hopes, vision, and values" (Saleebey, 2013, p. 19), you should not simply disregard a person's dream as unrealistic. Moreover, many strengths-based social workers insist that clients have the right to fail in the pursuit of their dreams, just like the rest of us. The issue is difficult to resolve and will undoubtedly take on a different look in different contexts.

One option for approaching such a situation is to help clients determine the steps involved in achieving their goals. For example, if you are the high school

EXHIBIT 4.3

Responding with Honesty

During an early meeting with Jasmine Johnson, the social worker learns that Jasmine routinely uses physical punishment to respond to what she perceives as her son's disrespectful behavior. Following is an example of the conversation that the social worker is legally and ethically obligated to have with the client. In the conversation, the social worker balances the goal of honesty with the client about her legal obligation, with an expressed desire to keep working together:

Social Worker: "Jasmine, it is my understanding from another source that when you feel Devon is being disrespectful to you, you hit him. Is that correct?"

Jasmine: "Yes, I just get so mad and he won't listen to me when I tell him he can't talk to me like that. I would never really hurt him though. I just slap him, just to get him to stop mouthing off to me. I haven't even left any bruises or anything."

Social Worker: "I want to thank you for your honesty in confirming the report that I hear about the way in which you respond to Devon when you feel he is being disrespectful to you. Having a complete picture of the relationship you have with him and the way in which you use discipline can help me to help you find alternative strategies for responding to his behaviors."

"Before we can move on to talk about other ways to handle your situation, I would like to talk about the issue of using physical punishment. We have to be certain that the form of punishment that you are using does not constitute physical abuse. I want to make sure that you know that if I have reason to believe that you are physically abusing him, I am required by state law and my social work ethical code to report that to Child Protection Services."

"Once a report is made, the agency will send a social worker to investigate the allegation of physical abuse. If that report is substantiated or found to be true, the agency will decide if they want to recommend that the court remove Devon from your care and place him in foster care. They could also recommend that because you and I are already working together, that they will not pursue removal but they will monitor the situation and get regular reports from me. The other possibility is that they will not substantiate the report and no case will be opened. I want to be very certain that you and I are both clear and comfortable with all this information and the possible outcomes that might occur. So, I will want to talk with Devon about those times when the two of you are having trouble. Using the information I have from the both of you, I will make a decision about making a report."

"I understand that you may be angry with me for even raising these issues and angrier still if I do feel I have to make a report, but I want you to know that I am legally and ethically obligated to ensure the safety of children. I hope that we can continue to work together, but I understand if you don't feel you will be able to continue working with me."

Jasmine: "I know you just got to do what you got to do. I don't think you will find that I have been abusing him, so go ahead and talk with him. If he says I have been, then we'll just deal with that."

social worker and your client states that she wants to be a physician, you and she can consider together what level of education she would require, what it might cost, and how long it would take. You should not use this kind of step-by-step discussion to discourage her, but rather to provide information that she can consider. You may then help her consider alternatives that could serve as stepping-stones. Breaking down goals into workable and achievable sub-goals, or procedural goals, can aid client systems in developing clarity regarding the practical side of the dream and how far they want to take it. This may or may not alter their enthusiasm, but it is an honest response to a sometimes troubling scenario. In this example, you may consider saying to the client whose goal is to be a physician: "Let's put together a plan for you to achieve your goal of becoming a physician. The plan might include gathering information on the educational and training requirements for going to medical school and completing a residency program. Once we know the requirements, we can create a plan that includes strategies for the steps you will need to take to reach each phase of the goal—what you can do while you are still in high school, then the major and experiences you may want to consider for college, and then finally what you can do to make yourself a strong candidate for medical school. What do you think about this plan?"

Using Mapping Skills to Enhance the Dialogue

As you have seen, assessment on any level is influenced by your interpretation of the client's story. This interpretation is shaped by both your own and the client's culture, ethnicity, class, gender, sexual orientation, and age, as well as evidence-based theoretical perspectives, practice skills, and professional knowledge and values. Human stories often have many layers, however, and can be difficult to grasp. Frequently, a visual picture can make the information more accessible. **Mapping** is a technique that represents complex phenomena visually so that they can be absorbed perceptually rather than linguistically. Two fundamental types of mapping—genograms and ecomaps—are discussed next.

Genograms   Technically a family tree, the **genogram** usually represents at least three generations and indicates various aspects of the relationships of the individuals included. Exhibit 4.4 presents a sampling of the conventions for indicating these relationships and Exhibit 4.5 is a genogram completed for the Sanchez family interactive case (see routledgesw.com). As you can see from the Sanchez family genogram, when used flexibly, a genogram can portray a broad range of issues and family patterns, such as strained relationships as well as stable marital relationships that can affect individual and family functioning. A review of the family history through the use of a genogram can add important information to the individual assessment, which may inform your work together on a contract. By adding labels to the map that indicate such attributes as life accomplishments or substance uses/

abuses and addictions, and health issues, genograms can be used in a more focused way to address a single issue. For example, a client struggling with a gambling addiction may be able to gain insight into her or his own challenges by identifying an intergenerational pattern of addictions that exists within the family. Patterns can indicate strengths of the family as well (e.g., educational achievements or career choices and longevity).

With client systems that are vulnerable, genograms should be introduced carefully and with sensitivity. The memories evoked and the visual nature of a genogram can result in more intense emotions (e.g., a sense of loss or regret) than mere words can. For example, a couple who review a genogram may become so distraught at the memory of a relative who died by suicide that each may abruptly leave their meeting with their social worker.

Additional concerns about using genograms are that they are frozen in time (that is, they represent relationships at the moment of the map's creation), and that they are sometimes subject to privileged interpretation, making it important to remember that the map represents one person's perception. Most importantly, they may seem deterministic, especially when they exhibit family patterns. For example, if a multigenerational exploration reveals that all or nearly all young male members on one side of a family have had major substance use problems; it may be tempting to see such a pattern as inevitable. When discussing the genogram, the social worker can give clients hope and the assurance that they can change. Therefore, family pattern exploration should never suggest that people are doomed by their familial history. Rather, a family history provides context and can help you locate your most effective intervention points.

| **EXHIBIT 4.4**

Genogram Symbols | **People in a genogram are represented by the following symbols:**
• Males = □ (occasionally represented by the medical symbol of ♂)
• Females = ○ (occasionally represented by the medical symbol of ♀)
• Unborn or aborted children are triangles (△)
• X through a person symbol indicates death (usually accompanied by age or year of death)

Relationships are represented by the following symbols:
• Strong, solid black line = strong, positive relationship _____
• Dotted line = tenuous relationship - - - - - - - - -
• Railroad track line = contentious or strained /-/-/-/-/-/-/-/-/-/
• Slanting vertical lines = separation or divorce (on marriage line) //
• Circle enclosing people indicates household
• Arrows along connecting lines indicate the flow of energy, reciprocity ↔
• Jagged line indicates a conflicted relationship with possible estrangement |

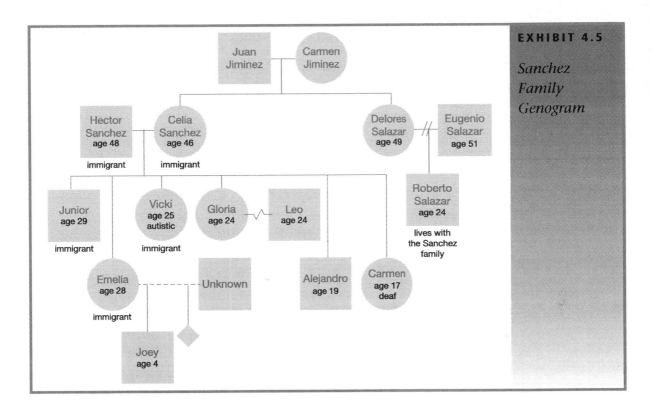

EXHIBIT 4.5

Sanchez Family Genogram

Ecomaps   A diagrammatic representation of the client's world that illustrates the client's levels of connection to such institutions as schools, religious centers or spiritual practices, the workplace, extended family, friends, and recreation is known as an **ecomap**. An ecomap helps clients make sense of their experience by showing them how their day-to-day world looks and the resources and strengths that exist within that world. Ecomaps can also focus on a particular aspect of the client's life. For example, a social worker who is involved with children with special health needs can use an ecomap of medical and social supports relating to the child's health status to help shape the intervention. Some ecomaps include a miniature genogram.

Like all mapping techniques, the ecomap enables your client to explore patterns of everyday living that may or may not be initially accessible in verbal form. For example, your client may complain about being lonely and estranged from his community. An ecomap could reveal that he has very few supports in the community—no satisfying work life, only one friend, no spiritual connections, and no outlet for recreation. Such a diagram would suggest that you expand your dialogue with him into these areas. Here, as always, it is important not to interpret and make conclusions directly from the map without exploring the meaning of the indicators with the client. Consider the Sanchez family ecomap as depicted in Exhibit 4.6 (refer to the Sanchez family case at routledgesw.com for more information). Can you identify the strengths, resources and areas for potential intervention that you

EXHIBIT 4.6

Sanchez Family EcoMap

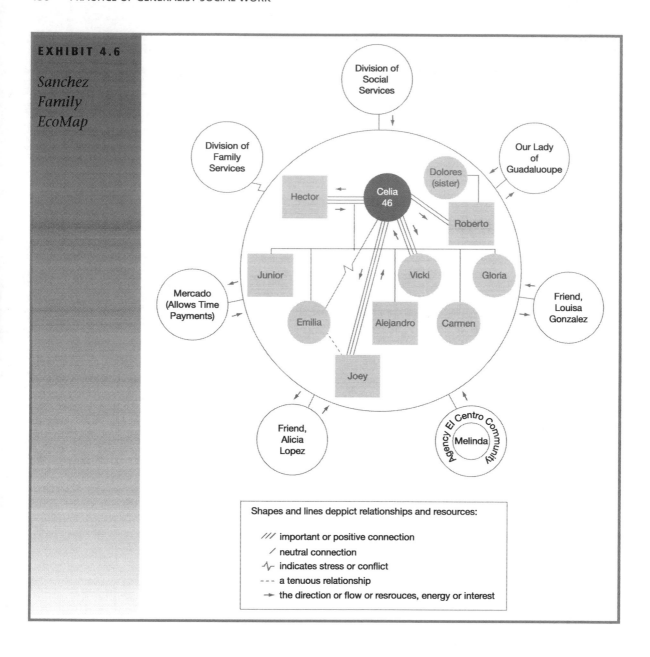

Shapes and lines deppict relationships and resources:

/// important or positive connection
/ neutral connection
-\\- indicates stress or conflict
- - - a tenuous relationship
→ the direction or flow or resrouces, energy or interest

see present in this ecomap? In reviewing the ecomap, you will note that there is a stressful relationship reported between Celia and Emilia, with Celia perceiving that the relationship is not reciprocal (i.e., Celia is devoting more energy to the relationship than she is receiving from the relationship). To fully understand the meaning of that statement, you, as the social worker, can invite Celia to share with you the reasons behind the statement, her emotional response to the relationship, and the implications of her feelings for her relationship with Emilia and others in the family.

Social workers occasionally use a sequence of ecomaps to evaluate the effectiveness and progress (or lack) of the work. For example, if the client mentioned above who is experiencing loneliness wants to expand his social world but feels fearful of that prospect, your work with him might be to develop safe connections, which could be indicated on subsequent ecomaps.

SKILLS FOR ASSESSING RESOURCES

By developing individual skills, such as creating a shared vision, becoming culturally competent, and using social mapping, social workers can work toward an effective assessment. Nevertheless, even the most skilled workers can rarely provide all the assistance that a client needs. Social workers should also familiarize themselves with resources that are available in the larger environment. There are several dimensions to assessing resources that are useful in reaching client goals, from their type and source to their availability.

Formal and Informal Resources

External, structured opportunities for assistance, which usually take the form of services, are **formal resources**. Training in parent education, financial assistance, caregiver respite, and support groups for children in schools are all examples of the hundreds of services you might be able to name that are helpful to clients.

Not all resources are formal, however. **Informal resources** may be people, your own creativity, or naturally occurring social networks, like church groups, neighbors, friends, or families. The informal, internal, or community-based assets can also be mobilized to better serve client systems. Within the practice context, natural helping networks, such as those available from individuals, groups, families, and communities, can be the most important. These resources may endure long past the time that formal resources are involved with a client system. For example, many children benefit from regular contact with an older child who befriends them, understands their troubles, or is just supportive of their abilities. An older adult may benefit from creating networks of friends that are similar age through attendance at a senior center or through volunteer work. It is not necessary in all such situations to hire someone or obtain a service. Some of the most meaningful relationships emerge from informal contexts of community life.

Assessment When Resources Are Available or Unavailable

Your familiarity with both formal and informal resources is an enormous benefit to your clients. Simply locating resources is not adequate. You also need to know enough about the client system and the resources to estimate the fit and likelihood of a successful interaction between them. For instance, your client asks you, as a family support worker, for help in managing an ill infant, and you agree that some supports would help her become more like the effective parent she wants to be. You might recommend some regular time away from the demands of the child and also some assistance in managing the child's challenging behaviors. You would probably then explore the availability of both formal and informal resources. For example, you could investigate formal parent support programs, focusing on the collaborative capacity of the trainers, as well as informal parent support resources such as a child care cooperative and family and friends, and include consideration of concrete issues such as cost and schedule.

Unfortunately, even the more careful social workers cannot locate necessary resources in all situations. At some point, most will have the experience of looking for a particular way to help a client and finding that it does not exist. Even in these situations, you are not powerless. You have at least two options when adequate resources are not available: You can advocate for a change in formal resources, policies, or programs that are not responding adequately (e.g., an older adult day center), or you can create resources where none have been developed. For example, you might assist single parents in a housing complex to organize a network of child care where none existed, or help a community write a grant proposal to fund summer youth employment for neighborhood teens.

These points suggest that social workers do not simply accept, for example, that an uncooperative agency does not respond appropriately to their client, or that a housing project has no child care facilities. Social workers need to remain hopeful and energetic in identifying and developing resources but must also be realistic and careful not to promise more than they can reasonably deliver. In a client's world, where many promises may have been broken, social workers need to maintain trust through an honest approach about the possibilities.

Social Action When Resources Are Inadequate

Many of the cases in which the resources you think necessary are inadequate, nonexistent, or discriminatory are political situations that call for policy practice approaches. Action on behalf of clients is one possible response to an unjust situation. **Social action** is a policy practice method in which the social worker generally aims to shift power structures in order to change an institutional response. The efforts come in many forms and may involve varying degrees of confrontation. For example, if the public health department is slow to investigate a situation of child lead poisoning at a local school, you might assist your client through their formal grievance process, and/or organize a protest with other concerned parents if the health department does not respond to the initial inquiry. Both the actions on behalf of one client and on behalf of many clients are forms of social action. However, the latter is clearly more confrontational than the former. In situations in which this kind of action is contemplated, the assessment will take on the additional dimension of an analysis of the actions that actually should or will happen, who will do it, and what the potential consequences will be. In addition, the assessment must address the client's interest in, and capacity to engage in, the action indicated. The social worker must be completely honest and sensitive to the possible consequences of political action for the client or the client's family. In many ways this last requirement can be considered a dimension of ethics. Read how this might play out in the tenants' story in Exhibit 4.7.

Despite the risks that may be involved, when an understanding is truly reached in partnership, few experiences are more positive or powerful than a successful social action based in client–worker advocacy. Even if the results are somewhat less than hoped for, the process itself can be remarkably empowering for both social worker and client.

Although this scenario expands into the phase of intervention, it points up the importance of assessment throughout the client relationship. The assessment drives the action stage of the work, and therefore, must also address the action inherent in it.

PLANNING

When you and your client have developed a shared vision and specified the goals, means, and end points you envision for the work, you will have addressed the components of a solid, detailed plan. This section will revisit these briefly and then consider the potential impact of two insidious influences—oppression and the emotional impact of change—on the planning process in particular and on the assessment in general.

The key components that influence the concrete plan for work include:

- Setting and prioritizing goals

- Identifying methods of reaching goals

- Developing a clear understanding of responsibilities

- Setting time frames

- Recognizing when an alternative plan is necessary

- Identifying resources

- Identifying an end point for the work.

Incorporating these elements into a plan can enhance the client's understanding of what will happen. The elements then provide a framework that guides the development of the intervention plan that enables the social worker and the client to maintain focus on the identified concerns, priorities, and available and needed resources. Having a structure for creating an intervention provides an opportunity for the social worker and client to collaboratively develop goals and action steps. These

EXHIBIT 4.7

Social Action: The Tenants' Story

In a privately run, not-for-profit housing project in a midsize city, there are several families with young children and some with grandparents. In the past, the housing was kept in marginally acceptable repair, but the residents stayed there because the rent is fixed and relatively low. A change in ownership leads to a distinct deterioration in maintenance, adequate and timely repairs, and the amount of heat allotted to each apartment. Toilets are clogged, rats are visible in the basement, and lead-based paint is beginning to crumble from the walls. The residents are unsure how they can respond to this deteriorating situation so they request that you, as a social worker from the city's housing authority, advocate for them. After discussing the situation with client spokespeople and then asking each client for approval, you sponsor a meeting for residents with the landlord's representatives. Although you have been careful to hear the landlord's side and have used your best engagement skills, the meeting results in a deadlock, and no progress is made.

Convinced that you have exhausted all of the available approaches, you consult with peers who affirm your position. At that point you consider the possibility of organizing your tenants to withhold their rent. You feel for many reasons that this is an appropriate and effective strategy to elicit a more satisfactory response from the landlord. At the same time, however, you know from prior experience that this particular landlord has a history of retaliation against complaining tenants.

You fully recognize the need to collaborate with tenants about the potential plan of withholding their rent. There is some hesitation among them because they too are concerned with the likely consequences of riling a powerful landlord who could throw them out. Some of the residents are people of color, and they are especially concerned with the availability and accessibility of other housing.

Let's consider the following ways the story might play out:

- *Plotline A:* You are so convincing that this is the way to deal with the problem that you are able to engage the tenants in a full-blown movement to withhold their rent. Your clients nervously await the landlord's response.
- *Plotline B:* You hear those who have specific concerns and play out alternative responses. You locate other sources of housing, and you identify legal advocates for those interested in pursuing new housing. You arrange for press coverage and obtain the full support of your agency and supervisor in taking this action.

Social Action: The Moral of the Story

You may fill in the ending of this story. It could be empowering or disastrous or a lot of things in between. The point is that when others live the results of your advocacy attempts, they need to have all the available information about possible consequences, the likelihood of certain outcomes, and the opportunity to choose which path they want to take. You, as the social worker, need to be very clear about who will pay the price of any repercussions. Because the "who" is often the client, the client should enter into such an agreement only when she or he clearly understands the risks and is motivated to follow through with the action consequences. Can you identify an alternative Plotline C?

components have been negotiated as an ongoing feature of your work together and constitute the mutually developed plan.

Although the ideal is that social workers and clients collaborate to plan the process, even the most careful approaches cannot account for all of the potential obstacles that might arise. Many clients face real disabilities. They may have emotional, cognitive, or physical challenges that make it almost impossible for them to participate in planning. Although social workers should never minimize these obstacles, they need to work directly with the client to the greatest extent possible, while bearing in mind that many clients who have experienced lifelong challenges have never been considered as adequate participants in service provision planning. Consequently, clients may be hesitant or even afraid to participate, even though they have the ability to make substantial contributions. Many others who have experienced racial or ethnic oppression will find it difficult to trust the process of mutual assessment and planning.

The transgression of excluding clients from this process has generated two negative consequences. First, service providers have often lost, dismissed, or ignored helpful information that only the person experiencing the context could provide. Second, many of these clients have internalized this kind of oppression and actually have come to believe that they have no right or capacity to participate. The **internalization of oppression**—the process by which individuals come to believe that the external judgments are valid, thus resulting in a devaluing of one's self—is one of the most sinister aspects in the oppression and domination dynamic (Van Soest, 2008) and clearly undermines the spirit of human rights.

Another component of assessment and planning that social workers must keep in mind relates to the impact of emotions. In their zeal to develop a plan, address the issues, mobilize the resources, and engage the contract, social workers may sometimes forget that certain elements of the work (e.g., emotional responses to loss and grief) are not as visible or as easy to categorize as others. By the time social workers have listened to the story and established a preferred reality, they will probably have heard and felt a lot of the emotional content that is related to the client's experience. They may not as easily, however, recognize the power that complicated feelings about change can generate.

For example, whole families are systems that can be organized around one member's addiction to alcohol. In some families, everyone knows how to respond if Dad is drunk and has passed out on the couch. John, the eldest son, may be the one to carry him to bed; everyone else might ignore Dad, step around him, and pretend his drinking does not happen. Mom takes this opportunity to make decisions about the family finances that Dad has always made. The specific patterns may not be as important as the idea of changing them. Although Mom has come to your agency to get help with Dad's addiction because it is obviously destroying his health, their marriage, and their family life, the social worker must be aware that there is a cost to change. Mom may feel uniquely competent when she is so clearly needed to make major decisions, or John may take great pride in being able to manage the chaos. Again, the patterns are not as important as the disruption of them. If Dad successfully withdraws from alcohol, Mom may actually have some regrets because she will have to develop a new role when her husband is more present. John may lose his place as competent caretaker, and all members of the family will need to relate to Dad and each other differently. This need for renegotiating family roles can have some disturbing emotional consequences.

Although there is a risk of over-interpretation, any change that the client is genuinely seeking is embedded in a social context and therefore can generate a powerful emotional response. A change in one part of the system has the potential to change the entire system. This response may take the form of reluctance to engage in ongoing collaboration or hesitation when making progress on specific goals. In such cases, consider engaging in a dialogue that goes beyond the technical or mechanical processes of goal setting and contracting. Specifically, it will be helpful to explore the meaning of the change itself in terms of emotional and logistical outcomes as well as the impact of the change on roles, functioning, and patterns.

STRAIGHT TALK ABOUT ASSESSMENT AND PLANNING: THE AGENCY, THE CLIENT, AND THE SOCIAL WORKER

You have seen that several agency, client system, and professional issues influence your work with clients. From those perspectives, the following discussion will address a range of issues that you will likely encounter as a social worker. Having insights in advance of such encounters can arm you with the knowledge and skills to respond competently and ethically.

The Agency Perspective

From the agency perspective, work methods, schedules, length of the work, documentation, and agency resources all need to be recognized and discussed openly with the client. For example, you need to provide information to the client regarding your schedule, appropriate ways to contact you, ways in which you will be documenting your interactions with her or him, and agency resources that are or are not available to her or him. The social worker needs to inform clients about their work processes from the beginning so that there is no confusion or frustration as the relationship evolves.

Administrative Tasks   With an increasing emphasis being placed on brief, effective, and accountable treatment for individuals, families, and groups, managed care organizations (MCOs) have influenced the assessment and planning process in significant ways (Hopson & Wodarski, 2009). You may be required to complete your assessment, planning, intervention, termination, and evaluation processes within a limited time frame. Standardized, rapid assessment measurement tools have become a mechanism for practitioners to complete assessments in a short period of time while maintaining optimal accuracy. In order to work within the current service delivery environment, your agency may opt to use a rapid assessment process. With numerous rapid assessment instruments (RAIs) available to choose from, the key is to ensure the instrument relates to the issues being presented by the client system and has been empirically validated with that population (Hopson & Wodarski, 2009). In order to have confidence in the efficacy of a standardized assessment tool, the social worker must have evidence that the assessment instrument has been psychometrically and clinically validated on the population with whom it is being administered. Utilizing any assessment tool on a population for which the instrument has not been shown to be effective can produce misleading outcomes in the areas of age, gender, race, ethnicity, and culture. Engaging in such evidence-based practice helps to enhance the potential effectiveness of the intervention.

Other formalities about which to inform the client relate to administrative tasks or financial coverage. For example, if you are required to submit monthly reports to the court, you should discuss this stipulation with your client. Even when clients are not happy with such an idea, they are likely to respect your communicating such information from the outset of the relationship. For another example, if your client will have to file a Medicaid application in order to continue working with you, explain the need for action as early as possible. For many clients, filing for financial assistance for services may be challenging, because it may mean to them that they are admitting defeat, disability, or helplessness. Being sensitive to your client's perspective about the meaning of these aspects of your work together is important.

Another mandate that is becoming increasingly prevalent in community agencies is the requirement for diagnosis as part of the assessment and planning process. As discussed earlier, many institutions now demand a *DSM* (from the American Psychological Association's *Diagnostic and Statistical Manual of Mental Disorders*), diagnosis in order to bill for third-party payment. This mandate can provide uncomfortable moments for social workers who believe strongly in helping clients to identify and use their strengths. The attempt to provide a psychiatric diagnosis requires a focus on deficiencies and can create a stigma. For example, a school social worker who is providing support to a five-year-old whose parents are addicted to heroin may struggle with diagnosing the child with an "adjustment disorder," which will then appear in school records, label the child for years to come, and may be transmitted to other institutions.

Obviously, if your agency requires social workers to make or use diagnoses, then you need to consider your feelings on this issue before you accept a position. Practicum students are not likely to formulate diagnoses, but the use of diagnosis inevitably categorizes clients, which can impact the working environment of a setting and some may object to working in such a setting. Others may find the process tolerable as long as it facilitates service provision. Still others may find the use of diagnosis helpful because they are able to use evidence-based practice to provide the best possible service to clients for their challenges, and collaborate easily with non-social work colleagues who also use the *DSM*. Ethical conduct obliges all social workers to inform their clients when a diagnosis is required to access services. The client can then decide if she or he wants to participate in that exchange. For some clients and social workers the diagnostic process may seem a small price to pay for receiving appropriate services and possibly medication, whereas for others, the benefits do not outweigh the costs involved in using the *DSM* with clients.

Documentation Recording information about your work with clients is both universal and unique. Social work encounters are recorded in virtually every setting, but the requirements and formats are typically specific to the setting. Social workers are required to have competencies to create a variety of documents aimed at diverse audiences to serve a range of purposes. In working with individuals, social workers will complete intake/psychosocial assessments, treatment plans, case notes, case studies, reports to external organizations (e.g., schools and courts) and compose other forms of communications (e.g., memos, letters, recommendations, etc). Social work documentation begins with the first encounter and continues through each phase of the planned change experience. Documentation is important for social work practitioners as recording information from the engagement, assessment, intervention, and termination and evaluation encounters serves a number of key functions, including accountability, supervisory and administrative purposes, practice enhancements, reimbursement, and planning (Kagle, 2008). Kagle and Kopels (2008) provide the profession with fifteen principles of recordkeeping that should be in the forefront of every social worker's practice. Quick Guide 10 provides a summary of those principles for practice.

Despite the need to learn the agency's preferred method for documentation, there are basic components that are likely to be found in the records developed by social workers. Kagle and Kopels (2008, pp. 38–40) suggests the following structure for creating a written record of the social work intervention:

1. opening summary;
2. data gathering and social history;
3. assessment;
4. decisions and actions resulting from initial assessment;
5. service planning, including service options and purpose, goals, and plans of service;
6. interim notes;
7. special materials (e.g., forms, consents, emergencies, and service reviews); and
8. closing summary.

QUICK GUIDE 10   PRINCIPLES OF GOOD RECORDS

- *Client goals* that demonstrate balance in competing goals (e.g., accountability, improving practice, efficiency, and client privacy)
- *Agency mission* guides the documentation
- *Management of risk* for the agency, social worker, and client
- *Accountability* for service decisions, compliance, and legal issues is evident
- *Abridgement*—only information that is needed is included in the record
- *Objectivity* includes observation, sources of information, criteria used in judgment, and appraisal
- *Client involvement* is evident
- *Sources* of information are cited in the record
- *Cultural context* factors are noted as each relates to the client's situation, service delivery, and outcomes
- *Usability* of the record includes organization, cross-referencing, and accessibility
- *Current information* is included in the record and summaries
- *Rationale* for service decisions and actions is evident
- *Urgent* situations are well documented
- Information *excluded* from the record includes information that is unsupported or not relevant to the client's service status

Source: Adapted from Kagle & Kopels, 2008, pp. 10–11

The opening summary portion of the documentation encompasses the engagement phase of your intervention. Information that is typically included in an opening summary relates to the client's demographic characteristics (i.e., name, gender, address, contact information, birth/age, family composition, employment, insurance coverage, etc.), reason for requesting or receiving services, and eligibility status. The assessment portion of the client record should specifically include the social worker's observations and descriptions, sources of the information, criteria used in making the judgments (e.g., formalized measures and information from past history with the agency or referring entity), and inferences and appraisals (Kagle & Kopels, 2008). This may be an appropriate place to record perceptions of strengths, resources, needs, cultural factors, and risks related to services that may be delivered.

While formats are agency or program-dependent, there are elements of social work recording that are common across the profession. Most agencies have a structured format for documenting client encounters or meetings, conducting an intake or eligibility interview, completing a social history, and a treatment/intervention plan. Examples of information to include in client records of these three documentation activities are presented here: (1) Exhibit 4.8 provides an example of information to obtain in a client assessment; (2) Exhibit 4.9 includes examples of information to include in a suicide risk and harm to others assessment; and (3) Exhibit 4.10 is an example of information to include in a treatment/intervention plan (adapted from

EXHIBIT 4.8

Information to Include in a Client Assessment

Demographic Data:
- Name
- Contact information
- Legal status
- All persons participating in assessment
- Presenting need
- Living situation (level of stability and safety)
- Social environment (level of activity, satisfaction, and relationships with others)
- Cultural environment (client satisfaction and view on helping-seeking, and cultural view of help-seeking)
- Religion/spirituality (statement of beliefs and levels of activity and satisfaction)
- Military experience (branch, time in service, discharge status, coping with experience, and view of experience)
- Childhood (supportive, strengths, and significant events, including trauma)
- Family (composition—parents, siblings, spouse/significant other(s), children, and others; level of support; and family history of mental illness)
- Sexual history (activity level, orientation, satisfaction, and concerns)
- Trauma history (physical, sexual, and/or emotional abuse or neglect and experience with perpetrator(s))
- Financial/employment circumstances (employment status, satisfaction, financial stability, areas of concern or change)
- Educational history (highest level achieved, performance, goals, and challenges)
- Legal needs (arrest/conviction history and current legal status)
- Substance use/abuse (history of addictive behaviors—alcohol, drugs, gambling, sexual, or other). Addiction screen (questions include: (1) Have you ever felt you should cut down on your drinking/drug use? (2) Have people annoyed you by criticizing your drinking/drug use? (3) Have you ever felt bad or guilty about your drinking/drug abuse? (4) Have you ever had a drink/used first thing in the morning or to steady your nerves or to get rid of a hangover (eye-opener))?

History of Emotional/Behavioral Functioning:
For each of the following areas, gather information regarding: current status (current, previous, or denies history); description of behavior; onset and duration; and frequency

- Self-mutilation
- Hallucinations
- Delusions or paranoia
- Mood swings
- Recurrent or intrusive recollections of past events
- Lack of interest or pleasure
- Feelings of sadness, hopelessness, isolation or withdrawal
- Decreased concentration, energy, or motivation
- Anxiety
- Crying spells

EXHIBIT 4.8

Continued

- Appetite changes
- Sleep changes
- Inability to function at school or work
- Inability to control thoughts or behaviors (impulses)
- Irritability or agitation
- Reckless behavior, fighting, or fire setting
- Stealing, shoplifting, or lying
- Cruelty to animals
- Aggression

Behavioral Health Treatment History:
- Date
- Program or facility
- Provider
- Response to treatment

Mental Status Exam:
- Attention (rate on scale of: good, fair, easily distracted, or highly distractible; and describe behavior)
- Affect (rate on scale of: appropriate, labile, expansive, constrictive, or blunted; and describe behavior)
- Mood (rate on scale of: normal, depressed, anxious, or euphoric; and describe behavior)
- Appearance (rate on scale of: well groomed, disheveled, bizarre, or inappropriate; and describe behavior)
- Motor activity (rate on scale of: calm, hyperactive, agitated, tremors, tics, or muscle spasms; and describe behavior)
- Thought process (rate on scale of: intact, circumstantial, tangential, flight of ideas, or loose associations; and describe behavior)
- Thought content (note: normal, grandiose, phobic, reality, organization, worthless, obsessive, compulsion, guilt, delusional, paranoid, ideas of reference, and hallucinations; and describe behavior)
- Memory (note: normal, recent (good or impaired), past (good or impaired); and describe behavior)
- Intellect (note: normal, above, below, or poor abstraction; and describe behavior)
- Orientation (note: person, place, situation, and time; and describe behavior)
- Judgment and Insight (rate on scale of: good, fair, or poor; and describe behavior)
- Current providers (including psychiatrist, primary care physician, therapist, caseworker, etc.)
- Community resources being used (including support groups, religious, spiritual, other)
- Client goal(s) for treatment
- Summary of social worker's observations and impressions

Adapted from St. Anthony's Medical Center, St. Louis, Missouri

EXHIBIT 4.9

Information to Include in Assessment of Suicide Risk and Harm to Others

Suicide Risk Assessment:

- Name, date, and time of assessment
- Clinical assessment (include current suicidal thoughts, obsessions with death or indications of putting one's affairs in order, even with no specific plan). If yes to any items, follow-up questions include: is there a plan, is the plan lethal, is there potential access to plan?
- Frequency of thoughts
- Intensity of thought (rate on scale of: (1) no pressure to (5) high pressure)
- Risk level (rate on scale of: negligible, mild, moderate, or severe)

Harm to Others Assessment:

- Name, date, and time of assessment
- Clinical assessment (current thoughts of harming another person). If yes, follow-up questions include: is there a plan, is there a target/victim, and is there potential access to plan?
- Frequency of thoughts
- Intensity of thought (rate on scale of: (1) no pressure to (5) high pressure; and describe plan and victim)
- Risk level (rate on scale of: negligible, mild, moderate, or severe)
- Notify physician
- Complete duty to warn protocol

Adapted from St. Anthony's Medical Center, St. Louis, Missouri

EXHIBIT 4.10

Information to Include in a Treatment/ Intervention Plan

Intervention/Treatment Plan:

- Preliminary assessment/diagnosis
- Preliminary plan for intervention/treatment (to be developed at first visit)
- Interventions for emergency/safety need
- Other interventions needed
- Needs (include date, identified need, status (active, inactive, deferred, or referred), and reason for deferral or referral)
- Strengths
- Facilitating factors for intervention/treatment
- Limitations
- Barriers to intervention/treatment
- Other care providers/referrals and purpose (including plan for service coordination)
- Plan for family involvement (note if client opts for no family involvement)
- Termination criteria/plan
- Planned frequency and duration of intervention/treatment

Adapted from St. Anthony's Medical Center, St. Louis, Missouri

St. Anthony's Medical Center, St. Louis, Missouri). A typical treatment plan will include (Sormanti, 2012):

- brief description of the issues to be address, including client strengths, resources, and challenges;

- theoretical orientation in which the intervention is grounded (may be optional);

- goals for intervention;

- specific and measurable objectives;

- estimated timeline for the intervention;

- planned intervention;

- plans for evaluation and review of progress (p. 117).

A second form of documenting practice that social workers find helpful is referred to as narrative recording. Narrative recording may or may not be a central focus of the agency recordkeeping process but can allow the practitioner to document client-specific process-related information (Kagel & Kopels, 2008). A narrative report might include: nature of the client situation; purpose of the service; service-related decisions and actions; process of the service delivery, and service impact (Kagel & Kopels, 2008, p. 125).

The Client Perspective

Just as agency issues can present challenges for the social worker, there are client situations that can prove to be difficult to address in the assessment phase of work. Client issues that require specialized practice skills can include working with client system situations that involve: involuntary, mandated, and non-voluntary clients; violent situations, including suicidal clients; and the need for crisis intervention skills.

Involuntary, Mandated, and Non-voluntary Clients   When you first considered social work practice, you may have assumed that people would come to you because they wanted your services and believed in the possibility, at least, that you would be helpful. The principles of self-determination and working toward the preferred reality seem the opposite of coercion, and some social workers question the appropriateness of this kind of practice. Yet, because of the constant tension between social work as an agent of social control and an agent of change, a significant proportion of clients (in some settings, *all clients*) are involuntary.

Involuntary has several different meanings. In some sense, all clients are involuntary in that few are happy to receive services as it likely indicates they are experiencing a crisis in their lives (e.g., loss, illness, etc.). Certainly those who seek public sector welfare services are likely to be involuntarily in their current situation. Thus, the distinction between voluntary and involuntary is not always clear cut, but three ways of distinguishing this type of client are accepted in the profession.

Clients are **involuntary** if they have been compelled to receive services, including mandated and non-voluntary client systems (Barker, 2003). **Mandated clients** are typically required by an authority (e.g., legal system, employer, etc.) to receive services in order, for example, to reclaim children, escape criminal charges, or avoid institutionalization. Such mandates occur most often in the systems of care that are heavily shaped and sanctioned by the law in, such as child protection, mental health, and criminal justice. **Non-voluntary** clients are not formally or legally obliged to participate in services, but are pressured into receiving them. For example, persons with substance abuse issues may be "strongly encouraged" by their employers to seek help for their addiction, or a parent "takes" an adolescent to family counseling. In these situations the client is in some way persuaded that she or he needs to get services in order, for example, to keep the peace, or remain married, or stay employed.

There are many, and often unpredictable, patterns of engagement with involuntary clients. Some angry mandated clients become convinced that they can benefit from genuine involvement, whereas others simply wait out their time. Non-voluntary clients may perfunctorily go through the motions to satisfy someone else, or they may work for real change. The whole notion of involuntary clients presents several challenges as well as opportunities.

Challenges in Working with Involuntary Clients   For many social workers, whether it is students or more seasoned practitioners, the idea of working with a client who does not want to be there is uncomfortable, if not daunting. In some cases, the issues relate to the interface the social worker needs to maintain with the mandating agency, which may seem rigid or overly authoritarian. In others, social workers might dislike the involuntary client's behavior (for example, negative feelings regarding the client's engagement in child abuse, criminal activity, or substance abuse). In still others, social workers are unsure how to build a relationship in such circumstances. These are understandable concerns, but they put many more limits on the situation than are necessary.

ENGAGEMENT WITH INVOLUNTARY CLIENTS   Being well-prepared for your first visit with an involuntary client is the first step. First, prepare *yourself*. In general, you can expect that your client may be angry, hostile, or fearful. If you are assaulted by spiteful or even hateful remarks at the beginning of a meeting, you may feel surprised and hurt, so it is helpful to know how you may react and to think ahead about how to respond. Striving to remain calm and non-reactive are goals for an encounter such as this one. Clients who are emotionally agitated or distressed may seek to shock or outrage the social worker. Having a professional response to the client's outbursts can serve to defuse the potentially volatile situation.

Involuntary clients often take on a perception that they need nothing you have to offer and they may feel they have a valid reason for believing what they believe (De Jong & Berg, 2013). As a result, they do not truly engage with you. Although this seems challenging, it can actually assist you in refraining from taking any angry remarks personally, because the client makes few role demands on you. Another way to prepare for the resentment and negativity of some involuntary clients is to place yourself in the client's position. Remember the times you have felt coerced, invisible, unrecognized, or ignored. You can connect with some part of that scenario that can make you more empathic and help you to understand the client's disinclination to trust anyone, including you.

Not all clients, of course, will barge in with this level of antagonism. Some will be overly polite or seem controlled. The client may perceive that if she or he maintains a complacent or controlled demeanor, you will not be able to break through the exterior to touch them in any significantly emotional way. Some may actually view meeting with you as an opportunity to think about making changes. Since the tendency may be to decide the client is resistant before you start the work, consider remaining as "unknowing" as you can. In any case, the most helpful approach is to listen, as it is in any client situation, to the story. Find out how the client sees the situation. Ask the client how things could have been different, respect the client's reality without challenging it, and assume the client has both strengths and competence in spite of the current predicament. Find out what is important to the client and how she or he wants things to change. This is often an excellent situation for questions that probe for the client's perspective because they can reveal that there is more to the client than the involuntary situation, and that she or he is not just "bad," or a personified crime—for example, "a B & E" (breaking and entering) or a "shoplifter." If the adolescent who is substance-using feels like a "loser" and worthless, you might ask, "What would your best friend say is your greatest strength?" or, "What would your favorite teacher say you're good at?" (De Jong & Berg, 2013). For more on useful approaches with involuntary clients, see Quick Guide 11.

In most cases of mandated services, you will have some externally structured pattern for your work. This can be a set number of sessions (such as 12) or a curriculum to follow. You may need to explain this to your client in order to encourage the engagement process. One of the major ways you can assist your client is to give her or him as much control as possible by emphasizing any choices that are real (for example, you may be able to meet once or twice a week, on Tuesdays or Fridays) and by being as clear as possible regarding requirements. If the court requires you to notify authorities if your client misses a single session, inform her or him of that. Consider that mandated clients typically feel as if control and choice have been taken away from them, and stay attuned to those issues. You will want to provide all the information you can about the contingencies as a respectful gesture of recognition and acknowledge to the client that you are aware that she or he does have choices (e.g., to work with you or not work with you), even if they do not feel they do.

QUICK GUIDE 11   INVOLUNTARY CLIENTS

Guidelines for Interviewing Involuntary Clients

- Assume you will be interviewing someone who probably will start out not wanting anything you might have to offer.
- Assume the client has good reason to think and act as he or she does.
- Suspend your judgment and agree with the client's perceptions that stand behind his or her cautious, protective posture.
- Listen for who and what are important to the client, including when the client is angry and critical.
- When clients are openly angry or critical, ask what the offending person or agency could have done differently to be more useful to them.
- Be sure to ask for the client's perception of what is in his or her best interest; that is, ask for what the client might want.
- Listen for and reflect the client's use of language into your next questions and responses.
- Bring the client's context into the interview by asking questions about her or his relationships with others.
- Respectfully provide information about any nonnegotiable requirements and immediately ask for the client's perceptions regarding these.
- Always stay "not knowing" (i.e., social worker's focus is on the client's frame of reference, not her or his own).

Source: De Jong & Berg, 2013, p. 184

LEGAL ISSUES   In spite of your best efforts to engage your client, mandated or not, you may find yourself having to take a position in the best interests of the client, including children and older adults. This position can become difficult when you have been trying to respect the client's situation and work with their goals and wishes. For example, do not decide prematurely that the single parent in a child protection scenario is not ready to have her children back or that your older adult client cannot live safely in the community. You want to keep working toward the goals of the client system to arrive at a point at which her or his preferred reality can be realized.

Nevertheless, there are times when you may have to take a conscience-driven stand that may seem to work against the client system. For example, if an older adult experiencing dementia has progressed to the point that she is unable to maintain her activities of daily living, including cooking, shopping, and bathing, she may not be able to continue living alone in her home. Your recommendation to the court may be to seek care for your client in a residential setting. The client may be upset about your recommendation and feel you are not supporting her. In such cases, inform the client(s) before any court hearing or other meeting. Avoid any surprises for your client that would further violate trust. For example, initiating an honest dialogue (e.g., "this is the way I see it") conveys respect even if the client does not agree with the assessment. Another way to help the client understand the perspective of others is to ask perspectival questions, such as "What do you think the judge might think about the times when you left the baby alone?" (De Jong & Berg, 2013).

Opportunities in Working with Involuntary Client Systems   Although involuntary situations are not ideal, they are seldom disastrous. Such interactions can present the opportunity to engage people who might never undertake any change without being forced into seeking services. These challenges also present social workers with an opportunity to practice their most cherished social work skills. By hearing and respecting clients' perceptions, assuming they can grow, and believing that their situation is workable, you have an opportunity to engage them in ways that can potentially transform their lives.

INDIVIDUAL SCENARIOS   Consider the man who was adjudicated by the court as a perpetrator of domestic violence, but because he also had a substantial history as a client of mental health services and had just been discharged from a psychiatric facility, you were asked by your supervisor to meet with him individually. This departure from the agency's standard intervention (which was group work for men) was made on the grounds that the client would not respond well in the potential confrontation of a group. He was painfully embarrassed at being singled out as inappropriate for a group process, angry that he was treated like a "common criminal," and exceedingly hostile about meeting with you.

Using a basic approach of respect, a willingness to hear him out, and a position of wanting to understand his life, you can connect with him after several sessions. He then confided that he had never had the opportunity to talk about the challenging issues in his family and that he felt humiliated and stigmatized in receiving mental health services. By the end of the twelve mandated sessions, he wanted to engage in further work, and he had come to believe that he could change his violent behaviors by responding differently to the stresses he felt. This is an example of the way clients respond to an affirming approach that recognizes their strengths even though the client was not initially involved in his treatment decisions.

POWER ISSUES   There is a clear association with power, specifically coercive power, in any kind of involuntary and especially mandated service. Regardless of the relationship or work that transpires, you have considerable power in determining the course of the client's life. In these situations, you need to recognize the power you have, be as comfortable with it as possible, articulate your understanding of the power to your client system and elicit the client's understanding. Transparency about power creates a rich opportunity to demonstrate your genuineness and trustworthiness to clients, and enables the opportunity to discuss power and its meaning to your client. You are presented with an opportunity to struggle with the tension between your role as a person of social control and your role as a person who assists client system who are caught in social control.

Focusing on power provides a context for discussing your client's goals and strategies. For the client, the goal might be as simple as getting authorities "off my back," or it may involve much more complex efforts to change a system that she or he experiences as oppressive. For example, the woman trying to regain custody of her children who has to see you, get clean, get a job, and find a home with no instrumental supports might, for example, want to join with others in an empowerment-focused approach to influence the system in helping her meet her mandates.

Violence   None of us can remain oblivious to the increasing violence in our culture. Violence looms in almost every newspaper, television news story, and on all too many urban street corners and rural crossroads across the country and beyond. From the wars waged overseas to violence in school settings, violence is expressed in ideological as well as interpersonal terms, and is a common theme everywhere. Social workers are frequently in the position of working with the effects of a violent culture and are not immune from direct exposure to violent threats in the workplace. The escalation of violence in our lives has become an insidious component of contemporary experience and should be addressed on personal, agency, and policy levels.

Safety in Social Work Practice   Although it can be comforting to think of violent acts, particularly those related to social work practice, as discrete, chance occurrences, there is often a difficult-to-deny connection between violence and many historical, political, and cultural processes that reflect a cycle of escalation. Violence in any setting breeds negative, destructive attitudes and relationships that are reinforced by periodic eruptions in social contexts, but social workers must be knowledgeable about and prepared for violence within any practice setting. School bullying, interracial taunting, and sexual harassment are just a few pervasive examples that are translated into interpersonal persecution that may occur in social work practice. Occasionally these situations escalate into physical violence on an interpersonal or social level, as witnessed by isolated attacks, school murders, or race-related altercations. Some people resort to violence to get their basic needs met, and our society struggles to eliminate the barriers that prevent people from doing so.

As evidenced by the NASW-issued policy statement calling for the development of organizational policies to address employee safety, the social work profession is committed to minimizing violence in the workplace (NASW, 2012–2014b). Social workers can observe the growing disparities between rich and poor in the United States, consider the wider political components that seem to spur people to hate one another, and renew their efforts to work for a just society. Building on the profession's systemic perspective, social workers can address racist and economic policies of welfare reform that punish people of color and maintain poverty, and can analyze, for example, legal approaches to limiting weapons, their commitment to antiviolence principles, and their positions on war. All of these issues are at the heart of people's lives and as such are in the realm of social justice-oriented social work practice from the most local to the most global levels.

Considering that practitioners are very likely to work with people who are the most affected by the quality of this cultural fabric, it is not surprising that, as symbols of the dominant culture and agents of social control, social workers are increasingly the objects of violence themselves (Jayartne, Croxton, & Mattison, 2004). As early as 1995, Newhill documented that "physical and emotional violence by clients toward social workers is increasing in all settings" (1995, p. 631). A recent survey of NASW members (Whitaker & Arrington, 2008) details specific areas of incidence and risk. Forty-two percent of social workers working in mental health services identify violence as a safety concern, while 43 percent of mental health social workers and 30 percent of social workers working in child welfare settings have actually experienced violence from adult clients in the workplace. Unfortunately, the trend toward violence against human service workers is not abating as workers report being subjected to physical assault, and "all workers were routinely subjected to psychological aggression" (Shields & Kiser, 2003, p. 13), such as outbursts of anger, profanity, or intimidation. As startling as these reports are, social workers can work together to ensure preparedness.

SKILLS FOR WORKING WITH CLIENTS WHO ARE ANGRY As a social worker in a context of potential violence, you can take reasonable precautions based on common sense and effective communication. See Exhibit 4.11 to consider these strategies. The single most useful tool for working with a client who is angry or hostile may possibly be empathy. Using your knowledge of basic social work engagement skills, you can also speak in a calm, quiet, and impartial manner while maintaining a physical and non-threatening distance (Burry, 2002). Empathic responses to clients demonstrate that you recognize they are upset, would like to understand the reasons for their distress, and want to assist them. In turn, such responses can defuse clients' anger. In addition to this overall compassionate stance, you will want to observe the following:

- Recognize your own tension. Be alert for feelings of defensiveness, and prepare yourself to avoid returning angry or hostile comments.

- Acknowledge the client's strengths as you listen, and include them in your responses when they can be heard as genuine (rather than patronizing).

- Focus on positive and current alternatives that are realistic and open to the client.

- Avoid moralizing or lecturing, no matter how destructive you think the client's conduct has been.

- When necessary, focus on keeping your own control rather than on the client's anger.

WHAT AGENCIES CAN DO   In situations of potential violence, agencies need to assume some of the responsibility to avert danger. Employers can recognize and validate the hazards to which workers are exposed, so that they will report their experiences openly. Policies for handling incidents should be clear and worker-focused. Opportunities for processing and group consultation should be frequent and responsive. Preparation for workers entering possibly dangerous situations and debriefing for those who have experienced aggressive or violent clients should be standard practice and should not require that the worker initiate a formal request. Larger agencies should consider establishing a management team to address violence and should offer training on physical safety and verbal de-escalation procedures that meet worker needs and interests. The majority of social workers surveyed by NASW believe their employers appropriately respond to issues of safety in the workplace (Whitaker & Arrington, 2008). Feeling safe in one's place of work is just one aspect of the social worker as an effective practitioner. The social work profession is committed to ensuring the safety of its members in their places of employment and has a work underway to develop guidelines for social worker safety in the workplace (NASW, 2013a). Read on about issues of self-care and ethical practice.

Crisis Intervention   Crises are an expected part of the social work professional's practice life; all social workers must develop competencies for responding to the range of crises that will present themselves. Events occur in individuals' lives that bring them to social workers as mandated, involuntary, frightened, or hysterical clients. While a life crisis may be the trigger that brings the client system to the social worker, each crisis is uniquely experienced by the client system and should not be considered routine. Crises occur when an intense and stressful event creates disruption in one's life that cannot be resolved using one's usual coping mechanisms, the precipitating event is perceived as meaningful or threatening, or the individual experiences fear, tension, confusion, and/or subjective discomfort following a period of disequilibrium (Roberts, 2008; 2005, p. 13).

Regardless of the origins of the crisis (e.g., forces over which the client has no control or the result of a poor life choice by the client), the trauma felt by the client can be real and devastating. The social worker's ability to quickly and accurately conduct an assessment can support and empower the client in her or his response to the crisis situation. Crisis situations provide the opportunity for the client system (and the social worker) to grow and change, but it is incumbent on the social worker to have a repertoire of practice behaviors that can help the client view what is likely perceived to be a painful experience as the opportunity to strengthen her or his coping skills and even quality of life.

Social workers can benefit from having a theoretically and evidence-based model for responding to crises. Two models that lend themselves particularly well to assessing and intervening in crisis situations are Roberts' seven-stage model of crisis and the solution-focused approach (Roberts, 2005; 2008). Roberts' (2005; 2008) model emphasizes the importance of assessment in the ability to facilitate an

What Social Workers Can Do in the Context of Violence

EXHIBIT 4.11

Violence

- Always inform your client of the time you expect to make a home visit. Keep to that schedule as closely as possible. This is not only respectful, but you are more likely to be safe if you do not surprise anyone with your appearance.
- Consult with others (especially a supervisor) before entering a situation you think may be dangerous. If your client has a history of violence and is highly stressed or is known to have firearms or other weapons, do not make a home visit or an office visit after hours without talking it over with others who may have more experience and can consult with you about your decision.
- When you leave the office to make home visits, always inform someone about your schedule. Check by phone in worrisome (or all) situations.
- Pay attention to your surroundings. If you are on a home visit and you hear fighting or crying from outside the residence, reassess the timing of your visit. If there is activity that seems suspicious and/or is not what you expected, avoid confrontation.
- Do not put yourself physically in a position to be trapped in a potentially violent client's home. Keep a clear path to the door, so you can leave quickly if necessary.
- Make connections with the local police if you work in a dangerous neighborhood, or with local shop owners or residents in rural areas. Alert them of your plans when circumstances warrant and to the extent that confidentiality permits. Know when police should accompany you, and do not hesitate to ask them when appropriate.
- Do not challenge an angry client with rebuttals or consequences. A calm, kind, and reflective presentation can encourage de-escalation.
- If you sense that something is wrong and you are at risk, even if you cannot tell exactly what it is, leave. There is time for analysis later, and if you overreacted, you can explore that in a safe environment.
- Report incidents of any kind to your supervisor, or use an agency-designated process if there is one. This both helps you work through your own reactions and skills and facilitates the agency's effective response to its workers.
- Recognize your own tension. Be alert for feelings of defensiveness, and prepare yourself to avoid returning angry or hostile comments.
- Acknowledge the client's strengths as you listen, and include them in your responses when they can be heard as genuine (rather than patronizing).
- Focus on positive and current alternatives that are realistic and open to the client.
- Avoid moralizing or lecturing, no matter how destructive you think the client's conduct has been.
- When necessary, focus on keeping your own control rather than on the client's anger.

effective intervention plan. Specifically, initiating a rapid and timely response to the client's crisis is a critical first step in working with the client's situation. In reviewing the seven stages of crisis intervention presented here, note the prominent role that assessment plays in the professional's response both in the initial steps and throughout the intervention (Eaton & Roberts, 2009, pp. 210–212):

1. Plan and conduct a crisis assessment including lethality, dangerousness to self or others, and immediate psychosocial needs.
2. Rapidly establish rapport and the therapeutic relationship.
3. Identify the issues pertinent to the client and any precipitants to the client's crisis contact.
4. Deal with feelings and emotions by effectively using active listening skills.
5. Generate and explore alternatives by identifying the strengths of the client as well as previous successful coping mechanisms.
6. Implement the action plan.
7. Establish a follow-up plan and agreement.

Through its emphasis on strengths, solutions, the present and future (as opposed to the past), and short-term intervention, the solution-focused approach is applicable to crisis intervention. Blending well with Roberts' model described here, a solution-based assessment and intervention homes in on the assumption that crisis provides the opportunity for clients to embark on changes; in particular, changes they may have been reluctant or unable to make previously (Roberts, 2005).

Suicide   Whether it is the client who is threatening or has attempted to intentionally take her or his life, the family coping with the suicidal death of a loved one, or the client experiencing depression that you fear is at risk for suicide, the social worker is faced with an individual or a group in crisis. Just as with the other forms of client crises, social workers are ethically bound to identify and assess clients who may be at risk for suicide and take appropriate actions to prevent suicidal attempts.

As with crisis intervention, the key to effectiveness is rapid and timely assessment. An important first step in preventing a suicide attempt or completion is a suicide risk assessment (Freedenthal, 2008). A suicide risk assessment includes questions focused on any thoughts the client may be having about death and suicide. The social worker then inquires if the client has developed a plan, a time frame, and a means for carrying out the suicide. The social worker must then assess the lethality of the client's plan. For example, does the client have access to the means of suicide that she/he has stated will be used, is the plan well formulated, or does the client talk about giving away personal possessions? These are indicators that the client is serious about ending her or his life. Also important is the client or family history with suicide attempts. Exhibit 4.12 provides an example of one agency's suicide risk assessment that you may find helpful in assessing a client's risk for suicide.

When assessing individuals from vulnerable populations for suicidal ideation (e.g., veterans, persons subjected to bullying, and LGBTQ youth), the social worker must have a specialized set of skills to competently and sensitively intervene. Considerations for assessing vulnerable populations include the following (also refer to Exhibit 4.13 for a sample dialogue between a social worker and a youth who expresses suicidal thoughts):

- Recognize that your client is likely to have experienced trauma or victimization for the short-term (veterans) or on a longer term basis (LGBTQ youth), so ensure that you do not re-victimize or traumatize them in your encounter.

- Allow the client to determine the pace of the assessment—they may want to talk or they may not wish to talk during the first meeting.

- Tap into the client's previous coping strategies that she or he feels have been successful in helping them to feel better.

- Understand your client's experience to the extent that generalized information applies or that your client shares, but do not make assumptions or presume to understand their experience even if you have lives something similar.

- Acknowledge that you may have limited time for intervening.

- Do not be overly directive as this may be part of the client's history that was traumatizing.

- The client needs to be in control as the crisis that brought on suicidal thoughts has likely robbed her or him of the sense of control.

- Do not minimize the crisis or offer positive prediction—you cannot know the experience or its outcomes.

- While some clients may find it helpful to share with others who have similar experiences or families, others may not feel comfortable being in a group environment, so do not force the issue.

- Be knowledgeable about the services and resources for clients by familiarizing yourself with evidence-based intervention practices and best practices for intervening with vulnerable groups.

- Above all else, maintain a non-judgmental status.

Critically important for the social worker is knowledge of agency and legal procedures for immediately and appropriately responding to suicidal ideations, threats, or attempts. Mobilizing an emergency response is imperative. Responses may include negotiating a contract in which the client agrees not to attempt suicide, involving members of the client's support network or community, facilitating the prescription of antidepressant medication, and/or seeking hospitalization for the client. Regardless of your response, a thorough and immediate assessment is critically important for your client.

An issue that may arise is the ethical commitment the social work professional makes to honor the client's right to self-determination. However, while advocating for the client's ability to determine her or his own actions is a social work value, preserving life takes priority in the case of threatened or attempted suicide (Freedenthal, 2008). In the practice context, this means that the social worker intervenes to prevent any threat of suicide by contacting authorities or negotiating a no-suicide contract.

The Social Worker Perspective: The Social Worker as a Whole Person

Just as it is necessary to be attuned to safety issues, social workers must be sensitized to their own physical and emotional health. As you have seen, several areas of challenge in social work practice might lead to increasing discouragement and

---

EXHIBIT 4.12

Suicide Risk Assessment

CRISIS WORKER_____   DATE_____   TIME_____

Life Crisis Services, St. Louis   *Suicide Risk Assessment*
Caller Name_____   Phone#_____

1. Are you thinking of suicide? (If yes go immediately to assessment)	2. Have you thought about suicide in the last two months?	3. Have you ever attempted suicide?
Y N	Y N	Y N

RISK FACTORS	RISK LEVELS	LOW	MEDIUM	HIGH	IMMED HIGH

INTENT: *expressed intent to die; availability of means to and opportunity for attempt; specificity of plan; and preparations for attempt.*

	LOW	MEDIUM	HIGH	IMMED HIGH
Attempt in Progress	no	no	no	yes
Plan to hurt self/other	unclear	some plan	well thought out	
– time frame	in the future	>24hrs.	<24hrs	0–12hrs
– method	unclear	some plan	well thought out	
– location	unplanned	unsure	known	
Preparatory steps taken (giving items away, goodbyes, etc.)	none	some	many	
Expressed Intent to die (on a scale of 1–4, with 1 being low and 4 high, when you think about killing yourself, how much do you really want to die?)	1	2	3	4

Details of method:_____

_____

EXHIBIT 4.12

Continued

DESIRE: *no reason for living, wish to die, wish not to carry on, desire for a suicide attempt*

Wants to hurt self/others	no	sometimes	most of the time	now
Hopelessness	none	sometimes	most of the time	always
Helplessness	none	sometimes	most of the time	always
Perceived burden	none	sometimes	most of the time	always
Feeling intolerably alone	not at all	sometimes	most of the time	always
Feeling trapped, no escape	not at all	sometimes	most of the time	always
Psychological pain	**1**	**2**	**3**	**4**

(on a scale of 1–4, how much hurt, anguish or misery are you feeling right now?)

CAPABILITY: *a sense of fearlessness to make an attempt; a sense of competence to make an attempt*

Means available	have to get	close access	have with	used
History of attempts	none	one	2–3 times	>3
Exposure to suicide	no	know someone	close relative/friend	
Hx of violence to others	none	occasionally	frequently	
Alcohol/drug use/abuse	none	average use	excessive use/abuse	
Currently intoxicated/ impaired				yes
Recent dramatic mood swings	none	within last month	within last week	daily
Increased anxiety	none	monthly	weekly	daily
Decreased sleep	none	monthly	weekly	daily
Out of touch with reality	not at all	occasionally	frequently	now
Recent acts/threats of aggression	none	some	many	daily

BUFFERS/CONNECTEDNESS: *how many connectors to meaningful components of life are present?*

Immediate support present	available now	possibly available quickly	not possible
Other social supports	available now	available occasionally	no support
Future plans goals	concrete plans	some plans	no plans
Purpose in life	strong purpose	some purpose	no purpose
Ambivalence about death	much	a little	none
Beliefs about suicide	against belief sys	belief system ambiguous	no belief system
Rapport with Crisis Worker	very engaged	somewhat engaged	not at all engaged

We've been talking for a while now . . . can I ask you about how you're feeling now? On a scale from 1–4, how likely are you to kill yourself?

	1	**2**	**3**	**4**
ENDING RISK	low	medium	high	Imm-high

____follow-up scheduled   ____Pager called   ____traced call   ____Police called

Source: Life Crisis Services, St. Louis, MO

EXHIBIT 4.13

Suicide Risk Assessment with Vulnerable Populations: A Dialogue with Christopher

You are a social worker working in a multi-service not-for-profit agency that serves youth who are at risk. Your organization provides crisis intervention, emergency shelter, transitional and independent living, mobile street outreach, educational and employment programs, and individual and family treatment. Christopher arrived at the emergency shelter the previous night and you are conducting an initial intake interview and assessment the following morning. Follow along with the interview below to learn more about assessing a client who is particularly vulnerable:

Social Worker: Christopher, it's good to meet you. What brings you to our shelter?

Christopher: My parents kicked me out a couple of weeks ago. I've been staying with different people, just sleeping on their couches.

Social Worker: How are you feeling today?

Christopher: Ok.

Social Worker: Do you want to talk about your parents kicking you out of the house?

Christopher: They don't like my lifestyle, so they wanted me out of their house.

Social Worker: What is it about your lifestyle that you think they don't like?

Christopher: They found something and flipped out.

Social Worker: What did they find?

Christopher: Well, actually, they saw an email between me and my boyfriend.

Social Worker: What was it about the email they had a problem with?

Christopher: It's that I have a boyfriend. Turns out they didn't know I was gay.

Social Worker: Sounds like they didn't take it well?

Christopher: You could say that. . . .

Social Worker: How are you coping?

Christopher: So, my boyfriend breaks up with me because my parents called his parents and now his parents won't let him see me anymore. They say I'm a bad influence. I'm just done. If my parents don't want me and my boyfriend doesn't want me and I don't have anywhere to go, then I'm just outta here.

Social Worker: When you say you're done, what do you mean?

Christopher: I mean . . . what's the point?

Social Worker: Again, let me ask what you mean.

Christopher: What's the point of living if I don't have a place to stay and nobody cares what happens to me?

Social worker: It's your mention of "point of living" that concerns me. Are you saying you would think of ending your life?

Christopher: Yeah, I guess that's what I'm saying. I don't have anything to live for, do I?

Social Worker: Have you thought how you would do that?

Christopher:	Yeah, I guess I have. There's a lot of times lately that I've sat at the Metro stop and thought how easy it would be to just step in front of that train. It would just end all the pain, you know, for everybody.
Social Worker:	Do you have a time frame for when you thought you might do that?
Christopher:	No, just been thinking about it sometimes.
Social Worker:	What's keeping you from doing it?
Christopher:	My friend, he saw this flyer at the Metro stop about you guys and he said I should call. So, I did. I don't know what you can do to help though.
Social Worker:	That was a good idea. That took courage for you to call. I'm glad that you came in. Let's talk about how we can help you to deal with some of these feelings you have without you hurting yourself.
Christopher:	{Silence . . . . he begins to cry}.
Social Worker:	I see this is emotional for you. Can you share what's going on?
Christopher:	I don't really want to die. It just seems like it's the easiest choice.
Social Worker:	Let's talk about something that would help you to be able to not follow through with this plan. Would that be ok?
Christopher:	Maybe.
Social Worker:	Let's talk about an agreement that includes some alternatives for when you are feeling bad.
Christopher:	Like what?
Social Worker:	We can both come up with ideas, write them down, and even sign them . . . just like a contract.
Christopher:	What kind of things?
Social Worker:	For instance, what makes you feel good?
Christopher:	Listening to my music.
Social Worker:	What is it about the music that makes you feel better?
Christopher:	I just connect with it.
Social Worker:	How about if you agree that when you are feeling bad, you will listen to your favorite band?
Christopher:	I guess maybe I could agree to that.
Social Worker:	What else do you do that makes you feel good?
Christopher:	Getting high.

At this point, the social worker and the client go on to discuss health and non-healthy coping strategies and long-term consequences of each. Eventually, they agreed on a list of alternatives to suicide which they write down. Each signs the contract, copies are made, and they agree to re-visit the contract within within the week. The social worker offers shelter and makes an appointment to meet within the new few days. Lastly, Christopher agrees to talk with a shelter staff member if he begins to have feelings of harming himself.

EXHIBIT 4.13

Continued

ultimately to the emotional and physical exhaustion associated with burnout (Nissly, Barak, & Levin, 2005). Working with clients who experience oppression and trying to help with inadequate resources, engaging involuntary clients who may challenge your capacities, confronting a violent culture and avoiding its direct expression against you, coping with the bureaucratic tensions of managed care—all of these may seem to confound and certainly challenge your best intentions in the practice of social work. Social workers report that the lack of time to complete job tasks is the major stressor, followed by heavy workloads and working with difficult and challenging client situations (Arrington, 2008). When added to the pressure that can be induced by organizational policies and practices (discussed in Chapters 10 and 11), social workers can experience frustration and alienation day after day if effective self-care strategies are not developed.

Therefore, it is important to address two phenomena that are difficult to manage: the incidence of painful events and the occurrence of personal triggers. Social workers need self-care strategies for dealing with the stresses that lead to **secondary trauma**, **burnout**, and **compassion fatigue**. Secondary trauma can occur when helping professions react to pain experienced by client systems, while burnout is a stressful response to the work. When both secondary trauma and burnout occur, the social worker may be experiencing compassion fatigue (Wharton, 2008). Given the intensity of social worker-client relationships and demanding work responsibilities, social workers, particularly in the early stages of their careers, are at risk for negative outcomes, including physical health problems, and should commit to a proactive approach to self-care (Kim, Ji, & Kao, 2011).

Painful Events   Although most of social work practice deals with struggles experienced by client systems, and does so as a routine dimension of the territory, many (probably most) practitioners at one time or another experience a particularly jolting event that shakes their confidence and makes them question their capacity or commitment. This event is often a crisis—for example, a client suicide, a murder, an unspeakable case of child abuse, or an annihilating fire, to name a few. Such events can be experienced as trauma that splits the social worker's world wide open, taking a generally well-balanced human being off guard. Because these occurrences are a departure from the social worker's usual ability to cope, she or he may be reluctant to recognize or acknowledge an unusual reaction. Ignoring the signs can be dangerous for the social worker's own stability and ability to bounce back. Social workers need to allow themselves to be human in this profession, and the experience of overload, contrary to negative connotations, can reflect the traits of a caring, human person.

When isolated situations such as these occur, the social worker should seek and receive as much support as possible. Time away from work, debriefing sessions, or a shift in responsibility may be indicated. The social worker's response should be normalized (i.e., assurance given that her or his response is appropriate to the situation). With appropriate supports and adequate time, most workers will work through such situations and return to their practice as committed as they were before.

Personal Triggers   In a related scenario, the social worker may have some unresolved or not-quite-resolved personal issues that affect her or his capacity to carry out the work of the agency. For example, a social worker who experienced an abusive father in childhood may harbor a great deal of rage. While anger may be understandable and normal, it can be a serious problem if, for example, the worker verbally or physically attacks a client who may be suspected of child abuse. In such cases the agency response is likely to be different from that for a painful event. The event is more likely to become a supervisory issue that is addressed administratively or with corrective action. Here, the social worker can engage in a process (i.e., therapy or education) to effectively change her or his behavior. An extreme reaction can be a difficult experience for someone who is committed to the profession and has the capacity to make a solid contribution. The intensity of the trigger situation can lead a social worker to assess herself or himself as inadequate and not suited to the profession. However, with appropriate consultation with a qualified professional (e.g., supervisor, mentor, or mental health practitioner), many social workers who experience an intense experience can work through the emotional aspects in order to reconcile the personal and the professional. As all helping professionals have areas that can be triggered, this phenomenon can be somewhat normalized. While no professional can claim perfect balance, social workers are obligated to identify and respond to those triggers that may impact personal and professional well-being.

Self-Care   In carrying out the work of the profession, social workers need to learn to care for themselves, even as they care for clients. As in the case of safety, the profession, through a policy statement, has deemed self-care a priority for all social workers from students to individuals to administrators (NASW, 2012–2014b). Although this process will take on different dimensions in different people, strategies that may prove useful include:

- *Learn to use yourself as a resource:* Knowing your own responses, biases, and limits in various processes of your work is a first step in self-care. Self-awareness is critical not only in ethical, culturally sensitive practice but also in taking care of yourself. Appreciating your strengths and accepting your vulnerabilities will help you make good practice connections with clients as well as avoid expecting too much of yourself, which frequently leads to feeling disheartened. In the face of extensive client need, you may expect to save the world and inspire clients to love you while you do it. When you know that about yourself, you can laugh at your own grandiosity, let go of the need for all clients to like you, and continue to strive for competent practice.

● *Understand shared power:* Within a model of planned change that includes engagement, assessment, intervention, and termination/evaluation, the use of the strengths perspective assumes that clients have the ability to control their lives and do not need you to do it for them. Such an approach carries an unexpected benefit of reducing the pressure to be all-knowing and all-delivering, a stance that can be burdensome to clients and professionals as well as harmful to clients. In essence, you do not have to assume responsibility for clients' behaviors; you cannot be responsible for any behavior other than your own.

● *Focus on practical goals:* Related to the impulse to rid the world of all evil, you will always be disappointed and will continue to fail. When you concentrate on the strengths clients have to achieve their own goals and when you stress social justice and human rights in ways the client can relate to, you are more likely to see joint successes more frequently. This in turn will bolster your commitment.

● *Find your own systems for support:* Social workers understand the importance of support for clients but may forget its value to themselves. Support, in and outside of the workplace, is critical for the connectedness discussed throughout this book. Developing a group of peers at work can be helpful and enjoyable. Such a collective can function as a peer supervisory/support network or a strictly social group. Friends outside of work and family are also critical. Spiritual support in the way of religious affiliation or a more informal connection with the dimensions that seem most important in life can help put the occasional but inevitable disappointments and frustrations of the work into a perspective that helps keep you from feeling overwhelmed.

● *Live healthy:* Social work jobs are demanding physically and emotionally. Ensuring that you are caring for your physical health through exercise, adequate sleep, and balanced nutrition is critically important as well (Wharton, 2008).

Although social work will never be an easy job, it need not be overwhelming. Recalling the joys and connections of working with people can fill out the compensation side of the balance sheet. Taking control of your own reactions and caring for yourself will enhance your capacity to stay committed.

Just as social workers emphasize the whole client in context, they need to think of themselves as whole people, too. This means that they are not just social workers. Like clients, they are children, partners, bicycle racers, amateur politicians, musicians, parents, belly dancers, and artists. They are good in some roles and need work in others. They may connect well with involuntary clients and yet steer away from children or older adults. They may be stimulated by institutional settings or find them hopelessly oppressive. They may need to work on how to change urban agency policies effectively or thrive in rural locations where the sole agency has no walls and policy is an on-the-go venture.

Social workers may also be subject to restrictions related to social justice themselves. Their human rights may be violated daily, and it is likely that they—although inadvertently—violate those of others daily. Their identities are privileged in some contexts and devalued in others. When you see yourself as a whole human, belonging in a context, you will be more likely to engage in your human work with enthusiasm and vigor.

Sustaining Ethical Practice in the Face of Challenges   Social workers are bound by the *Code of Ethics* (NASW, 2008) to care for oneself and practice ethically in spite of setbacks and situations that do not go as planned; to keep growing in response to new ideas, perceptions, and client-informed experience; and to keep a vision of what ethical social work practice can be and how you can contribute. The ethical aspects of practice can become obscured by the struggles workers experience. Just as practitioners need to guard against rationalizing funding cuts or avoiding client contact, they need to keep the ethical considerations of omission as much in mind as those of commission. For example, not working for active reform of harmful systems (omission) is as neglectful as committing an outright violation of the *Code of Ethics*. If workers increasingly withdraw, defending themselves against the challenges of dehumanizing contexts, they can slowly lose the spirit of ethical practice without even realizing it.

One of the most effective ways to negotiate troubling practice contexts is to engage fully in a positive, supportive supervisory relationship (Smith, 2005), in which both the supervisor and the social worker can grow. As the social worker, you can contribute to that relationship by bringing something to the process; supervision is not a commodity that is given to you (or done to you), but rather it is an interactive relationship. The term *supervision* conjures up images of a hierarchy. Literally interpreted as "watching from above," it suggests an authoritarian relationship in which one member judges and corrects the other. Fortunately, social work supervision is not limited to that configuration, in spite of the inevitable evaluative nature of the term.

Supervision, then, in all its forms—individual, group, ad hoc, formal case, and peer—has the potential to contribute a great deal to sustaining ethical practice in challenging contexts. Hearing others' beliefs about the issues involved in a challenging practice situation, for example, can expand your thinking and help you work through the places in which you are immobilized or perplexed. Such a dialogic process encompasses a genuine interchange regarding the values and ethics of a case situation. Effective, collaborative supervision can provide technical support, in that it has the potential to increase and improve your practice responses; and emotional support, in that it reduces isolation and increases hopefulness. See Exhibit 4.14 for tips on effectively engaging in supervision.

EXHIBIT 4.14

Tips for Using Supervision Effectively

- Be prepared for each session by reviewing your work and identifying the issues you want to discuss.
- Demonstrate a genuine eagerness about learning more and expanding your knowledge and experience base in practice.
- Take responsibility for your work, your thinking, and your reactions.
- Trust in so far as possible in the supervisory relationship so that you do not need to cover mistakes or deny any struggles you have with the work.
- Assume a willingness to take thoughtful risks.
- Understand the parameters of your work and the expectations your agency/supervisor has of you.
- Respect the difference between the focus of supervision (how an issue affects your work) and the focus of psychotherapy (how an issue affects your emotional life).
- Remain open and non-defensive if/when your supervisor suggests you do things differently.
- Demonstrate a respectful and professional approach to relationships with all colleagues, including your supervisor.

CONCLUSION

In an effort to help you experience the contemporary climate, this chapter has addressed a range of practice behaviors related to assessment (including the very idea of assessment) along with a variety of difficult and challenging issues that exist in social work practice today. Clearly, the process of assessment involves a major combination of efforts and tools. While there are literally hundreds of instruments that social workers can use depending on the practice setting, type of client served, and the range of issues, your agency will use a few instruments that are consistent with the mission and focus of the agency's services.

Assessment is viewed as an integrated activity that arises out of an effective engagement with the client system and progresses into the action-oriented phase of the work. The tone and focus should be consistent with the practice process as a whole. Finally, assessment continues throughout the practice process, sometimes shifting the work slightly and sometimes significantly. It is a fluid activity that fits into an ever-evolving integrated whole.

MAIN POINTS

- Assessment involves dialogue to discover the goals and aspirations of the client system.

● The theoretical perspective used by the social worker has significant implications for the assessment and the intervention. Five theoretical frameworks discussed in this chapter are psychoanalytic, attachment, cognitive, strengths-based perspective, narrative, and solution-focused.

● The social worker develops a shared vision with the client system by respecting the client's preferred reality and responding honestly to it.

● Mapping is a useful addition to verbal assessment. Two types of maps that are widely used within the profession are genograms and ecomaps.

● Assessment evaluates the types of resources, both formal and informal, that the client and the client's environment can bring to bear.

● When resources are not present or adequate or available to client systems, social workers need to respond creatively, appropriately, possibly with social action.

● Assessment and planning move from a shared vision to specific details regarding the intervention. Internalized oppression and the emotional impact of change can influence the entire assessment process and the client's capacity for participation in planning.

● Social workers are ethically bound to be clear and honest with clients about specific agency constraints and requirements that will influence the client's experience of the work.

● The phenomenon of involuntary/mandated/non-voluntary clients creates challenges for social workers that can be met effectively if framed adequately and met with a basic approach of respect and willingness to listen to the client.

● Violence is a pervasive quality in U.S. culture and influences both the work of the social work practitioner and the potential for her or his personal safety. Social workers can engage in social justice-oriented practice to try and minimize societal violence, and they and their agencies can develop skills and policies to deal with situations of potential violence associated with clients.

● Secondary trauma, burnout, and compassion fatigue reflect physical and emotional exhaustion in a complex practice context; it may arise out of painful events and personal triggers and can be mitigated through appropriate agency and social worker response.

● Several aspects of holistic self-care can carry you through difficult moments and are consistent with ethical practice. As in your work with clients, think of yourself as a whole person, with strengths and vulnerabilities and a need for your own systems of support.

● Effective supervision is a useful interactive relationship for negotiating difficult practice contexts.

EXERCISES

a. Case Exercises

1. Go to www.routledgesw.com/cases and review the case file for Emilia Sanchez, including the genogram and ecomap of the family for both its content and form. Click on the Assess tab and view the tasks you will need to complete. To prepare for this exercise, also review "Focus on Strengths" in the Values Inventory.

    Using Emilia as the anchor family member, develop two genograms. The first should reflect her relationships prior to age 14 and the second should be an "update" to her current age of 24. Include as much information as possible while ensuring that the drawing is informative and clear. After completing the first genogram and before completing the second, develop an ecomap that represents your interpretation of the systems and networks supporting Emilia's "change" (her involvement in drugs) that might have led to her estrangement from her family.

    In class, partner with another student and exchange the three drawings. Are they similar? In what respects do they differ? What information is particularly helpful? What facets of Emilia's life are the most effectively represented in a social mapping format? Which ones are the most challenging?

2. Go to www.routledgesw.com/cases and review Emilia's video vignette. After viewing the vignette, complete the following exercise.

    Emilia Sanchez has come to you for help because she has decided she must conquer her substance addiction. Emilia's long history of substance abuse, her family's outrage regarding the out-of-wedlock birth of her son Joey, and the following abortion of another child resulted in her feeling discouraged. Specifically, she is doubtful about her ability to make a place for herself in the family again and to make the changes she wants to make. Using the strengths perspective the social worker can identify and assess Emilia's strengths. Respond to the following:

    a. What was the most challenging aspect of identifying the strengths of someone who has had the number of challenges experienced by Emilia?

    b. How do you as the social worker encourage Emilia to recognize her strengths?

    c. Identify the strengths pointed out by the social worker in the vignette.

3. Go to www.routledgesw.com/cases and review the case of Carla Washburn. Click on Phase 2: Assess the Client System. Review the goals for this phase and complete Tasks 1, 2, and 3.

4. Go to www.routledgesw.com/cases and review the case of Hudson City. Address each of the following areas:

    a. Click on Engage and Discover. Select Case Files and review Your History, Your Concerns, and Your Goals. Respond to the questions posed under Your Goals.

INDIVIDUALS: ASSESSMENT AND PLANNING   **171**

b. Continuing with the engagement phase, click on Critical Thinking Questions. View Your Questions and respond to each of the questions presented.

c. Moving to the assessment and planning phase, click on Assess the Situation. Select Biopsychosocial Perspectives. Review the biological psychological, social, and spiritual lenses and respond to the questions related to each area.

5. Go to www.routledgesw.com/cases and review the Brickville case. Address each of the following areas:

a. Click on Engage and Discover and complete the Tasks for this section with specific emphasis on strategies for engaging Virginia and other key persons who are relevant in her life.

b. Review the ecomap centered on Virginia Stone and her family. Identify all relevant linkages Virginia has within her family and community, the character of those relationships, and strengths and areas for changes.

c. Review the Town Map and develop list of Virginia's needs on the individual, family, group, and community levels.

6. Go to www.routledgesw.com/cases and review the case of Hudson City. Two professional issues are relevant for the social worker depicted in this case: dual relationship and self-care (compassion fatigue). After reviewing the case materials, conduct a literature search regarding one or both of these issues and develop practice strategies that are appropriate for your future as a social worker.

b. Other exercises

7. To begin the process of developing engagement and assessment practice behaviors, complete the following role-playing exercise. Begin by partnering with two other students.

Using Confrontation: Engage in brief re-enactments of the following scenarios with each student assuming the role of the social worker, the client, or the observer. Use the engagement and assessment practice behaviors highlighted in this chapter to role-play the beginning phases of work with clients in these situations. Upon completion of the role-play, each member of the triad will provide balanced feedback regarding the others' performance of skills. Select from the following list of potential client scenarios:

a. Client sporadically attends scheduled sessions.

b. Client reports that when she was angry with her 10-year-old son at the mall she spanked him in public.

c. Client is having difficulty obtaining employment. You recognize that the client's style of dress and hygiene may be a concern for employers.

d. Client continues to use language that you find offensive.

e. Your colleague is not completing tasks and you are experiencing negative consequences.

8. Review and discuss with other students the following situations. Evaluate your level of comfort in them. What makes you comfortable or uncomfortable? Be specific. Upon completion of the discussion, brainstorm with other students strategies for maintaining your safety.
   a. Conducting a home visit
   b. Working after dark at your agency
   c. Driving a client to an appointment
   d. Working with young males (if you are female) or females (if you are male)
   e. Working with individuals with mental illness
   f. Working with individuals with substance abuse issues
   g. Having an initial meeting with a client who is unknown to you
   h. Working with a client with a criminal record

9. In writing, reflect on situations in which you have been involved that resulted in a confrontation between you and another person. If you cannot recall such a situation, remember a situation in which confrontation may have been appropriate, but did not occur. Imagine in that situation that you were confronted by a caring individual in your life. What was your reaction to being confronted? How did you receive feedback? How do you give feedback? What is your level of comfort ability with confronting others? What were the benefits to being confronted?

10. Quick Guide 7 provides questions to consider regarding the cultural heritage journey that each of us takes. To gain insight into your own cultural heritage journey, reflect on the items listed:
    The cultural heritage journey of both clients and social workers impact their perspectives of life and work. Consider the following:
    - Where do my family's roots begin?
    - If my family emigrated to the United States, how long has each parent been living here?
    - If I am in an adoptive family, what do I know about my biological family's journey to the U.S.?
    - What are my family's traditions that have been passed through the generations?
    - Do my family's traditions relate to religion or spirituality, holidays, rituals, and or significant events for the family?
    - What values have I learned from my family—may be related to ethics, wealth, religion and spirituality, race, ethnicity, health, education, sexual orientation, and image?
    - Have I challenged any of my family traditions or values? If so, what challenges have presented themselves and how have I responded?
    - What significant life experiences have shaped my view of the world, specifically people who are different from me?
    - How will my cultural heritage impact my social work practice?
    - Are there changes or additions that I would like to make related to my traditions? How might I start?

CHAPTER 5

Social Work Practice with Individuals: Intervention, Termination, and Evaluation

Where after all, do human rights begin? In small places, close to home—so close and so small that they cannot be seen on any map of the world. Yet, they're the world of individual persons: the neighborhood he lives in; the school or college he attends; the factory, farm, or office where he works.

Eleanor Roosevelt (1958)

Key Questions for Chapter 5

(1) How can I prepare for intervening with and empowering individual client systems? (EPAS 2.1.10(c))

(2) How can I use the strengths-based perspective to guide the development of intervention, termination, and evaluation strategies with individuals? (EPAS 2.1.10 a-d)

(3) What are the social work roles that enable me to effectively intervene, terminate, and evaluate with individuals? (EPAS 2.1.10(c))

(4) How do I determine the appropriate evaluation tool(s) for use with interventions with individual client systems? (EPAS 2.1.10(d))

THIS BOOK EMPHASIZES THE CONTEXT OF THE SOCIAL work intervention as it shapes the meaning of experience for clients and for social workers as they move through the social work intervention. In this chapter, we will explore the impact of context on all aspects of practice. Context both shapes what you do as well as is shaped by what you do. Whether you are working with someone recently diagnosed with a chronic illness such as multiple sclerosis, an adult child who is experiencing stress from caring for an aging parent, or an immigrant who is seeking U.S. citizenship, you interact with the choices and capacities of individuals, the systems in which people are embedded, and the accessibility of existing and potential resources. All of these dimensions contribute to the totality of the social work practice context.

There are multiple ways to intervene. Viewed through a critical constructionist lens (i.e., knowledge is not transmitted, but created, acquired, processed, and relative) (Barker, 2003, p. 93), a useful, direct intervention for one person will not necessarily meet the needs of another person in a similar situation. For example, a single woman with a six-month-old baby may identify job training and finding suitable child care as the goal of the intervention. In contrast, another woman in seemingly the same circumstances may need assistance with locating housing, handling roller-coaster emotions, and keeping a safe environment for her infant. Yet another single mother will need intensive advocacy efforts in order to deal with discriminatory practices in her employment. Social workers must expect differences and advocate against the one-size-fits-all interventions that are sometimes considered optimal by social service organizations and/or public policy.

In recognition of individuals' unique circumstances, you will locate the starting point of your work on the most pressing issue, as defined by each client system. This decision should reflect the priorities for action that are agreed upon in the assessment and planning phase. Each client is the chief negotiator of her or his journey; therefore the work will first assume and then reflect the strengths and capacities to be successful in that journey. From this starting point, the chapter examines the generalist practice competencies and practice behaviors of supporting client strengths within client environments. We will also look at the variety of social work roles and methods that support client–worker relationships, including strengths, narrative, and solution-focused approaches. This chapter also includes a case example focused on Thomas and his road to empowerment that will allow students to bridge the gaps between strengths-based, environmental, social justice, human rights, and social construction perspectives. As do social work interventions, this chapter will end with a discussion of termination and evaluation of the planned change intervention, focusing on strengths-based strategies to help the client to maintain goals they have accomplished.

SUPPORTING CLIENTS' STRENGTHS IN DEVELOPING INTERVENTIONS

Through its theoretical lens, the strengths perspective offers a focused, committed effort for identifying, expanding, and sustaining the resilience and assets that clients bring. Using a strengths-based method suggests that the worker will not only seek out and identify client strengths in the assessment (see Chapter 4) but will also support and maximize them throughout the working relationship. For many practitioners this effort will be the major focus of the work. The strengths-based approach views supporting clients as central to the worker's action. Social workers may need to remind themselves of the client's strengths and centrality as our culture, even for the most strengths-oriented practitioners, focuses on "what needs fixing." Strengths, like all human dimensions, need recognition, validation, and nourishment to remain vital. The following discussion will highlight intervention from the perspectives of strengths, narrative, and solution-focused approaches.

Strengths-Based Perspectives and Intervention

Consider four elements of the strengths-based approach as they apply to the practice context and the action of intervention. Saleebey (2013, pp. 109–111) identifies these as follows:

- *The social worker identifies hints and murmurs of strength even in the struggle:* As the client relates her or his current situation, typically focused on the challenges and stresses, the social worker can "listen" for the strengths and healthy aspects which every client possesses.

- *The social worker stimulates the discourse and narratives of resilience and strength:* Most of us have difficulty recognizing positive aspects about ourselves. The role of the strengths-based social worker is to identify the strengths, affirm positives that are offered by the client, suggest the possibilities that exist for the client, and be grounded in the client's daily life.

- *The social worker acts in context:* Through listening for the strengths as the client shares her or his concerns, the client's competencies become apparent. These competencies can then become the focus of the assessment and intervention process and they guide the development of goals. This "project" results from the social worker and the client collaborating to identify the strengths and plan of action.

- *The social worker can work with the client system to move toward normalizing and capitalizing on one's strengths:* The product of a strengths-based assessment is the intervention in which the articulated strengths and resources are brought together, normalized, and shared with others.

Acting in Context   The social worker's activity centers on helping the client use the strengths she or he is beginning to recognize, as well as those already discovered, and to link them with her or his goals and dreams. That kind of effort might lead, for example, to social worker support for more independence or more assertiveness. Inherent within the strengths-based perspective is the belief that social justice is the "ultimate goal" which requires a commitment to viewing the client-social worker relationship as one of equality (Gray, 2011, p. 8).

Consider a client who has experienced trouble interacting with her landlord and now feels that she cannot negotiate a lease arrangement. The social worker can help her consider the advantages of establishing a positive rental history as they work together to identify appropriate options. The social worker can also encourage the client to recognize her capacity to understand the business-focused details of renting and to negotiate for those specific items or conditions that she feels are important and fair. For example, if she is willing to paint the living room walls, she can ask the landlord to provide the paint and reduce her rent for one month. This endeavor involves some risk taking to stretch beyond her usual comfort zone to initiate change, but the action also offers an unusual opportunity to decide and act on her own behalf. When the venture is successful—the landlord agrees to provide paint and a rental rebate in exchange for painting the living room walls—the client experiences the benefit of adding another competency to her growing list. Conversely, when the effort does not go as planned, she becomes more aware of the specific areas in which she wants to direct her energies to make changes.

Capitalizing on Strengths   To focus on strengths is the consolidation step in which the social worker and client together recognize the successes they have had in establishing and stabilizing the client's competencies. Keeping with the same example, assume that the client found an apartment she likes at a rate she can afford. The social worker would then affirm the skills the client demonstrated, encourage her to generalize those particular skills into other arenas, and support her ability to

© Alexander Raths

generate new skills. The social worker can also help the client recognize that she is building useful relationships in the community. For example, when she has established a record of reliability in paying her rent on time and maintaining her apartment, her landlord is likely to become a resource if she needs a reference for a job or wants to find a larger apartment across town. In order to help the client grow, the social worker's efforts may include education, advocacy, and support for the client's capacity. Such efforts can enable the client to develop a network that links her accomplishments to her goals. By normalizing the task, in this case, of securing housing, the social worker not only demonstrates the client's capacities, but also reinforces the availability of community resources, and establishes that disengagement, or ending the work, is appropriate once the goal has been achieved. The ultimate normalization is the client's continued development and the recognition that she or he can successfully manage the everyday tasks of living.

Narrative Intervention

Building on the deconstruction of client perceptions that occurred in the assessment process, a narrative-oriented intervention remains focused on the client, her or his strengths, and the meaning that is assigned to current and future realities. Questions posed by the social worker are central to narrative-based intervention. As opposed to eliciting information and interpretation, intervention stage questioning is aimed at helping the client to move toward implementing changes in order to reach her or his goals. Return to Chapter 4 for a review of narrative-focused questioning strategies.

Once the original concern or issue is deconstructed and a plan is conceived, the social worker continues to use strategies (e.g., questions) intended to externalize the client from *being* the problem to having a relationship *with* the problem (Nichols, 2011). As the intervention is underway, the social worker continues to help the client view her or himself from a position of strength and working toward re-authoring the original issue. Upon achieving success, the client can celebrate and share her or his new perceptions by sharing the news about changes with people who are important to her or him. As with many interventions, the path from beginning to end may not be linear. The client-centered nature of a narrative approach may require both the client and the social worker to return periodically to the early discussions of the problem and reconstruct it in light of new insights that unfold throughout the intervention.

Solution-Focused Intervention

Using a solution-focused approach to the intervention provides the client and the social worker with a clear-cut strategy for arriving at the client's goals in a relatively brief, time-limited manner. As you recall from the discussion in Chapter 4 on solution-focused assessment and planning, the client engages in a self-evaluative process (evaluative questions) to learn how she or he views the situation. The client has also envisioned times when the problem did not (exception questions) and will not (miracle questions) exist, thus lending insight into possible strategies for resolving the problem using the client's existing and created strengths and resources. In the intervention phase of change, these questions continue to be important as reminders of the goal and connections to the client's past and future life.

Also used in the assessment and planning phases, scaling questions continue to be a critical aspect of the solution-focused intervention. Prompting the client to regularly assign a numeric value to a particular issue, experience, or behavior can empower her or him to continue to engage in self-evaluation and mutual feedback with you (Lee, 2009). While this practice strategy is evaluative in terms of progress (or lack of), it can also serve to motivate and empower the client and inspire confidence that change can, in fact, be a reality. Being able to acknowledge small, concrete gains can be a powerful motivator for continued work (for both client and social worker). Exhibit 5.1 provides examples of scaling questions and strategies for monitoring scaling questions over the course of the social work intervention.

Due to the emphasis on concrete solutions, the solution-focused intervention can be impactful and empowering for the client. As a social worker using this approach, as the sole approach or in concert with another intervention model (as discussed in Chapter 4), your attention will be directed toward maintaining the client's motivation and focus on solutions and change along with reinforcing your respect and confidence in the client's capacity to reach the desired outcome (Lee, 2009).

EXHIBIT 5.1

Integrating Scaling Questions into a Social Work Intervention: Virginia Stone

Virginia Stone of the Brickville neighborhood (www.routledgesw.com/cases) has identified a number of stressors in her life, including caregiving responsibilities for her mother and two grandchildren, redevelopment efforts in her neighborhood that may result in her family losing their home, and unresolved grief over the loss of her daughter and three nieces/nephews in a fire twenty years earlier. Scaling questions can be an effective strategy to help Virginia identify and prioritize areas of concern and strength. Scaling questions can also be infused into the intervention process to frame goals and track and empower Virginia's progress toward her desired goals.

During the engagement and assessment phases of work with Virginia, she identified that she was feeling particularly stressed about the potential loss of her family's home. The following dialogue provides examples of the use of scaling as an interventive strategy during several sessions:

Session #1—Engagement/Assessment phase:
Social Worker: "Virginia, on a scale of 1 to 10 with 1 being the worst and 10 being the best, how would you rate stress today regarding your housing situation?"

Virginia: "I would say my stress level is a 1. I am so worried that we're not going to be able to prove the house belongs to my Grandma Stella. Then, we won't even be able to sell it to that developer if he does move in here and take over."

EXHIBIT 5.1

Continued

The social worker and Virginia worked together to develop a goal to reduce her stress by taking action toward obtaining proof of family ownership of the house. They mapped out strategies which the social worker documented both in Virginia's case file and the client contract.

Session #3—Intervention phase:

Social Worker: "Virginia, since we last reviewed your progress toward your goal of reducing your stress, you have been to Legal Services to ask for help in locating a deed to determine ownership of the house and they told you that they believed you have a strong case. Using the same 1-10 scale, how would you rate your stress today related to this issue?"

Virginia: "I am feeling a whole lot better about things today. I would have to say I'm at a 5 right now. It seems there is a chance we will be able to prove Grandma Stella was the rightful owner and that it passes to my mom when Grandma died. I'm still worried, though, that something will go wrong."

The social worker asked Virginia to consider other actions that she could take at this point while she was waiting on the outcome of the legal investigation. Virginia acknowledged that there was little she could actually do, but wait. Once she realized that much was out of her control, she acknowledged that she was feeling better about things. The social worker documented the events and Virginia's perception of her current stress level.

Session #8—Intervention phase (three weeks later):

Virginia: "I heard from Legal Services and they were able to find proof that Grandma Stella is the rightful owner of the house! I can't tell you how much better I feel about this thing."

Social Worker: "Using that 1 to 10 scale, we have used before, where would you rate your stress level on the housing issue today?"

Virginia: "On this one issue, I would say I'm at 8 today! I'm still worried about losing the house to that developer fellow, but at least now, I know that if we do have to sell out, we will get paid a fair price for it so hopefully we can find another place for all of us to live."

With this particular stressor significantly improved for Virginia, the social worker and Virginia were able to turn their full attention to addressing other goals. While the client can certainly pose scaling questions to themselves, doing this exercise with the client may help her to engage in self-assessment and realize the differences even before the social worker has the opportunity to acknowledge the change, leading to a sense of empowerment. Despite the fact that Virginia had been working toward other goals, this stressor was overshadowing her motivation to fully engage in working on the less tangible issues of caregiver stress and unresolved grief.

Greene and Lee (2011) offer the following suggestions when using a solution-oriented approach:

- Use questions to identify solution-focused patterns that guide the development of client-focused goals.

- Use solution-oriented tasks between meetings to amplify patterns and promote stability and change.

- For clients who are motivated, direct behavioral tasks are developed for completion between meetings (e.g., engaging in a specific activity with client's family members). For clients who acknowledge a problem, but see the cause being external to themselves, observational tasks are best for between-meeting activities (e.g., attending a presentation on parenting).

- At the end of each meeting, provide an end-of-meeting message to affirm the client's work.

- Use client's beliefs, language, metaphors, and figures of speech to guide the tasks to be accomplished.

- Remember, clients have strengths, competencies, and resources to make change (pp. 126–127).

Cognitive Behavioral-Focused Interventions

Considered to be an effective modality for intervening with clients experiencing a range of mental health challenges, cognitive behavioral therapy (CBT) is another approach available to and used by many social workers working in clinical arenas. CBT is based on the premise that thoughts drive one's feelings and actions and changing one's thoughts can lead to changes in one's behaviors (National Association of Cognitive-Behavioral Therapists (NACBT), 2013). CBT has been well-studied over the past four decades and the evidence suggests that this approach is particularly effective with such illnesses as depression, generalized anxiety, panic and phobic disorders, and posttraumatic stress disorder (Butler, Chapman, Forman, & Beck, 2006). The meta-analysis conducted by Butler and colleagues (2006) also determined that CBT has moderate effectiveness within interventions addressing marital stress, anger, childhood somatic conditions, and chronic pain.

As will be evident, CBT shares similarities with other therapeutic approaches. The primary components of CBT include (NABCT, 2013):

Brief and Time-limited format—with a focus on behavior change, the number of meetings with the client can be limited to a specific number with the client continuing to work on change following the formal relationship.

Collaborative relationship—the social worker and client work together to identify and implement goals. The social worker's role is to listen, teach, and support, while the client's focus is on expression and learning.

Structure, Direction, and Education—facilitating change is completed within the context of a structure that includes the social worker creating an agenda with specific techniques being taught during the session and in the form of homework to be completed between sessions that emphasizes learning as well as unlearning behaviors.

While the delivery of CBT requires advanced training and experience, generalist social workers can incorporate aspects of behavioral-oriented strategies into their interventions as well as identify those clients who may benefit from working with a CBT-trained clinician.

Strengths-Oriented Practice Skills and Behaviors

When used effectively, many social work practice skills and behaviors can affirm client strengths and support their underlying position in the work. For example, the familiar "reflection of empathy" has remained a time-honored practice behavior because it mirrors the respect and regard for human beings that is inherent in social work. The practice skills and behaviors discussed in the following sections have also earned their place in social work practice history because they are consistent with the values of social justice and respect and dignity.

Supporting Diversity   As discussed in Chapters 3 and 4, a client's cultural background is a potential source of strength rather than an obstacle or deficit. As you strive to become a culturally competent social work practitioner, there are four assumptions on which you can build your skills: 1) reality is social constructed; 2) diverse worldviews need to be appreciated; 3) multiple realities affect individual personalities; and 4) diversity education has a positive impact on the journey to cultural competence (Kohli, Huber, & Faul, 2010, p. 266). Within the social, cultural, political, and historical context, these assumptions can be used to inform the way in which you view not only your client, but yourself.

An appropriate practice behavior is thus to use the traditions or aspects of your client's culture that nourish and give meaning to life. For example, many cultures demonstrate far greater respect for authority or the wisdom of age than U.S. culture has traditionally done. When you work with clients from such a culture, learning the practice behaviors that will support those values is critical because they are assets and can be incorporated as such into the intervention phase of your work. Although you should always confer with your clients regarding your actions, remain especially sensitized to the "fit" of your approach with your client's culturally influenced sense of propriety. Would it be appropriate, for instance, to encourage the client from such a culture to participate in a rent strike aimed at a prominent elder statesman who owns the property, if such an action would likely become highly adversarial?

SUPPORTING CLIENTS' ENVIRONMENTS

Having confirmed the importance of recognizing and supporting client strengths and cultural influences in the intervention process, we will focus now on identifying and intensifying the strengths of clients' environments. This dual emphasis on person and environment is consistent with the traditions of the social work profession. Accordingly, social workers often direct their efforts toward making clients' environments more responsive to client needs. This approach looks to the institutions and policies outside of the client rather than viewing the client's difficulties as a symptom of inner pathology. In maintaining this focus on improving environmental response, the work reflects an underlying commitment to a social justice framework. In the process, clients are assumed to possess the capacity both to identify their difficulties and to participate in the work necessary to alleviate them.

Principles for Taking Environments into Account

In a classic application of a systemic intervention that embraces the client's environment, Wood and Tully (2006) identify six major principles. Consistent with the basic tenets of the social work profession, these principles strongly emphasize a social justice orientation.

The Social Worker Should Be Accountable to the Client System   In being accountable to the client system, the social worker responds to the client's perception of the problem or situation despite her or his perception of its accuracy or clarity. At this point in the intervention, the client's perception is the reality that she or he is working with and, as we know, a basic tenet of the social work profession that we "start where the client is." If, for example, an individual client is experiencing substantial difficulty in accessing certain public assistance benefits because he does not speak English fluently, the social worker will direct energies to addressing that need. Strategies may include arranging for an interpreter, requesting an agency worker who is fluent in the client's native language, or simply accompanying the client to advocate for appropriate and equitable treatment.

The Social Worker Should Follow the Demands of the Client Task   Creating a "Frame of Reference" is a strategy for conceptualizing the demands of the client task (Wood & Tully, 2006). The Frame of Reference can be illustrated through a quadrant diagram covering the parameters of the social worker's tasks (see Quick Guide 12). This scheme suggests that practitioners may (A) work with client systems on their own behalf, (B) work with client systems on behalf of themselves and others like them, (C) work with others (non-clients) on behalf of client systems, and (D) work with others (non-clients) on behalf of categories of persons at risk. The principle of following the demands of the client stresses that the social worker will look beyond the individual client in order to see if there are others in the same situation. If so, the worker needs to adapt by moving from one quadrant to another or by functioning in multiple quadrants.

To learn how to apply the Frame of Reference into your work, consider the case of a social work practitioner working with the mother of a child with disabilities who has no viable transportation to school. The social worker, with the mother's agreement, can locate a driver, or a person to accompany the child on the regular school bus, or request extended school hours so that the mother can pick up the child after work (working with a client on behalf of the client, **Quadrant A**). However, if the social worker discovers that there are four other children with disabilities in the school who lack adequate transportation, she or he, still working with the mother, might organize and coordinate a team of drivers for all five children in the school (working with the client on behalf of the client and others like her, **Quadrant B**). If the social worker's efforts were directed at enlisting the aid of a community group to help locate and fund a small van that could hold six children, the work would fall in different quadrant (working with others out of concern for specific clients, **Quadrant C**). Finally, if the worker proposes statewide transportation funding to the legislature on behalf of a category of clients—in this case, children with disabilities—the work falls into **Quadrant D**, working with others on behalf of a category of clients. Although the worker's efforts may vary, move from one quadrant to another, and operate simultaneously, the work is always performed on the behalf of clients.

QUICK GUIDE 12 FRAME OF REFERENCE FOR STRUCTURAL APPROACH TO SOCIAL WORK TASKS

INTENDED BENEFICIARY		
FOCUS OF INTERVENTION	SINGLE	MULTIPLE
Clients	Work with clients on their own behalf	Work with clients on behalf of themselves and others like them
	Quadrant A	**Quadrant B**
Others	Work with others (nonclients) on behalf of clients	Work with others, such as (nonclients), on behalf of a category of persons at risk
	Quadrant C	**Quadrant D**

Adapted from Wood & Tully, 2006

The Social Worker Should Maximize the Potential Supports in the Client System's Environment   This generic principle requires the social worker to identify, access, and, where necessary, modify or even create the supports that a client system needs. These supports might include such resources as temporary housing for a family who is homeless, hospice services for a family dealing with an older member's imminent death, and child care services for a young mother who is employed outside her home. Chapter 3 discussed this aspect of resource development; it remains a central focus for this work phase as well.

The Social Worker Should Proceed from the Assumption of "Least Contest"   This tenet holds that the social worker should use the minimal amount of pressure that is required to meet a client's need. For example, a social worker can explore the availability of affordable child care for a client before pressuring the mayor to initiate a city-sponsored child care system. If these early attempts fail, and there is no adequate, affordable child care, the social worker can shift her or his focus to mobilize an effort to facilitate change at the community level.

In a different situation, the "least contest" approach might encourage the social work practitioner to contact another service before engaging in advocacy. For example, a practitioner working with a young woman who was not accepted into a local parent support group might make a referral to a different group. If a series of appropriate referrals does not yield the desired results, the social worker can consider other approaches of greater contest, such as advocating with the person who organized the original group.

The Social Worker Must Help the Client Deconstruct Oppressive Cultural Discourse and Reinterpret Experience from Alternative Perspectives   Embracing this principle guides the social worker and client to discuss the larger cultural and societal issues that impact the ways in which the client experiences the world. Exploring the oppression, discrimination, and/or violence that may be a part of the client's life experience can help the client gain insight into the origins of these negative forces and construct alternative perspectives. Creating a different view of oneself can free the individual to develop a new, healthier perspective. Consider the older adult refugee who recently came to the U.S. with her large extended family after fleeing her home country and living for several years in a refugee camp. This client brings with her a lifetime of oppression and discrimination, and she and her family have been victims of violence in their home country. As a social worker, your role can be to help her to create a new perspective on the world in which she now lives. By examining her previous life experiences within the context of her new life, you can work with her to develop an alternative to her earlier life of fear and dread.

The Social Worker Should Identify, Reinforce, and/or Increase the Client System's Repertoire of Strategic Behavior for Minimizing Pain and Maximizing Positive Outcomes and Satisfaction   Also known as the **"minimax"** principle, this principle cautions the social worker against holding the client responsible for a lack of community responsiveness in the development of resources or support. The principle recognizes, however, that the client may need assistance in developing behaviors that are likely to elicit cooperation from the community. For example, clients who have been excluded from the community on various levels might present as eager to the point of aggressiveness, and demonstrate a lack of interest in using the grievance process of an organization to file a complaint when they have been treated unfairly. This group of clients will likely improve their chances of being accepted if they curb their enthusiasm somewhat, by speaking in a softer voice and initially using a grievance process, for example.

In applying this principle, social workers are reminded that our culture has led many people to blame themselves for the violence or oppression that they have. Therefore, it is important that social workers, as well as client systems, do not attribute difficulties in accessing resources to our/their own failures by "blaming the victim." Social workers whose clients have adopted this perspective can employ consciousness-raising techniques to challenge these beliefs and transform personal issues such as child abuse and violence against women into political issues. Returning to the earlier example of the older adult woman who has recently arrived in the U.S., a focus of your work with her can be directed at examining the violence that she experienced in her home country. Within the social work intervention, you can help her to recognize that the perpetrators of the violence were responsible for the violence and, despite the pain and guilt she suffers, she and her family are not responsible. Freeing the client from the guilt can aid her in maximizing her opportunity to create a new life with her family in the U.S.

Social Workers Should Apply the Principles to Themselves As well as directing these principles toward the client intervention, social workers can benefit from ensuring they are practicing the same principles they promote to clients. Accordingly, social workers make clear, accountable service and/or therapeutic contracts and explore common practice issues with other social workers. Social workers are also encouraged to apply the principles to their work in agency settings.

Environment-Sensitive Processes and Skills

The groups of skills and behaviors discussed earlier can also be applied to the environment when relevant to supporting clients' effective use of environmental supports. Following is a discussion of practice behaviors that can be used with individual.

Providing Information Client systems often want, need, and ask for information about the environment, which may be their most valuable resource. You may be uneasy that you will over-influence decisions, be too directive, or create dependency. While such concerns are legitimate, they also can be managed. When you consider that information in our society is clearly linked to power, you may become less reluctant to provide it to clients. The challenge with providing information arises when it is confused with advice or is strongly one-sided. This concern becomes more difficult to mitigate with the understanding that you are never able to have all the information related to a particular situation or circumstance. Still, you can offer what you know as a simple proposition, always framed by the limits of your knowledge. Consider again the situation of the older adult refugee with whom you are working on creating a new and healthier perspective. She is concerned about the impact of previous experiences of violence on her younger female family members. She is aware that you have been working with her daughter and granddaughters. She has asked you to share the information that you know about her daughter and granddaughters and to direct them to not discuss their experiences with anyone. While you are sensitive to her concerns, you are ethically bound not to share information about other clients without their consent. Moreover, your commitment to practicing social work in an ethical manner would not allow you to advise any client system in their beliefs or behaviors.

Be clear that you do not hold any expectations about what clients do with the information. Information is typically not yours to be given away only with certain restrictions. For example, you are a social worker working in child welfare helping a young couple to learn more effective methods to care for their three-year-old son who has been diagnosed on the autism spectrum. You may want to generate a compilation of resources you think would be helpful to the parents. You may believe that they would benefit from respite services because you suspect it will be helpful for the mother to develop some trust in a caretaker outside of the family. Therefore, you provide her with the names of several agencies that provide respite services. When she does not follow up on this information, you may become frustrated or even irritated. In this case, your "information" was interpreted by the client as "advice," and the mother is not at a point where she is able to or agrees to act on the information in the way you assumed she would. This simple example illustrates the problematic nature of an investment in giving information. Fortunately, you can process this kind of situation with the client so that your agenda becomes transparent and the client is free to accept or reject it.

Refocusing and Confronting   Bringing the client back to the original focus of the work is known as **refocusing** and can span several possible directions, including two discussed here: referring back to purpose and confronting clients' beliefs, plans, and/or behavior. Social workers frequently use refocusing to assist clients in taking advantage of the potential in their environments.

When clients begin to digress into arenas other than the agreed upon focus area of your work (based on the previously developed contract), a simple reference to the original agreements regarding the focus of the work during the action phase of the work is sometimes all that is necessary. Occasionally the client (or the social worker) will simply become distracted by other compelling issues. Although such diversion is understandable, you are responsible for using your time with the client productively. For example, the social worker may say the following to a client who she or he believes has lost focus on the previously agreed upon goals: "Let us revisit the goals that we developed when we created our contract for working together. I think it would be helpful for both of us to check in with the goals to determine if we are both staying focused on the plan that we laid out. If we feel that the goals are no longer realistic, we can discuss revising our original plans."

In other situations, the client may not carry out the plan or contract as agreed because the plan is not truly embraced by the client as a priority, is proving uncomfortable, or the client is unable to implement the plan for some reason. For example, your client, Jane, established a plan to seek counseling because she had been sexually abused and was experiencing many painful memories. However, initiating that kind of contact involved breaking the culturally and family imposed rule of conduct that she should never discuss anything so personal with an outsider. She finds the effort to initiate this counseling more difficult than she had imagined. An alternative to refocusing is to confront her very gently, even though she may simply need more informal support and acknowledgment from you to manage the unanticipated struggle. You may say to Jane, "Help me to understand your concerns about talking about your painful memories with an outsider." This development might also signal that the plan is not working for Jane and is not likely to lead to the outcome previously anticipated. In situations like these you need to return to the goals, review the client's experience, and make changes as necessary.

Significantly, clients often perceive gentle confrontation of their behavior as not only helpful but supportive. If the social worker carefully contrasts what a client said she or he would do with what she or he actually did, the client might experience this activity as respectful because it affirms her or his capacity to meet the agreed-upon commitments. For example, you may say to the client: "Your goal was to obtain employment and you did not apply for any jobs this week. Please help me understand how your actions will help you reach your goal." Such a process must be carried out sensitively, of course, to prevent the client from experiencing the confrontation as hostile and argumentative. Confrontation is certainly not the first approach to use in working with clients in such situations; rather, referring to purpose (refocusing) should be tried first. When it is used selectively and sensitively, however, gentle confrontation can convey hope and respect.

There will be situations, however, in which a digression from the intended focus of your work with the client system is warranted. You will be called upon to exercise your professional judgment to be able to recognize when a digression warrants a new focus. Returning to the example of Jane, should Jane come to your office with an eviction notice in her hand, the focus of your work related to her addressing the prior sexual abuse would be placed on hold until she could resolve her housing crisis.

Interpreting Client Behavior   Interpretation refers to the social worker's making sense of the client's behavior in ways the client may not perceive or acknowledge.

The goal of interpretation is to inspire the client to consider her or his situation in a new or different way. From a social constructionist view, interpreting can be one of the most challenging practice skills to master.

You may ascribe a different meaning to a client's behavior than the client. Still, there may be a place for interpreting when you are careful to test it out with clients and when you acknowledge that the meaning you take from the situation is only one possibility. Again, returning to the earlier example of Jane, you might, in helping her to use her environmental supports, say "It seems to me as if you do not feel ready to take on the kind of work that this type of counseling is likely to be. Is that correct?" Jane may respond with: "I don't know what you're talking about. You may respond with: "Each time we try to discuss your painful memories, you appear to become uncomfortable and change the subject. I think we might best use our time right now by talking about other aspects of your life. When you feel more comfortable with me, we can revisit your memories. How does this sound to you?"

Although many of us have encountered situations in which the thoughts of others about our behavior were helpful, having others interpret our experience can also be frustrating and alienating, particularly when the interpretation seems judgmental or "expert" (as if someone else possesses the secret of understanding our behavior). Interpreting is a widely applicable process but one that should be used tentatively (after you have established a trusting relationship with the client), checked for accuracy with the client, and with the use of ongoing supervision.

Mapping as an Intervention Strategy Recall from the discussion in Chapter 4 that incorporating the family and community system within which the client lives is an important aspect of the assessment phase of the social work intervention. Just as mapping the client's family constellation and current living and relationships profile can be a useful strategy for assessment, the same visual depictions of the client's environment can become a component of the intervention itself. As you know, the genogram is a tool used to enable the client system to gain insights into her or his family history and patterns of a wide array of behaviors and issues, including physical and mental health, substance use and abuse, relationship patterns, and estrangements. Ecomaps, on the other hand, provide a visual depiction of the client's current life situation, including relationships, resources, assets, and challenges.

In the development and implementation of the intervention, the same information gathered from both of these tools in the assessment phase can be incorporated into the client's goals and action steps for facilitating behavior and life changes. For example, consider the client who identifies in the course of completing her genogram a multi-generational pattern of intimate partner violence directed toward the women in her family. She has also experienced violence in her own relationships. Understanding the pattern of the abuse can aid your client in understanding the nature of her previous relationships and strategies she can take to enable herself to seek out healthier relationships. The genogram can later be used to help the client determine if the pattern is broken.

In the assessment phase of your work with client systems, the ecomap aided the client in identifying strengths, directionality of her or his energy and benefits, and areas for change. While this information is important for developing the intervention plan, the ecomap can be used as a visioning tool to enable the client to view her or his life after the intervention is implemented. Specific behavior and life changes can be determined as a result of the client's new perspective on her or his current life. Previously unrecognized resources can be mobilized, unhealthy behaviors can be addressed, and dysfunctional relationships can be targeted for change. The ecomap thus becomes a "work in progress" and serves as a mechanism for monitoring progress and outcomes.

TRADITIONAL SOCIAL WORK ROLES IN CONTEMPORARY SOCIAL WORK PRACTICE

In thinking about the preceding descriptions of social work skills and behaviors, consider the specific roles that workers assume in order to support client strengths and environments. As introduced in Chapter 1, social workers engage in practice behaviors associated with roles that can be isolated, highlighted, and deconstructed. The following discussion will focus on the assumptions that underlie these roles and the ways they are actualized in the context of social work practice. Specifically, these roles are:

- Case manager
- Counselor
- Broker
- Mediator
- Educator
- Client advocate
- Collaborator

Together they represent much of social work's activity across the dimensions of supporting both client strengths and client environments.

These seven roles are not mutually exclusive. Rather, there is considerable overlap among them—for example, educating the legislature regarding the needs of foster children may also pave the way for a future effort at advocacy. Such a blending and overlapping of social work roles is common in social work practice, and it can be helpful to be clear about the roles in which you are undertaking with particular practice behaviors. For example, if you were to assume an educator role in presenting useful and relatively unbiased information about adults who have experienced psychiatric hospitalization, you would want to be clear whether you were going to additionally use a client advocate role to advocate for a particular treatment. While education and advocacy are often interlinked, it is important to be clear when you are advocating for a client system and when you are providing education about the client system.

Case Manager

Case management is "a procedure to plan, seek, and monitor services from different social agencies and staff on behalf of a client" (Barker, 2003, p. 58). Clients with multiple challenges and needs particularly benefit from case management. For example, a client with a serious mental health issue, a back injury, and a housing issue may be in need of medication, a referral to vocational rehabilitation, and a referral to a housing resource. This client may be an appropriate candidate for case management. **Case managers** not only coordinate these services but are also responsible for monitoring the responsiveness of services to clients by holding providers accountable, ensuring client participation, and collaborating with others to raise awareness of unmet needs. As mentioned, a case manager may also collaborate with others to advocate for and build needed resources in a community. Case managers are found in settings such as: aging services, behavioral health, substance abuse treatment, child, youth, and family services, corrections, disabilities program, education, employee assistance, health care, housing, immigrant and refugee services, income support programs, military and veterans services, and tribal programs (NASW, 2013b).

Common Components of Case Management   While there are multiple case management models, steps that are typically included are: (1) accessing the client system by ensuring eligible people are informed of available services; (2) assessing a client's needs and strengths; (3) developing a plan for intervention; (4) identifying and designing an appropriate network of services to be used; (5) creating a written contract that includes achievable and measurable goals, time limits, agreed-upon actions, and consequences of failure to fulfill the contract (if any); (6) implementing the plan; (7) monitoring of the plan to determine progress or a need to re-evaluate the contract; (8) evaluating the outcomes of the intervention; (9) terminating the case management relationship; (10) following up on the client after termination to determine if the client has maintained the desired change (Roberts-DeGennaro, 2008). The case manager may also serve as an informal, personal support/contact person or even therapist. As the provision of therapy is an advanced-level intervention, the social worker is bound by the *Code of Ethics* (NASW, 2008) to possess an appropriate degree and training. In contrast to brokers, who may match a client to a single service, case managers take responsibility for assessing, monitoring, and evaluating the coordination of all services required by a client.

For more than three decades, the term case management has been used to describe this overall coordinating function, although many scholars believe that modern generalist practice is actually a form of case management. In more recent times, the component of cost containment has become crucial to case management. Accordingly, contemporary versions of case management have two primary—and often conflicting—purposes: to improve the quality of care through coordination of services, and to control the costs of care. While case management was originally conceived to address integration of system-level services, it has been shown to have become an effective practice approach with an array of populations as well as a cost-effective service delivery mechanism. Case management is an accepted strategy for working in family preservation, school attendance and performance, substance abuse treatment, corrections, and health care delivery systems; however, the quality of the services provided and the outcomes for the client systems must be given priority over cost-effectiveness (Rothman, 2009a).

The National Association of Social Workers (NASW) has developed a set of standards for the practice of social work case management. The NASW Standards (2013b) are intended to support the case manager in competently and efficiently serving her or his clients by:

- strengthening the developmental, problem solving, and coping capacities of clients;

- enhancing clients' ability to interact with and participate in their communities, with respect for each client's values and goals;

- linking people with systems that provide them with resources, services, and opportunities;

- increasing the scope and capacity of service delivery systems;

- creating and promoting the effective and humane operation of service systems;

- contributing to the development and improvement of social policy (p. 17).

Counselor

The term **counselor** is used in a variety of ways within the helping professions. Within the social work profession, counseling typically is a specialized clinical social work skill performed by social workers at the graduate level who have advanced training in working in health, mental health, and family service settings. A counselor provides services to individuals, families, groups, and communities that encompass the provision of suggestions and information along with establishing goals (Barker, 2003). Counseling is a term also used to describe volunteers and professionals or paraprofessionals who work in group settings (e.g., camps and residential settings).

While the counseling role can include an array of different activities, the social worker functioning as a counselor typically works with the client system around a specific issue or concern. Dependent on the expertise and training of the practitioner and the needs of the client system, different approaches and strategies are used.

Broker

Social workers typically act as **brokers** by linking clients to a needed service or resource. Needs may range from instrumental assistance (e.g., food, clothing, and children's toys) to intangible services such as counseling, support groups, and advocacy. Consider the functions and context of brokering, the process of building the necessary networks, and making the match of client to service.

Brokering Functions and Context   Five social work functions of brokering are:

- Assess client need

- Assess available resources

- Match and initiate referrals to appropriate services

- Link or network services

- Share information

The brokering role can also include modifying resources and creating new resources where none exist. All of these activities can strengthen the client's environment, or context, as well as to support her or his strengths. These activities also overlap, and they reflect much of what was discussed in Chapter 3 regarding assessment of clients and resources, and the fit between them.

One of the early roles associated with the social work profession, the broker role may have less status today than it once had because of the profession's increased interest in psychological and therapeutic roles and the availability of information through on-line sources. Brokering may also seem fairly simple at times. For example, if your client requests assistance with locating used children's furniture, you simply "refer" her or him to the appropriate agency. It is probably somewhat misleading, though, to think of the brokering role as simple, because it requires the social worker both to be familiar with the resources and to nurture and maintain a network of and relationships with such resources. In the complex world of contemporary, urban social services, the ability to know one service or resource from the next can itself be a feat. In rural areas, where the service system may be much less comprehensive, with fewer choices, knowing and maintaining effective relationships with the available resources becomes even more important.

Building and Maintaining Networks for Brokering   Developing a network will include such activities as making initial and subsequent contacts with appropriate providers in other agencies and organizations, discovering those who will help your clients most effectively, and becoming familiar with eligibility criteria and service elements involved in programs. Effective brokering also involves establishing stable working relationships with people, organizations, and systems that will help your clients. Building an effective working relationship with outside people and systems could involve developing a two-way relationship with network resources, so that you are able to provide referrals that best fit the system of the resource, and able to contribute energy to the relationship in other ways, such as offering to appropriately assist your networks in their service, advocacy, and/or fundraising efforts. For the social worker to build and maintain a network, key elements include reciprocity and open, two-way communication.

To build a network successfully, you must acquire knowledge of the informal aspects of the resource as well as the formal ones. For example, it will be helpful to understand the mission statement of a particular community hospice program—the program's goals and methods of service delivery. Other examples include learning that a program has received a large grant for an additional building, that the local university is about to place a field unit there for student training, or that the board of an affiliated organization has approved funding for a development director. Although you will not have access to all informal developments in the services you use, your attention to, and continuous connections with, both formal and informal aspects of the network you maintain will enrich your understanding of your community's resource environment.

Making the Match in Brokering   As discussed in Chapter 3, knowledge and skills are needed to facilitate the matching of clients and services as you maintain an ongoing and continually growing network of contacts. Developing and maintaining a resource network (or "service system linkage") nearly always requires you to use the skills you first learned about in Chapter 3 in connection with clients—that is, looking with planned emptiness and others. When you understand the purpose of the service, as well as the opportunities and challenges of the service, your ability to use the service for your client will increase.

Following up on your referrals is important, both to determine if the client is participating in and benefiting from the service and to gain insight into the provider's response to your referral. Remember that the match between clients and services is important from the service's point of view as well as from the client's perspective. When you have taken the time and care to learn about an effective and successful referral to a particular service, you will gain the trust of other providers who will respect your competence and skills. Further, when you develop a genuine understanding of, and appreciation for, the work that others do, you are likely to find your own work gratifying, and perhaps more to the point, you will find "the system" more responsive to your client.

Mediator

Mediation has both a formal and informal dimension in the delivery of human services. Mediation as a professional practice has its own identity and is often more associated with the legal system and public policy than it is with social work practice. There are, in fact, social workers who complete specialized mediation training and become mediators in such areas as divorce and child custody. Most social workers, however, engage in some aspects of mediation fairly frequently, if on a less formal basis. **Mediators**, as outsiders to a dispute, try to: (1) establish common ground between disputing parties; (2) help them understand each other's point of view; and (3) establish that each party has an interest in the relationship's stability past the current area of difference.

Finding Common Ground Locating the points of agreement in the midst of a dispute is a common skill in social work practice. Assume you are working in a youth agency and have just seen Josh, a 15-year-old who has recently run away from home for the first time. After conversation with him, you understand his challenge to be about the relationship between his mother and her new live-in boyfriend. He is upset and scared and is not certain about the ramifications of being on his own. He dislikes his mother's boyfriend and objects to the curfew and other house rules he imposes. Josh also says he misses his mother's companionship the way it was before "he" entered the scene. When you are satisfied that Josh is safe at home (that he is not being abused in any way), you ask him for permission to schedule a meeting with his mother, her boyfriend, and Josh to explore the issues among them. He somewhat reluctantly agrees.

When his mother enters the youth center, she seems exasperated with her son, but she is also relieved to see him. She begins to cry and hugs him. Her boyfriend remains quiet, but when he catches Josh's eye, he smiles at the boy just slightly.

Walking through It In exploring the process for one possible avenue of mediation in Josh's family, you might do the following:

- Represent yourself warmly and genuinely as one who wants to help resolve this issue without taking sides. Convey to the family that you trust the process of working through the issues and you will walk that journey with them. You might let the members of the family know that you are there to listen, explore options, facilitate, and possibly mediate, but not to direct or advise.

- Attempt to establish common ground. In this case, Josh's present well-being is the immediate point of common interest. You discover that Josh is 15 years old and is not prepared to support himself physically, financially, or emotionally (or legally in most places). He is scared and concerned about his future. At the same time, however, he is vocal about his freedom and does not want to be bound by all the rules imposed upon him by his mother's new boyfriend. His mother cares for him and wants him to be safe. Her boyfriend wants a peaceful household and likes Josh well enough, although he has no strong connection with him. He cares deeply for Josh's mother.

- Help Josh, his mother, and her boyfriend understand and appreciate one another's points of view. Assume that Josh's mother and her boyfriend may have different perspectives and opinions, and acknowledge all points of view. Facilitate their direct dialogue with one another. The skills of planned emptiness and looking from diverse angles will be beneficial to this interaction. A dialogue aimed at establishing a process in which each person is willing to listen to the others' points of view may focus on the following:

  > Social Worker: "I would like each of you to be able to share your perspective on the current situation. In order for each of you to be able to hear what the others are saying, it is important that we establish some ground rules for the discussion. For instance, I would like to ask that we each remain silent until the other person is finished speaking. If we have questions or thoughts to share, those should wait until the person has completed her or his statements. I will try to summarize what I am hearing each of you say so that you can confirm or modify our understanding."

- Establish with Josh, his mother, and her boyfriend that it is in the interests of all of them to work together during this interaction so that they can come to a resolution regarding Josh's living arrangements and his general safety.

- Help all parties recognize that each will benefit from an ongoing positive relationship in which they can settle differences that go beyond the current dispute. The quality of Josh's future may depend on their ability to work together, as well as the quality of the relationship between Josh's mother and her boyfriend. These are the abiding points of common interest.

This scenario represents just one way in which this situation might play out. You might make other arguments to reach the same goals of finding common ground and identifying solutions. You will notice that no *particular* solution is implied in this process. Josh, his mother, and her boyfriend might agree that Josh should return home, live with another family member, try to survive on his own on a trial basis, or any number of other possibilities. The major point is that the relationship between Josh and his mother (and her boyfriend) is collaborative and that each has a stake in working out a solution that affirms their mutual benefit.

Educator

There are many ways in which social workers function as educators. While the mission and scope of individual practice settings will determine the nature and level of such an activity, some form of education is commonplace in most social work settings. On an interpersonal and concrete level, as an **educator**, you may teach clients about a range of topics; for example, you might inform clients about services, new programs, completing an application form, or something as basic as the correct bus to take to travel uptown before noon. You may assure an adolescent client that she is indeed normal when she worries that her moodiness indicates she is not, or you might educate clients about the maximum allowable percentage of income charged for rent in public housing. Much of the direct service aspect of acting as educator relates to helping clients to access resources, as well as to make the behavioral changes they want to make.

Developing Client Skills   On a direct practice level, you, as the social worker, can assist clients in developing the skills they need to participate in the intervention and reach their goals. The motivation for this kind of change needs to generate from clients' vision change, rather than from your opinions about needed improvements. You can assist clients in understanding the steps needed to accomplish that change. For example, if your client, Ramon, feels intimidated by his co-workers or supervisor in his job and wants to increase his assertiveness in the workplace, you might help by demonstrating a more assertive (but respectful) stance in a relevant interchange, and/or role-play or assist him to practice the new behavior. In other settings, you can help older adult clients learn to organize their medications to increase their compliance. You can teach parents strategies for supporting their children through positive feedback. You can model clear communication and gradually assist your client in participating more effectively in problem situations. For example, if you are working with an adolescent who is frequently suspended from school for short periods because of her angry outbursts at teachers, you can work with her on appropriate strategies for expressing herself more effectively without alienating adults. Additionally, you might accompany her to her first meeting with the school principal on her return to classes. You can help her describe her situation and support a more focused, direct, and respectful level of discussion based on her goals. Outside of the meeting you can help her process the session, including evaluating what went well and what did not go well, and support her ability to negotiate these relationships with further practice. You may then role-play the next meeting rather than attend yourself to support her goal of self-sufficiency in these types of situations. At the same time, when you use skills with clients, you are modeling skills that they may use on their own in their relationships. In all of these situations the ultimate goal is a more positive sense of strength and agency—a change in clients' thinking about themselves and a greater integration of self and skills into the environment.

© Adam Gregor

Working with the Public   Social workers frequently are called upon to educate larger groups about issues that affect client systems. In some instances, these efforts take on aspects of primary prevention in that they are designed to prevent the development of a problem. For example, you may teach a parenting class or present a session on ways to support racial tolerance in the classroom to a preschool group. Other education efforts may involve providing testimony in a legislative hearing regarding the cultural needs of a group of refugee children, or speaking to a community group in response to their concerns about a new group home in their neighborhood for discharged clients with psychiatric histories.

Although the activities differ in nature, all of these scenarios involve working in an educator's role. Social work practitioners have a long history of acting as educators when they take on the position of field instructor for students in an academic program leading to a social work degree.

Client Advocate

Advocacy in social work practice is aimed at securing the rights and well-being of clients who are at risk for a negative outcome. The **advocate** may obtain resources, modify existing policies or practices, and promote new policies that will benefit clients. Regardless of the activities involved, advocacy represents a struggle over power, and by definition it seeks to obtain and ensure clients' access to resources.

This role of defending, championing, or otherwise speaking out for clients is one of the original cornerstones of the profession's commitment, although its intensity has varied somewhat according to the political climate (Haynes & Mickelson, 2010). Although advocacy often focuses on the political or civil rights of clients, as a social worker, you frequently have the opportunity and obligation to advocate in informal, everyday situations when a client is treated disrespectfully or inefficiently, or not provided needed services. Advocacy assumes an active role that is not always comfortable or popular with others.

Many social workers distinguish between **case (or client) advocacy**, defined as advocacy on behalf of an individual client or a single group of clients, and **cause (or class) advocacy**, which is initiated on behalf of a category of clients. **Legislative advocacy** is specialized version of cause advocacy in which some aspect of the law is addressed. We will take a brief look at each type.

Case Advocacy   Social workers usually practice case advocacy on an agency or organizational level. For example, if your client has been denied food stamps (Supplemental Nutrition Assistance Program (SNAP)) to which she or he is entitled, you can advocate with the public welfare income support organization that houses the SNAP services. To be an effective advocate, you will need to know a number of variables, such as eligibility requirements, agency appeal policies, regulations, and power structures, as well as contextual variables such as the way in which SNAP fits into the overall welfare organization. You will also want to recognize the situation as one of potential conflict and to enter at the point of least contest so that you do not inspire more resistance than necessary. Therefore, you can address the situation with the worker who originally denied the SNAP application and use the grievance process at the agency before taking any other measures, such as involving powerful outside parties in advocacy, calling the media, or organizing a protest.

Cause Advocacy   More political than case advocacy, cause advocacy involves both larger numbers of people and, by definition, a cause that affects them all, either by imposing obstacles to attaining resources or by directly depriving people of these resources. Cause advocacy involves speaking for a large number of people who are not your clients and who are not likely to be empowered to participate directly in forming either the goals or the preferred advocacy methods related to the cause. Nevertheless, cause advocacy can be effective when you can partner with other concerned organizations, and can spur change that positively impacts many people. It is also helpful when the issue is very clear and does not impose any solution on people who do not choose it. For example, if your community has no adequate facilities for child care after school, there may be agencies, churches, businesses, or other institutions that are concerned and want to develop a facility for parents in need of such a service. Assuming that you will require funding from local public sources, your coalition of interested partners (who are well organized and well-rehearsed) will be in a position to advocate for adequate child care facilities. As in case advocacy, you/your coalition will need to be skillful and to have researched background information regarding the number of children to be served, their needs, the estimated costs of providing the service, the likelihood of participation, the process of getting the issue on the city's agenda, and the benefits to all stakeholders.

Legislative Advocacy   This form of cause advocacy is devoted to adding, amending, changing, or eliminating legislation in order to benefit a large group of clients. For example, when a social worker advocates to lower the legal alcohol limit for driving a moterized vehicle from .08 blood alcohol content (BAC) to .06, that is a case of legislative advocacy designed to benefit a large category—drivers and passengers.

There are many other kinds of scenarios in which a large group of people would benefit from a change in legislation, because nearly all entitlement policies have limitations that may present obstacles to your clients. Controversial resources such as family planning and abortion clinics are likely to have more access constraints (such as age limits or pregnancy duration) than do other less disputed services such as food pantries.

A considerable amount of knowledge, insight, and organization are required to change a law, but it is well within the arena of social work practice to initiate or collaborate in such activity. With careful preparation and diligence in appropriate situations, legislative advocacy can produce the desired results. Legislative advocacy usually requires intensive, cooperative work with one or several organizations.

Thoughts about Power and Advocacy   Social work advocacy is frequently related to the social justice dimension of resource allocation; therefore, it is helpful to recall the assumptions about power. In general, advocates recognize that power is not easily relinquished, not equally distributed, involves conflict, and is necessary to make substantial change. These points may appear harsh if you have not considered power and advocacy within such a context. Familiarity with these concepts can prepare you to enter a more political arena and are beneficial in understanding the social locations of those who have been oppressed and have had little power to exercise in our culture.

Like work with individual clients, advocacy can be a highly rewarding undertaking (Abramowitz, 2005), but can be complex, frustrating, and mysterious. When not executed well, it can be costly to clients, as you will recall from the tenants' story in Chapter 4. See Exhibit 5.2 for a list of cautions and strategies regarding advocacy. More on advocacy and empowerment will be discussed later in this chapter.

Collaborator

Guided by the mission of the social work profession, collaboration is inherent in the daily practice of most social work practitioners. In addition to using collaboration skills in developing plans for change with clients, social workers engage in collaborative relationships with other professions in many different ways. For example, social workers work with health, social service, and legal professionals in the delivery of client services, development of programs and policies, advocacy work, and research. While the terms to describe cooperative practices are sometimes used interchangeably, they do have distinctive connotations. Multidisciplinary practice, for instance, refers to groups of professions working together toward a similar aim, while maintaining their individual interventions (Moxley, 2008). Interdisciplinary practice (also referred to as interprofessional collaboration, collaborative practice, and partnered practice), on the other hand, involves professionals from different disciplines integrating their professional knowledge to work together toward a common goal. In these situations, social workers engage in **interprofessional collaboration**.

Social workers often work in settings in which the primary mission is not the provision of social work services. Known as **host settings**, these include health care facilities and programs, educational institutions, law enforcement or legal systems, military programs, and even financial institutions. While it is essential for social workers working in host settings to understand the philosophies, professional cultures, and language of other professions, social workers will be collaborators in

EXHIBIT 5.2 *Strategies Regarding Advocacy*	• Enter any situation in which you want to advocate at the point of least contest (Wood & Tully, 2006).
	• When clients are directly involved, ensure they genuinely support your efforts and understand the possible repercussions.
	• Prepare yourself fully with the knowledge that is relevant to the situation (e.g., eligibility requirements, entitlement limitations, number of people in a given category, history of advocacy on this topic).
	• Be clear and specific about your goals; a complaint about a policy means little if there is no solution presented.
	• Begin with simple efforts at persuasion; assume first that there has been a mistake or an oversight.
	• Be clear when persuasion is not working with the targeted audience and you must enter a new level.
	• Assess that new level prior to action: Can you be successful? Are your clients still supportive? Do collaborative partners agree on the next step?
	• Use carefully cultivated social work skills in an advocacy effort, including listening, empathy, clarifying, and firmness.
	• Use the discourse of collaboration and common ground.
	• Seek supervision and consultation. You are as vulnerable to "not seeing" in thinking about cause advocacy as you might be in the most intense interpersonal work.
	• When entering into formal processes (e.g., legislative advocacy), become knowledgeable about the technicalities of the legislative process.

virtually every setting in which they are employed. While effective in enhancing the delivery of services or care, interprofessional collaborations can be complex and, sometimes, frustrating relationships in which to work (will be further explored in Chapter 10). Social workers may find that challenges occur in communicating with other professions who have their own professional "jargon," perceived hierarchy of professional influence, and conflicting opinions on approach, roles, implementation, and outcomes related to the work to be completed.

Social workers can use their training in collaboration and negotiation with client systems to prepare for becoming effective professional collaborators. Abramson (2009) suggests that social workers can become competent collaborators by learning about the other professions with which they will be engaged. Upon learning about your collaborators' professional socialization experience, language, and culture, you can then seek out common ground on which to begin the process of building a professional collaboration. Taking advantage of opportunities to gain interprofessional competence will serve you well as you move into any area of practice.

PUTTING IT ALL TOGETHER

Thus far, this chapter has explored various aspects of social work practice in action. Areas covered include supporting clients' strengths and their environments as well as various actions by workers. Empowerment practice will be used as a model for examining the possible points of integration for social work roles within the context of a strengths-based practice approach.

Empowerment Practice

Social work as a profession has been committed to empowering clients for much of its history. As one of the early scholars of empowerment practice, Simon (1990) asserted that, in its purest sense, empowerment cannot be given to someone else because empowerment is not ours to give. Empowerment resides within the individual and can only be encouraged or perhaps released, but not given, by others. The social worker may be in a position to promote or facilitate empowerment, but ultimately, the change comes from the client her or himself. Best practices in the application of empowerment theory suggest (Parsons, 2008, p. 124):

- A sociopolitical lens is used to frame situations experienced by the client system.

- Power within the helping relationship is couched within the context of client strengths, self-efficacy, and education as opposed to pathology.

- Informal social networks are integral to the empowerment of the client system.

- Collectivity is key to the intervention with a specific emphasis on support, mutual aid, validation, and promotion of social justice.

The practice of empowerment-focused social work establishes a clear theoretical connection among individual client strengths and their environment and the client's capacity to act in a way both to empower her or himself and liberate others. This connection in turn supports intervention in multiple and highly interconnected settings and recognizes the importance of unleashing or strengthening the client's sense of self and agency. Exhibit 5.3 provides an illustrative case example of an appropriate and sensitive empowerment approach.

The concept of empowerment has applicability within each of the roles that a social worker assumes when intervening with a client system. As the social worker serves in the roles of case manager, counselor, broker, mediator, education, client advocate, and collaborator, she or he has the opportunity to use the concepts of empowerment to carry out the tasks associated with each of these roles. Even in situations of crises or tragedies that impact entire communities, regions, or the country, the social worker, working at any level, can approach interventions with an empowerment perspective in mind. If you are working with individuals who have experienced violence (shooting or bombing), a natural disaster, or community-wide crisis, your role can be not only to meet their immediate needs, but to help them become empowered to survive, advocate for themselves or others, or change the system. As Kohli and colleagues (2010) write: "We should appreciate the strengths of individuals that help them to survive in the worst of situations and use their strengths to empower them" (p. 266). Consider the resilience and fortitude of a client who has survived a significant life event and begin the intervention at that starting point.

The social worker working to advocate on behalf of a client system can build the advocacy effort on the strengths of the client, use the client's self-efficacy and life experience in the advocacy, and help the client system become empowered to engage in self-advocacy. Advocacy work that elicits power from clients includes facilitating as strong a role as possible for clients to advocate for themselves. Activities to promote self-advocacy and empowerment include helping them to prepare, providing education and support, and accompanying them during the process. Consider the situation in Brickville in which Virginia Stone is fearful of losing her family's home to a developer. What strengths and experiences do Virginia or her family members possess that might be used in advocacy on her behalf?

Social workers need different strategies to initiate and facilitate a strengths-based intervention that empowers clients. For example, while the narrative approach does not provide specific therapeutic skills, the client can be empowered with the social worker's help in constructing, deconstructing, and finally reconstructing their conceptualizations (i.e., stories) that have previously been used to define their lives. The social worker's role is to help the clients determine if they wish to change their perceptions by challenging and broadening their thinking about their lives.

The story takes place in a small New England city in the 1970s. Thomas was born with a neurological disorder, cerebral palsy. His family had little idea of how to deal with his severe physical limitations and had three other children to rear as well. His parents cared about him and did what they could to learn ways to support him and his abilities, as well as cope with his disability. He received the standard medical care of the time and was sent to school with his age group.

Early on it was evident that Thomas's body did not reflect his aptitude for schoolwork. He was, in fact, intellectually capable and did exceedingly well in the subjects in which teachers supported him. His family did not understand, however, the ways in which Thomas's socialization was affected by his physical impairments or how he experienced his life. Although he had only a few friends, he attempted, for several years to maintain a positive outlook, even developing an excellent sense of humor. Nevertheless, he continued to feel excluded by peers and adults alike.

COLLIDING WITH THE WORLD

Over time he became hostile with teachers because he had to prove himself over and over again, every time he entered a new grade or school. Teachers and school administrators first assumed that he was unable to do grade-level work. One teacher questioned his very presence in the regular classroom. His pastor at church advised that he seek supported employment through a public vocational program. Everyone seemed to assume he was unable to do, know, or even feel anything. His medical treatment included the excruciating requirement to walk in physical therapy that was promoted in those days. No one noted Thomas's pain or responded to it.

By the time Thomas was an adolescent, his parents sent him to a psychotherapist to learn the reasons for his intense anger. To Thomas, this was yet another insult to someone who had already suffered so many. His therapist told him he needed to "get the chip off his shoulder" and tend to his schoolwork. There would be no money for college—he would have to earn scholarships if he wanted to do more than sit in the living room until he could be matched to a job he did not want.

MEETING MAURA

Finally, Thomas was in need of a new wheelchair. He was directed to a social worker for assistance when his parents could not pay the required deductible. The social worker, Maura, who worked in the primary care office of Thomas's physician, first spent some time getting to know him and hearing what he had to say about the wheelchair. She assisted him in getting the funding for his new chair and continued to ask about his overall experience. He began to talk about his needs to consider his options regarding school, his family, his anger, and his growing sense of estrangement from the world.

Maura helped Thomas by hearing with an openness and reflexivity that assumed he was the expert on his experience. She heard his story, took him seriously, and helped him look at what his disability meant to him by asking him to talk not only about the physical pain but also about the exclusion that he experienced. She helped him look at his resilience in the face of all he had been through. She did not try to challenge his perception

EXHIBIT 5.3

Thomas's Story

EXHIBIT 5.3

Continued

of his experience but was completely respectful of it. She asked him to articulate the way he wanted his life to be different in view of his strong capacities. She encouraged him to reflect on his own position and how he might address his goals. As time passed, Maura helped Thomas secure vocational rehabilitation funding for college and validated his, by then, strong commitment to working in human services on disability issues. She also met with Thomas's family to help them understand his choices and the impact these choices would have on the family.

CHANGING DIRECTION

Thomas went on to study for a master's degree in disabilities and is preparing for a career in which he can advocate for others with disabilities as well as himself. He is a full participant in school and in the surrounding context of his family, friends, and culture. He is still angry sometimes, but he is not fearful or alone.

Reading this (almost all) true story, you can see how Maura implemented the empowerment focus identified in the chapter: She assists in liberating Thomas's "potent self" and encourages his interpersonal connections. She supports his understanding of his environment and she helps connect him to a direction in which he can address the more political aspects of his experience. Consider now the connections between Thomas's experiences with other social work perspectives.

Social Justice: Maura works diligently to expand Thomas's access to the benefits of his society by helping him to find funding for education. She does not accept the status quo arrangement in which those with private resources are privileged and those without are not. To complete this role, Maura would work toward reallocation of educational funding for all people, not just Thomas.

Human Rights: Thomas's status as a person with a disability is not accepted as a rationale for discrimination in education, nor does his disability preclude the "full development of the personality" (Article 26, United Nations, 1948) through education. Maura recognizes him as a person who has both needs and abilities and the right to fulfill his life as he chooses.

Strengths Perspective: Maura recognizes Thomas's considerable strengths, validates them, and encourages linking their full expression to his goals. She sees him as a whole human being, with many talents to offer, not as a "victim of cerebral palsy."

Critical Social Construction: Maura sees Thomas's disability as a social construction that results from the prevailing collective meaning given to it by our culture. She questions the limitations imposed by that construction, and she believes Thomas can do what he sets out to do. Maura accepts that there are multiple realities and so honors his experience of exclusion and oppression. Because his experience has developed within the context of his social location, on which he is the expert, she makes no attempt to "correct" his understanding. Her effort goes into changing his future experience in order to make it more consistent with how he wants his life arranged.

Because the narrative approach is consistent with the social work values of strengths and collaboration, and the client as the expert on her or his life, the narrative approach can be integrated with other clinical approaches (solution-focused) and cognitive behavioral approaches (Parsons, 2008).

One practice behavior that can aid social workers in helping clients to redefine their perceptions of their lives is **motivational interviewing (MI)**. Initially developed for practice with clients who are not seeking services voluntarily, motivational interviewing builds on the client's strengths and right to self-determination to implement a plan for change she or he created (Miller & Rollnick, 2013). Aimed at helping clients who are ambivalent or reticent about change, MI is a collaborative process between the social worker and the client in which the practitioner aids the client in becoming more aware of the implications of her or his decision to engage in change process (Lundahl, Kunz, Brownell, Tollefson, & Burke, 2010). A more technical definition of MI can be described as: ". . . a collaborative, goal-oriented style of communication with particular attention to the language of change. It is designed to strengthen personal motivation for and commitment to a specific goal by eliciting and exploring the person's own reasons for change within an atmosphere of acceptance and compassion" (Miller & Rollnick, 2013, p. 29).

Currently used in a variety of settings with a range of client situations (e.g., child welfare, treatment of substance abuse and other additions, intimate partner violence, and family and group work), motivational interviewing is an evidence-based strategy in which the social worker uses the practice skills of empathy and reflective listening (Wahab, 2005). As the expert on her or his life, the client is responsible for articulating her or his story and developing a motivation for initiating life changes. The underlying perspective (or "spirit") of MI encompasses four interrelated components: partnership, acceptance, compassion, and evocation (Miller & Rollnick, 2013, p. 15). These components along with the principles of MI and skills and strategies for integrating motivational interviewing into practice are further delineated in Quick Guides 13 and 14, including the concepts of change talk, evoking change, and the four processes of MI, known as OARS (Open questions, Affirmation, Reflection, and Summary). MI will be again highlighted in use with families in Chapter 7.

Motivational interviewing is but one of the numerous techniques and skills available to the social work practitioner, but one that is built on a strengths-based, person-centered framework. In determining if MI is an appropriate intervention strategy, consider the following (Miller & Rollnick, 2013, p. 25):

1) Are conversations regarding change occurring?
2) Should conversation regarding change be initiated?
3) Will the client outcomes be influenced by client changes?
4) Is the client ambivalent about change?

The process of implementing an intervention using motivational interviewing encompasses four components (Miller & Rollnick, 2013, pp. 26–29):

- Engaging—establishing a helpful connection with the client to promote a working relationship;

- Focusing—developing a specific agenda for change;

- Evoking—drawing out the client's motivation for change and articulating arguments for making the change; and

- Planning—helping the client to transition from being motivated for change to being ready for change which could be evidenced by the client changing her or his conversation from the "ifs and whys" of change to the "where and how" of change.

The role of the social worker is to maintain a focus on the client and her or his desire to make a change. Specifically, the social worker is engaged in MI when she or he is listening and reflecting, affirming those behaviors/activities that the client is doing well, identifying change talk, drawing out the client's motivation, wisdom,

---

QUICK GUIDE 13   THE SPIRIT AND PRINCIPLES OF MOTIVATIONAL INTERVIEWING

The Spirit of Motivational Interviewing is intended to embrace the following concepts:

Partnership establishes that the client is the expert on her or his life and behavior change is more likely to result from the social worker and the client working together to determine the areas of and strategies for change.

Acceptance is comprised of the social worker's belief that the client has absolute worth, genuinely attempts to understand the client's perspective, respects the client's autonomy, and affirms those strengths and efforts demonstrated by the client.

Compassion is conveyed by the social worker to the client through actively promoting and prioritizing the client's needs.

Evocation is rooted in the idea that the client possesses strengths, wisdom about her or himself, reasons for acting as they have, and have the motivation and resources with which to make desired change.

Stemming from the spirit that guides Motivational Interviewing, Principles on which the strategy is built emphasize the social worker's ability to:

1) **Express empathy** to enhance client rapport and comfort and decrease client resistance.
2) **Develop discrepancy** to enable client to gain insight into the gaps between her or his value system and current behaviors.
3) **Avoid argumentation with client**
4) **Roll with client resistance** to convey to the client that her or his apparent resistance to change is respected and viewed as normal.
5) **Support client self-efficacy** to promote client's confidence in her or his ability to change behavior.

Source: Lundahl, Kunz, Brownell, Tollefson, & Burke; 2010 pp. 137; Miller & Rollnick, 2012, pp. 14–21.

> ## QUICK GUIDE 14   SKILLS AND STRATEGIES FOR MOTIVATIONAL INTERVIEWING
>
> Focused on the person and her or his strengths and resources, three areas of skill related to motivational interviewing include (Miller & Rollnick, 2013):
>
> - **Change Talk** occurs when the client makes statements that relate positively to making a change in her or his life and can be contrasted with **Sustain Talk** which refers to the statements which favor maintaining the status quo (p. 7).
> - **Evoking Change Talk** is based on the assumptions that people have wisdom and experience that includes ambivalence about change which can be elicited by the social worker, using a person-centered approach aimed initiating a desired change.
> - **OARS** is a pneumonic that embraces four core skills used to facilitate the four processes of MI and includes (pp. 32–34):
>   - **Open questions** are intended to engage the client in expanded and reflective discussion about feelings and potential change, such as "how are you feeling about your goal to lose weight?"
>   - **Affirmation** enables the social worker to offer positive statements regarding the client's intentions, strengths, efforts, resources, and courage related to the proposed change. For example, the social worker may say to the client who is striving to lose weight: "I admire your commitment to living a healthier life. You will be a good role model for your children."
>   - **Reflective Listening** provides the social worker and the client the opportunity to clarify, deepen, and explore the client's previous statements. In working with the client goal of weight loss, the social worker might choose to say: "It seems that you are able to maintain your weight loss plan well, but struggle with compliance when you are going through stressful periods, like the loss of your mother."
>   - **Summarizing** allows the social worker to expand on the previous reflections and transition to the next phase of work. Summaries bring together all the aspects of the conversation and aid in the planning of the change process as well as the evaluation of change that has occurred. Summarizing with the client working on weight loss can emphasize: "We have talked about your past efforts to lose weight, the strategies that you feel worked and did not work, and your commitment to being successful this time. We have also explored the triggers that make it more challenging for you to stay on your diet and exercise plan. We have brainstormed some strategies that you can put into place when you are feeling particularly stressed and wanting to revert to your old eating patterns."

and strengths, and resisting the need to offer the "righting reflex" (i.e., advice-giving, confronting, or arguing) (Miller & Rollnick, 2013, p. 324).

As with any practice approach, the social worker is ethically obligated to review the evidence to ensure the efficacy of the intervention. In a 2010 meta-analysis of 119 research studies using MI, Lundahl and colleagues concluded that MI can be effective with an array of issues presented by clients, including: addictions (e.g., substance abuse and gambling); high and low levels of distress; increasing healthy behaviors; decreasing unhealthy behaviors; and enhancing client participation in the intervention process.

STRAIGHT TALK ABOUT INTERVENTIONS: UNEXPECTED EVENTS AND ONGOING EVALUATION

In contrast to other ways to view social work practice, the intervention process may seem orderly. If you practice sensitive engagement, careful assessment of both the players and the environment, and plan consistent methods of taking action on behalf of your clients, you might be seduced into thinking that the work will always go smoothly. This is a mistake.

In the real world, not everything goes as planned. Because social work is completely immersed in people's lives, social workers have only limited control over their work with client systems and need to develop an acceptance (or at least tolerance) for the up-and-down nature of working with clients who are living within their environments. In this real world, people get sick, get into accidents, change their minds, get fired and laid off, become disheartened, move, their kids get into trouble, and they experience violence. On the positive side, people also get promoted, find their strengths, find their voices, fall in love, get jobs, read an inspiring book, discover a new friend, and develop insight. All of these factors and many more have the potential to interrupt, postpone, redirect, or even terminate your work together.

In some cases—for example, if the client becomes discouraged or seems simply to lose interest—you will want to inquire about the role you might have played in contributing to these developments. In many cases your responsibility will be to honor, acknowledge, and explore the meaning of the new situation to the client, and reconfigure your work when indicated. It will generally be helpful for you to *expect the unexpected*. Flexibility is one of the most critical of social work attributes.

Just as you have been consistent in evaluating the progress of your work with your client and others involved throughout the engagement and assessment processes, you will want to evaluate the direction of the intervention periodically as well, particularly if there have been substantial changes or unexpected events in the client's life. You may easily get carried away in the plan that you so carefully put together, even when it may no longer fit very well. Checking and re-checking in with the client will serve you well during the intervention process. In that way you can stay connected to your client throughout and provide more relevant service.

As you recall from Chapter 3, crisis intervention and suicide threats are expected, yet unexpected, aspects of social work practice. The social worker's ability to intervene effectively is contingent upon a timely response and accurate assessment. After ensuring during the engagement and assessment phase that the client is safe, the focus during the intervention phase is on empowering the client to resolve the crisis by reconstructing her or his perceptions of strengths, assets, and resources (Eaton & Roberts, 2009). The social worker and the client collaborate to devise a viable plan for addressing the root causes of the crisis and emphasize concrete action strategies and a clear, agreed-upon plan for follow-up and maintenance.

SUPPORTING CLIENTS' STRENGTHS IN TERMINATION AND EVALUATION

"Important things are almost never easy." This statement captures the real work of the termination phase of planned change. Even when all participants in a relationship agree that it is time to move on, ending can be a wrenching process, often punctuated with doubts about whether you have given or done enough, or wishing, in some vague way, to start over. This is a common experience that you have probably had yourself, perhaps when you left home or ended a significant relationship. Endings tend to raise ambivalence: On the one hand they may be sad, while on the other hand they represent a kind of freedom to be on your own, make a new start, and be who you want to be.

The endings between social workers and client systems often reflect these tensions. Some social workers, as well as clients, will be tempted to minimize any significance in saying goodbye and simply "slip out the back." Others will not even want to talk about it, finding it more comfortable just to be gone. Yet others tend to make scrupulous notes with contact information and schedules and agree to call, text, connect on social networking sites, or email each other. Clients may, in fact, demonstrate both positive and negative feelings about their termination process, viewing termination with a sense of accomplishment or as (possibly another) loss in their lives (Fortune, 2009). People tend to develop patterns about endings, and usually these serve to mitigate the loss that inevitably occurs in all significant relationships. The remainder of this chapter will address the processes of ending a professional relationship with your clients. We will explore strategies for evaluation of the practice intervention on multiple levels. While terminations that occur at the level of the family, group, organization, or community are unique and will be explored in the following chapters, this chapter will emphasize general issues related to terminating and evaluating the social work relationship with the individual client.

ENDINGS AND TERMINATION

There are considerable variations in the way endings occur in social work practice. Factors that play a role include age, gender, ethnicity, socioeconomic status, cultural experience, working style, and personality of both the client and the social worker. The social location of the work, its purpose, the agency, the perspective used, and the organizational pressures surrounding the work also have an impact. Even within this range, however, general planning is necessary and can be helpful to both the client and the social worker in bringing closure to the professional social work intervention.

Planning the Process: Overview

Begin by considering several tasks that relate to endings. These tasks will not apply to all social work relationships and their order need not be rigid. Your theoretical orientation and the specifics of the practice situation will guide much of the timing and ordering, but these tasks are common to a range of relationships and are consistent with ethical practice. They are:

- Negotiate the timing of the termination.

- Review the agreement for work.

- Process successes and shortcomings.

- Develop and clarify plans for termination and maintenance of change.

- Share responses to ending.

- Respect cultural consistency.

Negotiating the Timing Because termination is a goal that is established at the beginning of the relationship, it should always be before you as you progress through the intervention. In some circumstances your client and you have the opportunity to specify the number of sessions at the beginning of the relationship. The closing date can then be determined during the first meeting. Frequently the same is true of mandated arrangements or managed care situations in which the agency is bound by prearranged guidelines. All of these are, of course, artificial proclamations that the work is done, and they are all externally imposed. In some cases, clients may request an additional number of sessions (as in task-centered models), and social workers may petition for an extended number of sessions (as in managed care).

Predetermined boundaries tend to distract from the ideal timing for ending, which is when the social worker and client have reached the mutually formulated goals. The point at which the goals have been reached may not be clear, particularly if goals have not been clearly established. Even when goals are put in precise behavioral terms (for example, "the client will contact the school social worker"), there is some risk that the established goals and intervention plans may still not be realized. For example, your client may have unsuccessfully attempted to contact the school social worker, or made contact, but was not able to clearly articulate her message; therefore, meeting the goal technically does not meet the spirit of the issue. Nevertheless, you will need to make a reasoned judgment that the client no longer needs your services. Because you and the client may not necessarily agree on the exact moment when that occurs, in many cases you will need to negotiate the timing and criteria, remembering that you are each subject to your own foibles relating to ending relationships.

There are at least three areas for you to consider in your negotiations regarding the termination of the relationship. The first is your responsibility and relates to preventing an unanticipated ending. In many circumstances you will not have control over the timing, but in others you can anticipate, if not change, the conditions. You have a responsibility to provide full information to the client about the possible nature and timing of termination. For example, a state contract or managed care restrictions may determine an ending date that you think is inappropriate. In such a situation you can at least prepare the client for the possibility that your appeal for more sessions may be denied. Other examples of situations in which you need to be straightforward with clients and make the smoothest transition arrangements you can include a situation in which your agency is about to eliminate services or programs, your position is threatened, or you know you are leaving your job or internship on a particular date.

The second issue relates to maintaining an ongoing dialogue with your client regarding her or his status or progress to determine if your work together is helping, and what needs to happen for the client to know she or he has met the goals. Just as you have ongoing discussion with your client about goals, contracts, and progress, you should also discuss termination throughout your relationship. While the agency or the plan you established with the client may determine the ending point, discussion about termination can and should be integral to the work.

The final issue involves predetermined endings. Knowing that you will have a specified number of meetings with the client provides you with the opportunity to regularly check in with the client to ensure that you both share an understanding of the point at which the work will be completed. Some clients may assume you can or will extend the number of sessions as you prefer, and others might expect if they behave well (especially in mandated sessions) that they will be "dismissed" early. Always be as clear as possible about any limits that are imposed on you for the work and be as open as possible to discussing the meaning of those limits to your client. Even when both sides clearly understand the ending date, either the client or you may still experience difficult ending dynamics.

Reviewing the Agreement for Work   In the process of finishing your work together, you and your client should review the formal contract or informal agreement you made for work. You will want again to negotiate the understanding each of you has about the agreement, as it may have changed in view of your completed work, circumstances in the client's life or your life, or due to the agency's ability to offer services.

Processing Successes and Shortcomings   As you precede through the work together, you have discussed and processed the aspects that went well, the things that missed the mark, and any approaches that clearly headed in the wrong direction. This type of assessment should be ongoing and not just take place at the last meeting. Nevertheless, processing a summary of successes at the end is an appropriate strategy to gain perspective on the whole experience and gain insight into the client's view of the work. Sometimes, for example, an experience that seemed difficult at the time, in retrospect, is associated with personal growth (for your client and or you). Recall the example from earlier in the chapter of the client with mental illness who had struggled to locate housing. You may have suggested that she was ready to meet with the landlord on her own, which at the time seemed challenging to her. As your work together progressed, however, she may have seen that it was a useful push that helped her to recognize her strengths. Looking through an empowerment lens, this success suggests your client's ability to act on her own needs, and she no longer views herself as a victim, but as a survivor of the power imbalance experienced by people with labels, such as those given to people with mental illness. It is useful for clients and social workers to process successes based on power and human rights definitions.

At this point in the social work relationship, it is not uncommon for some clients to indicate that they have not made as much progress as you believe they have. The client may point to various criteria to demonstrate her or his own shortcomings or to register some verbal sense of being abandoned. This behavior can be complicated to deal with as it raises the possibility that ending would be premature and possibly harmful. Also possible, however, is that clients choose this means to make known their lack of self-confidence in making their way independently, or their reluctance to leave the relationship. One effective way to address such dynamics is to be transparent about the process and discuss issues in detail with the client. Whether or not you offer to continue to work with a client for a longer time is a matter of your professional judgment, your supervisor's judgment, and all the other constraints in the agency or funding sources, such as managed care.

A second possibility is for clients to threaten to leave the relationship early in anticipation of ending. This often seems like an "I'll fire you before you fire me" response to minimize a sense of abandonment and is sometimes called **flight**. In this case, you will want to make every effort to engage the client in the process for at least one more session in which you can address the issue of ending and strive for closure. Clients who pointedly (and physically) avoid the ending steps in the work, in some cases are the ones to benefit most in personal growth by a positive ending process.

The identification of shortcomings in your own work is another area in which you may get an unexpected challenge. When you ask for ideas about those aspects of the intervention that did not unfold as you or the client had hoped or expected, you must be truly open to receiving a critical answer. An account of your shortcomings or those of the process itself can come as a surprise to you, especially if you have not heard such expressions of dissatisfaction along the way. Although hearing critical comments can be difficult, you can choose to learn from them and remain open to the meanings your client has in making them.

Making and Clarifying Plans   Although there are times when the work is clearly finished and the client is ready to move ahead without your assistance, this does not always happen. You may determine the best course of action is to make a referral to a different type of service, or transfer your client to another social worker in your agency or to a different agency. For example, you have worked with a client named "Sam" throughout the school year in which you had your practicum at the agency where Sam receives services. As the month of May nears, you realize that it is not likely that he will complete the work he wanted to do before you leave your position. In negotiation with him, you will need to reassess, much as in the earlier phases, where he is in the agreed-to work, where he wants to go from here, and through what arrangement it is mostly likely he can continue his work effectively. You can help to clarify his options and assist him in assessing his own needs at the point in which you must terminate with him. Be aware that this stage represents a potential ending to services, which may either set him back toward reaching his goals or promote a continuation in his growth.

When you decide to refer the client to another agency, you will assume the role of broker. As discussed previously, the broker is responsible for establishing a relationship with a network of resources. Regardless of your experience or skill level and the success of your work with the client system, you have the potential to undo any client gains with an inappropriate or ill-conceived referral.

In those cases when the client is ready with some certainty to end professional services, a plan is needed for maintaining the gains and/or continuing further growth in the community. Some clients who no longer have the protective atmosphere of the client–worker relationship may find it difficult to maintain the changes they made without some clear way to reinforce them. You can help by identifying and exploring the situations that are likely to challenge the client's gains and consider strategies for addressing future stresses. You can also help by encouraging the client to identify a natural network of support before ending the work, or identifying other community resources that will be available. Consider again the termination with Sam. You and Sam may identify formal (e.g., agencies, services, organizations) and informal (e.g., individuals) resources that are available to him following the termination of his relationship with you. Most clients will have to return to the context in which their struggles arose, so a strategy to cope with that environment can be critical to maintaining the gains. In Sam's case, a "rehearsal" of his response to potential situations that may arise may be a strategy that he can rely on when you are no longer a resource for him.

Sharing Responses to Endings   Identifying and articulating one's feeling about ending may be the most sensitive of the dimensions of ending the work. You will need to anticipate, to the extent that you can, the way in which your client will respond and those issues that will require attention. Remember that many clients have endured a series of difficult endings in their lives. For example, they may have experienced relationships that abruptly fell apart without their having any control or understanding of them. They also may have been removed from abusive homes or witnessed violence that resulted in death or separation. The devastation of poverty and social exclusion frequently creates the chaotic climate that leads to turbulence and disconnection. If ending well with your client turns out to be the most positive aspect of your work together, that, in and of itself, will be work well done and could provide a model of a healthy ending to a relationship.

Like their clients, social workers themselves may have experienced similarly chaotic, sometimes traumatic, disruptions. It is crucial that social workers know and understand their own history with regard to disengaging from relationships. Many social workers struggle as much as or more than clients in termination, and they need to gain the insights that will allow them to facilitate a professional process that benefits the client. Many new social workers are surprised to realize how attached they become to a client, and find they are not prepared to respond appropriately to their feelings or the client. This understandable reaction is an issue for supervision. To facilitate this process, the social worker can recall that the relationship is for the clients' well-being, and not for her or his own well-being.

In discussing feelings about termination, you can let clients know about your ambivalent feeling about the ending: that you feel sad at ending the relationship; but that you have enjoyed knowing them, and that you have confidence that they will meet their own needs effectively in the future. It may be helpful to offer the possibility of future work together if the client needs assistance. It is generally not appropriate to express acute feelings of loss so that clients feel they must help you cope, and it is not appropriate to continue the relationship beyond whatever follow-up arrangements you might make without a new process. Exhibit 5.4 depicts two termination-related exchanges between Virginia and her social worker as they move toward ending their working relationship.

Over the course of a career, many client situations challenge these guidelines. There may be situations, particularly in rural social work practice, that involve the potential for **dual relationships**. A dual relationship exists when the social worker has both a professional and a personal relationship with the client. For example, a social worker would be in a dual relationship if she or he is working with the client in a social work capacity and purchases real estate from the client. Such situations have less than clear boundaries. In these cases, the social worker can seek dialogue with supervisors and peers to resolve issues for which no simple rule is adequate. The following describes another situation that may challenge these guidelines.

Respecting Cultural Consistency   Just as social workers are ethically bound to honor the cultural dimensions that the client brings to your work together, social workers

The two examples of an exchange between Virginia Stone (www.routledgesw.com/cases) and her social worker presented here demonstrate the power of an effectiveness of the termination process.

As you recall, Virginia has been working with a social worker on a number of challenging issues, including housing, caregiving, and grief and loss. Due to the number and intensity of the issues they worked on together, the relationship spanned over a year. The time has come for them to terminate the social work intervention. As you will learn in reviewing the two samples of dialogue presented here, terminations can be handled in different ways. Which of the two examples would you choose?

EXHIBIT 5.4

The Dialogue of Termination

Termination Option #1	Termination Option #2
Social Worker: "Virginia, as we discussed the last time we met, today will be our final meeting. I want to take this opportunity to review our progress."	Social Worker: "Virginia, I think today should be our final meeting. What do you think of this idea?"
Virginia: "I've been thinking a lot about this since last time and am feeling good about things."	Virginia: "What do you mean? I didn't know we had to stop. Did I do something wrong?"
Social Worker: "Looking at the contract that we developed early in our time together, let's take a look at each of the items on the list." "Before we get started looking over the list, I want to let you know that I have enjoyed working with you and seeing you make such positive changes in your life. I will miss our meetings. I want you to know that you can feel free to contact me if you feel any needs in the future."	Social Worker: "No, you haven't done anything wrong. I just decided that we have accomplished as much as we are probably going to. Let's do a quick re-cap. I think that will help you see what I mean."

must also anticipate and attend to the meanings they make of endings. Clients who identify as ethnically different from the dominant culture may have approaches and reactions to endings that are different from that of the social worker. The social worker is, thus, bound to incorporate cultural diversity into the termination process. Termination processes and rituals need to be culturally appropriate (Fortune, 2009). For example, it is not uncommon in some cultural groups for a child's schoolteacher to be invited to a family birthday party. If the client lives in a culture with tightly woven informal ties, a client may view the formal aspects of ending the work and cutting off the relationship as a reflection of your annoyance or rejection.

On the other hand, some clients may find it difficult to share their emotions verbally about ending because their cultural orientation encourages a restrained and private approach to sentiment. It may be more productive to emphasize the return of the client to her or his ethnic community, whose support and nurturance will help to consolidate the gains. This rejoining of the community is viewed as especially critical to communities of color (Lum, 2004). In some situations a client may bring you a gift, a gesture that represents a more comfortable way to express feelings, reflect celebration, and offer a suitable token of closure. In others you may want to maintain a more formal, businesslike relationship similar to the tone maintained throughout the work. Avoid objectifying or stereotyping any client by focusing solely on cultural difference (Diller, 2011). You will need to be mindful of your client as a whole person who brings idiosyncrasies as well as cultural traditions to the social work relationship. The dynamics of endings probably engender more commonalities than differences, and the person sitting across from you saying good-bye is, after all, the same person you have shared the work with all along.

Finally, your own cultural sensibilities about endings will influence the scenario as well. Anticipating your own patterns is helpful, particularly if they are likely to vary from those of your clients in any significant way. As mentioned, endings can evoke difficult feelings for all people, and some of us have developed more sophisticated ways than others to get around them. Even in the social work literature, there is more attention to beginnings than endings. Endings can represent a metaphorical death, and by recognizing your responses to that, you will be able to say goodbye to clients in an effective, positive, and professional way.

STRAIGHT TALK ABOUT TERMINATION AND ENDINGS

The termination processes discussed thus far have assumed conditions in which ending is planned. However, endings often are not planned and are not under your control and your role becomes to negotiate the unexpected. Earlier we explored scenarios in which the client leaves the relationship before you expect. In addition to life-changing events, clients terminate because they do not feel the relationship has been fully established. For example, when the client is not clear about expectations or does not share the social worker's understanding of the purpose, the meetings may seem unfocused and without direction. Social workers identify outcome goals that the client may not find relevant (even when clients "agree" to them). If there is not sufficient clarity about an ending point in the work, the client may not be committed to remaining engaged in the process. In such cases the work may continue, but the client grows discouraged.

There are countless reasons for the work to end prematurely, all of which present the social worker with dilemmas regarding the appropriate response and its meaning to the client system. Without pressuring the client unduly, the most useful strategy is to make contact, if possible, and encourage the client to return to the work, if only to end it in a more purposeful way. This approach conveys respect for the integrity of the work while leaving the client in control of the future of the work. This contact provides an opportunity to explore the dynamics of the relationship, which may provide important information to inform your future work.

In institutional settings in which the client has been transferred to another unit or service, you may want to petition your supervisor for a single meeting so that you can reconnect with your client, if only briefly, to review the work and acknowledge the shift in care. Although institutions tend to organize themselves in terms that meet the staff's, rather than clients', needs, make an effort to prioritize the latter whenever you can. When you have made every effort, either to continue the work with a client or simply to have one session to end your work together, and your client does not respond, you will need to respect the manner of ending she or he has chosen. It is the client's choice at that point, and you, representing the ideal of self-determination, will humbly understand the limits of your influence. You will most likely hope that your client has made a new plan.

An ending that will occur for most social workers at some point in their career is the termination that results from the worker leaving the organization. The already complex process of termination is made more challenging when the social worker informs the client that she or he is leaving the program or organization. When the social work relationship ending is worker-initiated, the client can feel a loss of control, disappointment, and/or anger, both over the ending and the future of her or his intervention (Siebold, 2007). While the social worker and client are best served by following a traditional termination process to the extent possible, another important task is to openly process feelings about the termination. For example, you may say to the client, "You may not have experienced the ending of a formal helping relationship before. What feelings are you having about our work coming to a close?"

FORMAL EVALUATIONS

The term evaluation raises anxiety in many practitioners, but most social workers can appreciate the importance of evaluating practice. Evaluation can provide feedback regarding goal achievement, the utility of a particular method and about the experience of the client in the work. Evaluation can be globally-focused, behaviorally-directed, and process-oriented. Evaluations can also reflect on individual progress, group development, and changes in power structures as it recognizes success. In short, evaluation is useful for practitioners when viewed as a tool for development rather than a threat.

The following discussion will highlight issues in evaluation as they increasingly occur in everyday practice.

Priorities in Evaluation

A longstanding debate continues regarding the types and relevance of various evaluative strategies that are most appropriate for social work practice. As you recall from Chapter 4, some experts maintain that social workers need empirical, evidenced-based practice to function in our contemporary society. Practice evaluation data can have multiple uses, such as informing and improving your practice assessment and interventions, providing data to incorporate into proposals for funding, and program planning. While considerable attention is devoted to justifying social work practice, social workers are ethically obligated to maintain their professional commitments and values. Bloom, Fischer, and Orme (2009, p. 15) suggest that, in order to be an evaluation-informed practitioner, one must strive to be a **scientific practitioner**. A scientific practitioner combines the use of evidence and evaluation without compromising the "art and creativity of practice," thus scientific practice includes: (1) incorporating research and evaluation to identify interventions that are known to be effective; (2) systematically monitoring and evaluating one's practice through single-system designs (described below); (3) being committed to ongoing learning to improve practice competencies; (4) approaching social work practice with a goal of problem-solving, investigation, and discovery; and (5) remaining committed to the values and ethics of social work practice (p. 15).

Social work ethics mandates that the provision of service obligations take precedence over data collection and evaluation efforts. However, evaluative strategies are an important component of effective practice. Evaluation of practice can provide feedback that is consistent with social work practice requirements, values, commitment to informed practice, and serve to empower both the client and the social worker. Imagine how empowered a client may feel when they realize she or he now has experience in successfully overcoming a challenge that has been plaguing her or him for years.

In the following two sections, two major methods of evaluation will be presented: empirical design processes and reflective assessment. These two evaluation methods serve different purposes, and they can help to balance the overall activity. First, they recognize the contemporary requirement for evidence-based practice, that is, practice guided by empirical, scientific evidence that the intervention has been successful. In addition, they can validate the contemplative, postmodern social worker's inclination to critique the traditional processes of practice and evaluation through focused critical reflection.

Quantitative and Empirical Processes: Evidence-Based Practice

Quantitative and empirical evaluation processes have two tools in common: single-subject design and goal attainment scaling (GAS).

Single-Subject Design Applicable with any theoretically guided intervention, the **single-subject design** (SSD) is a grouping of evaluative procedures based on an intuitive framework for examining changes in one client over time (Fischer & Orme, 2008). Usually, the social worker and client system complete repeated measurements of an action, feeling, or behavior over the course of their work together. For example, consider Virginia's return to her walking program. The two of you could use a single-system design approach to charting and monitoring her progress. The visual depiction of her activity provides you both with an accounting that can serve as the basis for discussion, revising the intervention plan, as well as evaluation of progress. Conversely, a social worker could follow one intervention over a series of clients. Single-system design evaluation plans serve to assess, monitor, and adapt to changes and compare the effectiveness of different interventions (Bloom et al., 2009, pp. 264–265).

In order to administer this design effectively, the client and social worker meet over a period of time to allow for repeated measurements. The social worker and client first agree on the behavior, attitude, or belief to be measured and the way in which this variable will be measured. The behavior(s) selected is one that should reflect(s) the goals of the work. Peripheral concerns, (e.g., if the client arrives on time) are not typically the focus of a single-subject design evaluation. There is no maximum number of attributes that the social worker and client can select, but in general you will want to limit them to a few after considering a wider range (Bloom et al., 2009).

Measurement can be attained by frequency (the number of occurrences) or through a standardized scale that yields a numerical result. The social worker and the client establish a relatively consistent interval for taking measurements throughout the course of the evaluation. For example, if you want to monitor the number of new social contacts your client initiates over time, consider the accumulated frequency over a standard period of time (e.g., every two weeks).

In most cases, the phases of your evaluation begin with a **baseline** (the rate at which or number of times the behavior occurred before the client came to you), intervention, **maintenance** (the period of stabilizing client gains), and follow-up. The measures can then be charted to depict a visual pattern. If a baseline cannot be established because your client needs intervention right away, improvements can still be plotted as they occur during the intervention (for example, after one month, then two months of working together).

If a baseline can be established without withholding the intervention, the differences between no intervention and intervention can be depicted. This provides a bit more evidence that the improvement is a result of your work as it assumes that the baseline measure would continue without your intervention (this is, of course, a big assumption). If the intervention is interrupted (e.g., the client or you go on vacation and no one fills in for you) and the frequency of your client's social initiations is reduced, you can see if restarting your work results in renewed improvement. If it does, the connection between your work and your client's improvement is strengthened, and the possibility that some other event brought on the improvement is reduced. However, such interruptions may not be in the client's best interests, and the course of the work should not be disrupted simply to demonstrate its success. The single-subject design is highly flexible and generally easy to use. Exhibit 5.5 provides an example of an SSD graph.

Goal Attainment Scaling   Used in a variety of settings, **goal attainment scaling** (GAS) represents a standardized framework that is customized by inserting individual goals. The first step of the multi-phase process is to identify two to five client goals and develop a scale for each of them based on the quality of the outcome. The following is a useful scale (Bloom et al., 2009):

0   Most unfavorable outcome thought likely.
1   Less than expected success.
2   Expected level of success.
3   More than expected success.
4   Most favorable outcome thought likely.

EXHIBIT 5.5

Simple Single-Subject Design

In this example of the single-subject design model, you can see that during the baseline period, before the work began the client initiated three, then two, then no new social contacts at intervals of two weeks. During the weeks of service, the numbers increased from two to seven, with a setback at week 8. After the work was over, the client initiated seven, then five, then six, and seven new contacts. This indicates improvement from the baseline period and extends over a four-week period following the work.

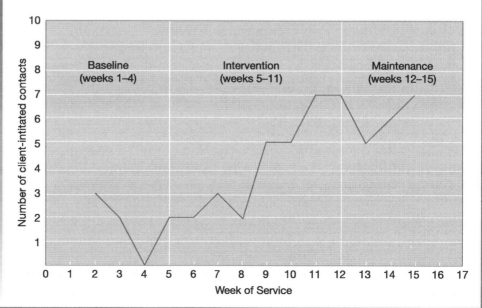

In this model, in a few words, the social worker and client describe the client's condition or status before the intervention on each goal, reflecting position 1 on the scale. The next step is to indicate the way in which deterioration may look at the lowest point (0). Finally, the social worker and client describe the best scenario (4), and they distinguish between levels 2 and 3. For example, if one of your client's goals is to attend parenting classes as part of a plan to regain custody of her children, you may agree on the statements depicted in Quick Guide 15. Consider that you have also agreed on two additional goals: attending GED classes and maintaining a clean apartment.

Following this process, you and your client can add further precision to measuring goal achievement by assigning a relative weight to each goal reflecting its importance in the individual case situation. As you can see in Quick Guide 15, the individual weight is 75 for attending parenting classes, 15 for keeping her apartment clean, and 10 for attending GED classes, for a total weight of 100. Your client's progress is tracked by placing a check mark in the cell that best describes the client's status at the point of entry into the social work relationship (acting as a baseline) and an X indicating the best description at the point of ending. You then calculate the weighted change score by subtracting the beginning score from the ending score and multiplying the difference by the weight.

The next step is to compute the percentage of possible change for each scaled goal. To calculate this percentage, you determine the highest possible mark on the scale and then divide it into the actual weighted change score. Finally, you calculate an overall score by summing all the possible scores for all the goals and dividing that number into the sum of the actual weighted change scores. Although this procedure may seem daunting and overly empirical, it becomes intuitive with practice.

Originally developed for use with individual clients, goal attainment scaling is well-suited to a range of social work interventions, including child and adult mental health, families, and organizations. GAS has also been shown to be an effective strategy for evaluating crisis interventions (Roberts, 2008). With crisis interventions and GAS both having an emphasis on short-term, structured and specific characteristics, the social worker engaged in crisis intervention work can easily and quickly use GAS to evaluate for both the client and her or himself the process and outcomes of their work together.

Other Forms of Evaluation   There are other quantitative measurement scales and other methods of attaining evaluative data that are not always quantitative. The latter includes such instruments as client satisfaction scales or client evaluations of the agency, the social worker, or both. Although these tools are often considered highly subjective and not particularly rigorous, they can frequently provide both the social worker and the agency with valuable information about the client's experience.

QUICK GUIDE 15   QUANTITATIVE AND EMPIRICAL PROCESSES

Client: _Clarissa_

Key: ✓ = Beginning level   × = Ending level

Attainment Grade	Task 1: Attend parenting classes	Task 2: Attend GED classes	Task 3: Keep apartment clean
0 Most unfavorable outcome likely	✓ No attendance at parenting classes	No attendance at GED classes	✓ No satisfactory ratings for cleanliness
1 Less than expected success	Attend less than 50% of the time	✓ Attend less than 40% of the time	Receive satisfactory ratings less than 60% of the time
2 Expected level of success	Attend 50–75% of the time	Attend 40–60% of the time	Receive satisfactory ratings 60–80% of the time
3 More than expected success	× Attend 76–95% of the time	Attend 61–85% of the time	Receive satisfactory ratings 81–95% of the time
4 Best anticipated success	Attend 96–100% of the time	× Attend 86–100% of the time	× Receive satisfactory ratings 96–100% of the time

Summary	Task 1	Task 2	Task 3	Total
Percent of goal	75	10	15	100
Change in score	3	3	4	
Total score	225	30	60	315
Possible total score	300	40	60	400
Percent of goal attained	75%	75%	100%	79%

Postmodern Views of Evaluation   In this world of managed care and increasing calls for evidence-based accountability, effective social workers must be knowledge-able about documenting, evaluating, and accounting for the usefulness of their efforts. To be able to choose the right evaluative method and be informed about your options, you should know the shortcomings and criticisms of each method.

Critics further charge that such models exclude the richer and nuanced contributions of aspects of client situations that are not easily quantified, as well as qualitative research, and overshadow the ethnographic forms of evaluation that question the assumptions of the everyday world.

These approaches have been based on expectation, researcher bias, and a political agenda that most social workers find unhelpful. Even when workers find themselves in a situation in which they must accommodate the system by demonstrating the value of their work through empirical evaluations, they cannot afford to de-emphasize the possibilities of vulnerabilities and opportunities for error in the evaluations themselves. This is a place for a form of bilingualism—that is, you will need to speak the language of contemporary demands, and at the same time keep alert to and keep up with the readings of their critiques.

There are social workers who find these empirical methods mechanical or lacking in substance, and want to examine their work on additional levels. Although the support for evidence-based practice in quantitative terms is strong in the profession, social workers are not restricted to these evaluative strategies. Social workers can also use the processes of reflection about their practice, as discussed below (see Lawler & Bilson, 2004).

Qualitative and Reflective Processes

Now we will look at other ways to evaluate social work practice. We will explore case studies, explorations of compatibility with theoretical perspectives, and quality of relationship. These methods have the power to expand your thinking about social work practice. These are not meant to compete with empirical processes, but to balance the experience of self-evaluation.

With any of these methods, many questions could be asked; we suggest some here. The most useful perspective is to focus your inquiry purposefully on a specific set of outcomes. If you want to take your inquiry further, consider engaging your supervisor as a second reader in a case study or your peers in a conversational group to explore particular issues. Such a focus is also a useful framework for ongoing staff meetings and professional development activities.

Case Studies   Like single-system designs, **case studies** involve intensive analysis of one individual, group, or family, and depend on accurate, careful, and detailed record keeping. A case study typically begins with the engagement of the client and social worker and continues throughout the course of the work. Case studies do not generate empirical data from planned comparisons because they are an accounting of the client's situation while working with you. However, case studies, still in use in contemporary practice, have a long history within the social work profession and provided the basis for the evolution of the formal evaluation process (Bloom et al., 2009). While space prohibits the presentation of a case study here, as they can be lengthy, you are encouraged to review the interactive cases found on the book's companion website for the Sanchez and Stone families or Carla Washburn, (www.routledgesw.com/cases). As you review the cases and complete the interactive assignments, you can document your thoughts, reactions, and plans in the online notebook. These notes can provide the foundation for a case study. Upon completion of the full planned change effort, you will have ample information to finalize a case study in which you can evaluate progress and efforts.

A full case study, though certainly not a new method, can provide you with a rich history for postmodern reflection. For example, in reviewing the initial contact, you might wonder if your sensitivity matched the client's need for validation, consider other ways you could have articulated your purpose, or reflect on the course of your work if you had taken another tack. Or perhaps you thought you were clear about the agency requirements, but looking back on the records, you can see where there might have been some confusion in the language. How might the work have gone if those requirements were clear to the client? There are hundreds of ways you can reflect on the work that may be helpful, both in supervision and on your own. A record of your observations, what you thought about it, and what the client said—all of these can provide material for analysis later. Case studies are particularly useful in noting your own growth as you learn from the experiences that your clients have with you and.

Inherent in the process of compiling a case study is the ability of the social worker to communicate clearly, succinctly, and professionally in creating an agency document. Recall from Chapter 4 the discussion related to general uses and components of documentation. Such guidelines continue to be essential during the intervention, termination, and evaluation phases. During these phases of social work practice, a well-organized and articulated social work record is essential for both the client and the social worker. To help you in developing your knowledge and skills regarding professional social work writing, Quick Guide 16 provides guiding principles for clinically-focused practice writing. Quick Guide 17 presents a sample case summary. With the integration of standardized forms and electronic records, case summaries may be abbreviated or encompassed within a "fillable" on-line form, but the data gathered and compiled continues to include key items of information related to the client's situation (identified need for services, assessment and planning, actions taken, and termination and evaluation notes).

Explorations of Compatibility with Theoretical Perspectives   Another strategy to evaluate your work is to explore its consistency with the perspectives you want to guide your work. This approach is especially helpful when those perspectives present challenges to you and stretch your thinking.

Social work practice has been placed with four frameworks thus far: social justice, human rights, the strengths perspective, and critical social construction. These theoretical frameworks introduce multiple perspectives, and therefore can be translated into broad criteria for evaluating the work you want to do. For example, are you consistently recognizing your clients' strengths, or do you tend to be pulled into the pathology orientation that historically dominated much of social work

QUICK GUIDE 16   GUIDING PRINCIPLES FOR CLINICAL WRITING

- Keep the values and ethics of the social work profession at the forefront of your writing. Just as with your practice, your commitment to client self-determination, strengths, empowerment, and cultural competence must be ever present as you write about clients in any form of documentation, including case records, case studies, letters, emails, and court reports.
- In keeping with your commitment to ethical practice, maintain your writing focus on words that do not label, denigrate, depersonalize, or marginalize the clients about whom you are writing (e.g., do not use words like "abusive parent," "handicapped boy," or "welfare mother").
- Extending the strengths perspective into your documentation requires the social worker to use strengths-focused language in order to provide "voice" to the client's narrative story (e.g., "In sharing her experiences of surviving a sexual assault, Mary focused on the coping strategies that she used to regain her self of safety and security.").
- Maintain your person-centered practice standards to emphasize the client's right in the areas of confidentiality and self-determination. Operationally, respectful writing means that you include only information that is relevant to the current situation. If you believe that collateral sources can benefit from additional information, you can consider an alternative form of communication (e.g., case conference, telephone contact, etc.) with a release of information signed by the client.
- Ensure your knowledge of and compliance with agency and state requirements regarding information that can be communicated about a client, including health or disability status and compliance with Health Insurance Portability and Accountability Act (HIPAA).
- Always contemplate individuals and groups who may have access to the writing that you have produced, including clients and their families, co-workers, court systems, or even media outlets and ensure that your writing is fact-based or that opinion-based content is clearly labeled as such.

Source: Sormanti, 2012, pp. 129–131.

Resources to Support Practice-Related Writing:

American Psychological Association. (2010). *Publication Manual of the American Psychological Association (6th ed)*. Washington, D.C.: American Psychological Association.
Green, W. & Simon, B. L. (2012). *The Columbia Guide to Social Work Writing*. NY: Columbia University Press.
Kagle, J. D. & Kopels, S. (2008). *Social work records (3rd ed)*. Long Grove, IL: Waveland Press, Inc.
Purdue Online Writing Lab (OWN).
Szuchman, L.T. & Thomlison, B. (2011). *Writing with style. APA style for social work*. Belmont, CA: Cengage Learning.

practice? Are you alert to social justice concerns when you meet with clients who seem unable to make their way in this culture? Can you truly remain open to the multiple realities that critical social construction emphasizes? Do you rationalize human rights violations because they are so common in our culture?

These considerations can also apply to more specific practice perspectives, such as feminist or narrative lenses. If, for example, you adopt a feminist theory that stresses the importance of power analysis, is your work consistent with that type of

QUICK GUIDE 17   CREATING A CASE SUMMARY

Effective and accurate documentation is essential for social work practice. While formats and content are agency-specific, the following example provides an abbreviated presentation of a typical case summary that you might prepare. For this example, let us return to Jasmine Johnson, a client discussed in Chapters 3 and 4.

Opening summary
Jasmine is a 33-year-old African American woman and a single parent for a 14-year-old son, Devon. Jasmine requested services from the agency to help her improve her relationship with her son and to find better strategies for disciplining him. She admits to having hit him when she believes he is being disrespectful to her but denies any current or past abuse or neglect.

Assessment
Family history: Jasmine was married at age 18 and gave birth to Devon when she was 19 years old. She and Devon's father divorced two years ago when Devon, aged 12, was starting middle school. Jasmine was awarded sole custody of her son and monthly child support, but the payments are sporadic. To save money, Jasmine and Devon moved out of the family home and live in a 1-bedroom apartment across town where Devon sleeps on the fold-out couch in the living room. The move required Devon to attend a new school.

Employment and financial situation: Jasmine has worked in the housekeeping department of a large hospital for the past eleven years. The hospital provides comprehensive benefits (e.g., health insurance, retirement, and educational support) but her salary is low and she often struggles to pay all her bills each month.

Social support and resources: Jasmine and her family did not have any family living in the community where she moved with her then-husband for his job. Her ex-husband remains in the area, but has remarried and is expecting a new child. He seems to have little time for Devon but does see him occasionally, usually when Jasmine contacts him and asks him to see his son.

Jasmine's supports in the community now include a tight-knit group of co-workers, a neighbor, and members of her church. Her co-workers provide emotional support and often invite the Johnsons for holidays. The neighbor frequently "cooks too much food" and brings dishes over for Jasmine and Devon. Her church friends have provided spiritual support which Jasmine finds comforting.

Jasmine and Devon have received support through the church's Christmas adopt-a-family program, the youth programs, and the food and clothing pantry. Devon is eligible for the breakfast and lunch program at his school.

Prioritized concerns: Jasmine's primary concern is her relationship with her son. While she wishes that he did not speak disrespectfully to her and violate the rules she has established, she recognizes that he is a teenage boy whose life was turned upside down by the divorce, the move to a smaller apartment and a new school, and the lack of support from and contact with his father. She wants to improve her relationship with Devon and find alternative ways to establish and maintain acceptable boundaries for his behavior.

Strengths and areas of challenge: Jasmine possesses a number of strengths, including: 1) desire to be a good parent and not use physical discipline; 2) history of consistent employment that provides benefits; 3) while not extensive, an active support system; 4) willingness to seek help for relationship challenges and use community resources; and 5) resilience to overcome adversity.

<u>Areas of challenge for Jasmine include</u>: 1) history of physically striking Devon and lack of knowledge about other parenting possibilities or strategies; 2) low self-esteem—Jasmine assumes her son's disrespect for her is her fault for having divorced his father and taken him away from Devon, moved him to a new area and school, and never having enough money to provide for anything beyond the necessities; and 2) stress related to her income, including sporadic child support payments from her ex-husband.

<u>Intervention plan</u>: With monthly reviews, a three-month contract was jointly developed by Jasmine and social worker, including:

Goal	Client Tasks/Timeline	Social Worker Tasks/Timeline	Follow-Up	Termination
1. *Improve my relationship with my son*	*Begin attending weekly family therapy with my son as soon as an appointment can be made.*	*Refer Jasmine and her son to a family therapist and communicate regularly with the therapist regarding progress (with Jasmine's informed consent).*	*At 3 months, Jasmine and Devon are attending family therapy.*	*At 6 months, social worker recommends case be closed as goals have been met.*
2. *Learn better strategies for disciplining my son, particularly when I am angry*	*Participate in weekly parents of teens class and support group (next group begins the first of next month).*	*Refer Jasmine to parenting class and support group and communicate regularly with the group facilitator regarding progress (with Jasmine's informed consent).*	*At 3 months, Jasmine is a regular member of the class/support group.*	*At 6 months, social worker recommends case be closed as goals have been met.*
3. *Get Devon's father to pay child support more consistently*	*As soon as possible, contact Legal Services Child Support Enforcement office to inquire if they can help.*	*Provide Jasmine with Legal Services contact information and eligibility requirements.*	*At 3 months, Jasmine has made an appointment with Legal Services*	*At 6 months, social work recommends case be closed as goal is in process of being met.*

Intervention (including service options and purpose, goals, and plans of service)
Goal #1: Jasmine and Devon attended eight sessions with the family therapist. They continue to attend, although Devon sometimes refuses to go with his mother. He has attended six of the eight sessions and agrees to "keep trying it for a while." The therapist reports to the social worker that she believes the Johnsons are making small strides toward improving their relationship. As suspected, Devon is extremely angry with his mother and does not understand why things had to change. Jasmine refuses to disclose to Devon that his father's drinking and infidelities are the reason for the divorce. Early on, there was an episode of physical contact in which Jasmine slapped Devon for calling her a name.

QUICK GUIDE 17   CONTINUED

Goal #2: Jasmine has been regularly attending the Single Parents of Teens classes and support group. Not only has she found the information provided very helpful, she has found a new group of friends who share her experiences, empathize with her, and offer helpful suggestions.

Goal #3: Jasmine was slow in contacting Legal Services, but once she got an appointment and completed the application, she was seen by an attorney and social worker. A letter has been sent to her ex-husband regarding his delinquent payments. He has yet to respond. Should he not respond, she will have to return to court and request enforcement.

Closing summary
During the past six months that the social worker and Jasmine have been working together, she has initiated work on each of the three goals outlined in the contract. Positive changes have been slowly occurring in her relationship. There have been no episodes of physical contact in five months. While they still occur, the number of episodes of Devon violating curfew and speaking disrespectfully to his mother have decreased somewhat.

Shortly after the implementation of the intervention, the social worker learned that an allegation of child abuse has been made against Jasmine. Upon further investigation, the social worker learned that the call to the child abuse hotline had been made by Devon months earlier after a particularly emotional confrontation with Jasmine. The report was unsubstantiated.

The social worker is recommending closure of this case as substantial progress is being made on all goals. Jasmine and the social worker will discuss termination and evaluation. Should Jasmine feel she is in need of additional services, she will be invited to request her case be re-opened.

Sample Case/Progress/Interim notes (brief excerpts from the case record):

October 1: Jasmine Johnson came to the agency seeking help with her 14-year-old son. She believes he is "out of control" and she is worried that she will become abusive. She has physically struck him on several occasions when he has violated her rules and or spoken disrespectfully to her. She feels she is at her "wit's end." Intake assessment forms completed. Social worker asked Jasmine to bring Devon to the next meeting so his perspective could be included.

November 28: Jasmine and Devon have attended two family therapy sessions. Devon was a reluctant participant but did agree to attend with his mother. Jasmine reports that he said very little in the first meeting, but opened up more in the second session. Her fears are confirmed—he is very angry with her. Jasmine has attended two Single Parents of Teens classes and support group meetings. She reports that she is getting a lot out of the sessions, particularly in terms of the tips she is getting for ways to interact with Devon. She has not yet contacted the Legal Services office.

analysis? Do you return to a more traditional perspective regarding the issues that your client brings by emphasizing her reluctance to leave an abusive relationship or her lack of self-esteem? Can you keep an analytical structure of gender relations in the forefront rather than falling back into our society's tendency to blame women for occupying a power-down position? Which aspects of feminist theory do you carry out well, and which seem to call for continued growth?

If your framework is narrative, are you completely open to the complexity of the client's story? Do you wish you had been more transparent in your responses to it? Do you recognize your client as expert, or is there a temptation to believe you know better? Which aspects of the perspective are troubling to you? Which seem to come naturally? These considerations, of course, do not yield empirical data; rather, they can help you decide to what degree you can work within the constraints of particular perspectives, and they can help you identify the ways in which you want to grow intellectually and skillfully.

Explorations of Quality of Relationship   The nature of the relationship you develop with your client offers another opportunity for exploration. Is the relationship consistent with your purpose in the work? Was openness a characteristic of the connection early on? Was openness difficult for either you or your client to establish? What can you learn from the client's struggle? What can you learn from your own? What cultural dimensions influenced the development of the relationship, and what was successful in working through those issues (see Maramaldi, Berkman, & Barusch, 2005)?

This might be a situation in which to look at your own idiosyncrasies. For example, do you respond more easily to people most like you? Do you wonder if you encouraged the client's dependence on you? Do you struggle with keeping useful boundaries between you and your client? Are you comfortable with the amount of self-disclosure you engage in? Are there some kinds of clients that are hard for you to be positive about?

Additional questions that can arise out of this kind of self-examination lead to more generic issues: What is the ideal relationship between client and worker? How does it look? Is it different in one setting from another? Is it likely that all your client relationships will fall into this range of ideal? How does the ideal relationship interface with the client's goal attainment?

STRAIGHT TALK ABOUT EVALUATION AND RECLAIMING OUR KNOWLEDGE

As the discussion on intervention, termination, and evaluation comes to a conclusion, consider the distinctiveness of social work as a profession and consider its greater purposes, as social workers are informed by social justice, human rights, strengths, and the value of multiple realities. This is a profession that lives through a combination of practical orientation and a strong sense of caring. As you struggle to account for yourself and your professional knowledge and competencies, remember social work's legacy of context and care. These are always difficult to measure, and as long as you are pressured to assign numerical indicators, some of your most important contributions could be missed. You will recall from Chapter 1, we used the term "practice wisdom." Practice wisdom is the social worker's application of her or his "accumulation of information, assumptions, ideologies, and judgments" (Barker, 2003, p. 334). Identified through reflecting on one's practice experiences, practice wisdom enables the social worker to translate empirical and conceptual knowledge into practice behaviors that are most appropriately suited to the client situation (Chu & Tsui, 2008). Social workers' distinctive knowledge, based on a practice wisdom of person-in-environment, strength and struggle, and heart and grit is the "humble stuff of lived experience and values and flies in the face of most current views about how social workers know what they know" (Weick, 1999, p. 327).

As an empowering practice, then, with real people and real misery as well as real joys, social work will ask other questions about its effectiveness, including:

- Who benefits from this work? If the goals of social work relate to helping clients find empowerment, how are they realized? If a client demonstrates improved capacity to manage a household budget, for example, is there a link to her experience of poverty? Is it identified and challenged? Will she quietly and skillfully manage on close to nothing (and is this progress?), or will the benefit go beyond, to others like her, and to challenge the structure that supports poverty?

- Whose values are most salient? Social work's values are at the heart of the work. Do they dominate the client relationship? How do social workers negotiate differing values with people who do not share theirs? This is one of social work's most challenging dilemmas.

- What changes have occurred in the power structure? Have social workers helped to raise consciousness about oppression and the internalization of it? Have clients joined with others to respond to oppressive imbalances in the structural arrangements of our culture? As work with clients ends, is some move toward empowerment visibly in progress? (Lee, 2001).

These questions lead us back to the beginning of our exploration of social work practice. As a social worker, you will likely find satisfaction in connecting with clients. You will struggle to meet the demands of their preferred realities and document them for those who require an accounting. You will experience frustrations and successes, answer many questions, and come up with many more. The world will change dramatically and will challenge you with its excitement and danger as well as its incredible potential. You will figure out how to respond to it in ways that are consistent with your values, your social work ethics, your sense of social justice and human rights, and your respect for different views and experience.

Your own story as a social work practitioner will be embedded in the hundreds of stories of your individual clients, groups, and families, the organizations and communities you serve, and in the story of global change. Your choices are legion and your opportunities enormous. In the end, these will return to the story of the profession. You as a member of the next generation of citizen/social workers will author the next chapter.

CONCLUSION

In this chapter we have looked into several dimensions of the intervention, termination, and evaluation actions that social workers take. The practice setting, theoretical perspectives, and perception of roles influence the ways in which you work to support clients' strengths and to support their environments. The overall fit of your activities with and your beliefs about people creates a sound backdrop for expanding your work into other system levels. The next chapters will integrate and extend these same processes to groups, families, organizations, and communities. This will serve both to consolidate the principles and skills in this first part of the book and to take them beyond, across system levels.

In the paradoxical way that some things constantly change while they remain the same, so endings and evaluations continue and evolve. All the same, we will all experience new ways to do things in the future of social work practice. Some of you will work in much more complex settings that integrate public and private sectors through partnerships and alliances. Some of you will choose more radical forms of practice in which you will challenge the historical and professional legacies you see as unjust or obsolete. Others will continue to practice with only incremental changes from the models under which you trained. All of these dimensions of practice will have implications for beginning and ending the social work relationship, for executing and evaluating the work, and for grappling with the obstacles that get in the way of the vision. There are no magic formulas for anticipating all of the implications of change for social work. Your flexibility, integrity, and penchant for the experiential context of your work, as well as your caring for people who struggle, will be your own best guides for your future practice.

MAIN POINTS

- The most pressing client issue from the client's perspective is the starting point for the intervention. This issue is defined by the context of the client's situation and by the actions agreed upon in the assessment and planning phase.

- To support clients' strengths, social workers act in context to normalize and capitalize on those strengths. Social workers engage in responding to feelings, determining their meaning, and supporting diversity.

- To support clients' environments, social workers are accountable to the client system; they should follow the demands of the client task; maximize the potential supports in the client's environments; identify, reinforce, and/or increase the client's repertoire of strategic behaviors; and they apply these principles to themselves.

- Examining traditional social work roles and their assumptions within the context of contemporary practice provides another perspective the context of the roles of case manager, counselor, broker, mediator, educator, client advocate, and collaborator.

- Exploring a case situation through the lens of empowerment perspectives demonstrates the work in context and how it fits with the perspectives emphasized in this book.

- The planned ending process with clients consists of several components that can benefit and empower clients by consolidating the gains of the work and the relationship. Social worker and client must have clarity regarding timing, original agreement, successes and failures, and responses to termination.

The same basic process for ending work, when flexibly used with different emphases, can be applied to families, groups, organizations, and communities with additional and specific components to consider with groups (to be discussed in Chapters 7, 9, 11, and 13).

- Unplanned endings pose a special challenge to both social workers and clients. Social workers can attempt to reconnect and end the work, but may have to acknowledge the limits of their influence and honor client self-determination.

- Quantitative and empirical evaluation processes are increasingly required in social work practice. Two options for evaluation are the single-subject design and the goal attainment scaling.

- Qualitative and reflective practices also constitute an important method for evaluation and professional growth as they relate to philosophical commitments, theoretical perspectives, and relationship building.

EXERCISES

a. Case Exercises

1. Go to www.routledgesw.com/cases, review the video vignette with Emilia Sanchez and respond to the following questions:
   a. What strengths does she possess?
   b. What thoughts and/or feelings resulted from watching the video that, if verbalized, would not reflect a strengths-based approach?
   c. As the social worker, what are your next steps with Emilia?
   d. Either with a group in or outside of class, develop a strengths-based intervention plan for Emilia.

2. Go to www.routledgesw.com/cases and click on Carla Washburn. Begin by reviewing the engagement and assessment phases. After your review of the first two phases, click on Phase 3: Intervention. Complete each of the tasks identified in the five steps (introduction, goals and needs, client tasks, social worker tasks, timeline, and coalitions).

3. Go to www.routledgesw.com/cases and click on Brickville (with your focus on Virginia Stone and her family). Begin by reviewing the engagement and assessment phases. After your review of the first two phases, click on Phase 3: Intervention. Complete each of the tasks identified in the five steps (introduction, goals and needs, client tasks, social worker tasks, timeline, and coalitions).

4. Go to www.routledgesw.com/cases and click on Brickville. With your focus on Virginia Stone and her family, review the four components of the social work intervention, including engagement, assessment and planning, intervention, and termination and evaluation. Upon completing the tasks identified in each section, develop a case summary using the guide included in Quick Guide 15.

5. Go to www.routledgesw.com/cases and click on RAINN. After reviewing the information, click on Engage, review the two client scenarios (Sarah and Alan), and respond to the three questions that are listed below the scenarios. In addition, develop a list of skills that you would need to develop in order to competently work with a survivor of sexual assault.

b. Other exercises

6. The admissions unit of a psychiatric care facility issues a daily report to all staff on a single sheet of paper that summarizes all the patient admissions, discharges, visits, legal proceedings, and other activities for the preceding 24-hour period. This summary is meant to convey useful information for all staff and it is viewed as helpful to social workers who needed to monitor client status. One of the categories on the form is entitled "body count" and is a tally of the number of patient/residents actually in the hospital at midnight. A social worker from a local community mental health agency was visiting her client who was in the hospital and she heard reference to this sheet. She was horrified at the language of "body count," to which you were by then accustomed.

   a. What do you think is the issue here? Why does it matter? Be as specific as you can.

   b. In what way would you address it? Would such an activity be consistent with your view of a social worker's role?

   c. Develop a plan that includes at least three steps you might take.
      Compare with other students and be prepared to discuss in class or write brief responses to each question to submit.

7. Tammy is a 35-year-old Caucasian female seeking treatment after her release from a 21-day residential drug/alcohol treatment facility. She is referred to your agency by the public child welfare agency. As a result of continued substance

abuse and an arrest and conviction for driving under the influence with her eight-year-old son, Jared, in the car, physical custody of Jared was granted to Tammy's mother.

Tammy has a 20-year history of drug and alcohol abuse and she has been diagnosed with Bipolar Disorder. She was recently prescribed medication by the staff psychiatrist at the residential treatment facility. It is a new medication and is not covered by her insurance. She has been non-compliant with medication in the past due to the side effects and her drug/alcohol use. She has had brief periods of sobriety but often relapses after a few weeks. This is the longest she has been sober since the birth of her son eight years earlier.

Tammy has never been married and has a difficult relationship with her family of origin. Her mother placed her in foster care at the age of eight due to Tammy's behavior. Tammy reports that her mother was physically and emotionally abusive to her and often would leave her with various relatives when her mother found a new boyfriend. Tammy's father is not involved. Tammy has few friends and little contact with her mother. She is angered that her mother has custody of her son. Tammy has been involved with her son's father sporadically for the past nine years. Currently, she names him as a source of support.

Tammy is not currently employed but receives public assistance in the form of Medicaid, disability assistance, and food stamps. She lives with her boyfriend but would like to have her own apartment. She has her high school diploma and is interested in continuing her education.

Partner with other students and complete the following exercises:

a. Using narrative interventions, reconstruct this case from a strengths-based perspective.

b. What roles would you play in helping Tammy?

c. Construct a strengths-based plan of intervention utilizing solution-focused interventions.

d. Share your findings with the class and compare plans of interventions.

8. Reflect on a time in your life when you contemplated making a change. What did you do? Was it successful? What factors led to the success? If not successful, what was missing?

9. Identify an area or behavior in your life that you would like to change (e.g., texting while driving, exercising, budgeting money, or time management). For one week, chart your journey in a journal, and note those factors that are helping you to maintain the change and those that are a negative influence on your efforts to maintain the change. At the end of the week, reflect on your progress or lack thereof. Consider your feelings before the change, during the change, and the results.

10. Write your own life story from a strengths-based perspective.

CHAPTER 6

Social Work Practice with Families: Engagement, Assessment, and Planning

The social work profession and the family have traveled a long distance together, sometimes in close companionship and sometimes on divergent paths, only to meet once again on the same road. Our profession began in the company of the family and has returned to it once again.

Ann Hartman and Joan Laird, 1983

Key Questions for Chapter 6

(1) What competencies do I need to engage with and assess families? (EPAS 2.1.10(a) & (b))

(2) What are the social work practice behaviors that enable me to effectively engage and assess families? (EPAS 2.1.10(a) & (b))

(3) How can I use evidence to practice research-informed practice and practice-informed research to guide the engagement and assessment with families? (EPAS 1.2.6)

(4) How do the perspectives and experiences from my present family and/or family of origin impact my engagement with and assessment of client families?

FAMILY IS THE EARLIEST, MOST BASIC, AND, SOME SAY, most challenging small group one can experience during a lifetime. It is also probably the most powerful in shaping who we become. For some, family means home, safety, and acceptance. For others, family means violence and danger. Some people may feel important and cherished with family or never quite good enough or even useless. Others may feel swallowed up in their family's dysfunction, or may long for, and bask in, the wholeness of unconditional support. Most of us experience a mix of feelings in between these. Family is a complicated enterprise, and many of us harbor some of its tensions that make us both joyful and troubled. Virtually all members of society have experienced some kind of family, and most have ideas (or dreams) of the qualities that an ideal family could or should possess.

This chapter explores the concept of family—definition, meaning, and place within the contemporary social context—and the process of engagement and assessment with families. We will also look at theoretical perspectives for working with traditional and contemporary family structures as well as dynamics, skills, and tools for working with the range of families. This chapter will also guide you through challenges and strategies for understanding the impact that family issues can have on your clients and you.

FAMILIAR PERSPECTIVES AND SOME ALTERNATIVES

The enormous and rapid changes in social rules in most Western countries have led many to contend that the "traditional" family has been lost. A traditional vision of the ideal U.S. family portrays a nuclear group consisting of two heterosexual adults and two or perhaps three children. The father supports the family economically, and the mother supports it emotionally. There are clear roles, rules, and jobs that each member undertakes. If the mother also works outside the home, she is still free to participate in the kindergarten car pool, do the laundry, entertain friends, and "be there" for her husband and children. A more contemporary version of the father makes him increasingly sensitive to the feelings and needs of his wife and children, but his work still takes priority. In contrast, the National Association of Social Workers (NASW) (2007a, para. 3) provides this definition of the contemporary family: "two or more people who assume obligations and responsibilities generally conducive to family life." This inclusive perspective on the family embraces parents who are divorced, separated, unmarried, grandparents, gay, lesbian, bisexual, transgender, or questioning, adoptive, and fostering, along with couples who have no children, partnered couples, and families caring for older adult members. Social workers must use practice approaches that recognize the diversity of family constellations (Hull & Mather, 2006). For example, to avoid assumptions and confusion, the social worker should ask the client to describe and define her or his family unit (Wood & Tully, 2006) rather than make assumptions, based on household composition. Given the changing nature of the family and the fluidity of membership in some families, it is important to ask each member of the family, "Who do you consider to be part of your family?" A follow-up question may then be, "Of these people, who is biologically and/ or legally related to you?" For instance, a client's "aunt" may not be a legal or biological family member but is considered by the client to be a part of her or his family.

Those with nostalgic older visions of family are appalled at what they see today as a lack of morality in many young families. There is distress about young adults who live together with no permanent commitment. Single women often choose to become parents and do not suffer the social stigma so prevalent only a few generations earlier. Parents who are gay, lesbian, bisexual, transgender and questioning (LGBTQ), who earlier had to conceal their sexual identities, are joyfully parenting children, and many are serving as foster and adoptive parents in conjunction with state child protection agencies. In the past, LGBTQ individuals were not even given the option to foster or adopt children; now they are recruited by these organizations. Although these developments in our U.S. cultural norms are lamented by some, fear of change may overshadow the research about new, more inclusive family and parenting arrangements. We will take a more in-depth look into this issue later in the chapter.

Nonetheless, there are real issues in the surrounding context of the contemporary family that create concerns for social workers. The U.S. divorce rate of 52 percent for men and 44 percent for women (U.S. Census, 2010b) is closely related to the ever-growing number of children who live in a single parent-headed household with incomes below the poverty level. The percentage of children who live in poverty with a single mother has reached 32 percent, while single father-headed households experience poverty rates of only 16 percent (U.S. Census, 2011a). Some believe that commitment to the stability and endurance seems to have vanished. While the number of unmarried teen parents is slowly declining (31.2/1000 teen females in 2010, down from 34.2/1,000 teen females in 2008) (Hamilton, Martin, & Ventura, 2012), the rate of births to teen parents of color continues to be high (36–47 births/1,000 teen mothers) (Hamilton et al., 2012). The associated legacies of perpetuated poverty, increased family violence, and health and mental health issues, and a sense that two generations—the child and the child's child—have sacrificed much of their potential, provide little hope for the future. Clearly, the change in predominant family compositions has required changes in service delivery, and impacts society. The widening of the income equality gap and economic pressures resulting from the recent economic recession has created additional challenges for families that impact their well-being. The result of such change has been serious societal issues with which society and social work must grapple.

There are other ways to look at family. In keeping with this book's focus on multiple realities, we will explore the experiences of an array of family constellations to expose illusions about the family and develop a more balanced perspective. This chapter will discuss the experiences of both traditional and nontraditional families. Consider Exhibit 6.1, which describes the experience of family for countless people who have been marginalized, that is, who have lived in the margins of our culture and others.

HISTORICAL ANTECEDENTS FOR FAMILY SOCIAL WORK

The social work profession has a long history of working with families, shaped by the context of the times. In community mental health centers and youth agencies, hospitals and schools, and welfare and child protection efforts, social workers work to strengthen families. Collins, Jordan, and Coleman (2013) encourage social work students interested in working with families to explore this area of practice using the following list of questions:

- What is the purpose of family social work?

- How does family social work differ from family therapy?

- How can I work effectively with families who are different from my own family?

- What is my role as a family social worker?

- How will I know what factors contribute to the family's difficulties?

- How should I work with an entire family and with all members in the same room at the same time?

- How will I know what questions to ask family members? What do I say to the family?

- How do I engage all the members of the family, particularly if they seem resistant, feeling blamed, uncommunicative, or overpowered by another member?

- What should I do if family members get angry at me or another member of the family?

- What do I need to know to help families change? What knowledge will help the families begin to change?

- What skills will I need when there are young children in the interview? Older children?

- What can I do to protect individual family members when the rest of the family is attacking or blaming?

- How do I help families that are paralyzed by a crisis to rise about it and solve their problems?

- What do I need to know about family social work when working with families of different ethnic, racial, or sexual orientation backgrounds?

- What skills are needed for prioritizing the family's problems and then for each of the phases of work with families (pp. 1–2).

Why is there such a concern for the maintenance of the family? What does our culture expect the family to do, and how do we think the family should work? Although there are many possible responses to these questions, two particular perspectives are examined that seem to have special relevance for social workers today—the family as a functioning unit and the family as a system.

EXHIBIT 6.1

Family: Views from the Margins

©: Harry Hu, courtesy of Shutterstock® images

- What we may consider now to be the ideal family has never actually flourished in any culture for any length of time: Children have historically been considered commodities that enhanced the economic status of their fathers, who owned them, and their value was often equated to the amount of work they did. Childhood as a time to be nourished and cherished is a relatively new and narrowly prescribed phenomenon of Western culture and some nations in the East. The United Nations instrument "The Rights of the Child" reflects the need for considering children as genuine people and not possessions. In contrast, in parts of the United States, children are prostitutes, drug dealers, and hired thieves. In much of the world they are all these and also soldiers.
- Ideal conceptions of the family have historically served to restrict and diminish the role of women as categorical caretakers. Many women throughout history have been required to abandon their dreams and have also been seen as property to be exploited. Men, too, are assumed to fit into tightly proscribed roles that may not be compatible with their identities or goals.
- In many families of the past, infant and childhood mortality was high and parents may have died by their early 40s. Such circumstances produced a crisis for remaining family members, resulting in placements with distant relatives, community members, or in orphanage care. The idealistic notion of earlier times places blinders on the realities of disease, early death, and other forms of danger that surely shaped the overall experience of family.

EXHIBIT 6.1

Continued

- It is still the family, even in contemporary U.S. society, that is the primary force for socialization and nurturing of children and support of the community. As an example, the families of children experiencing mental illnesses still provide most of the care and nurturing of their adult children, in spite of federal and state programs designed to assist them.
- The definition of family as rigidly restricted to biological and legal ties and hetero-sexual partners or adoption has always marginalized and scapegoated significant numbers of people who have meaningful and productive relationships and who make significant contributions to the community, the socialization of children, and the general economic and social order.
- More contemporary notions of family have liberated both men and women to develop and carry out the roles of child caretaking, economic provision, manage-ment, personal development, and health care in a way that does not deny their individual aspirations and talents; the same principles hold for gay, lesbian, bisexual, and transgender families with biological or adopted children.
- Contemporary families have greater biological control over the number and timing of pregnancies and can plan family composition in a way that is consistent with their financial capacities and other internal demands (for example, one partner is in school or another is committed to the care of a parent). Women are now more able to exer-cise control over their reproduction, and make decisions about their future.

Family as a Functioning Unit

A concrete way to think about the importance of the family is to explore the societal expectations of the family. The following functions are among those that are typi-cally viewed as critical to the contemporary maintenance of families:

- Provide the material and economic necessities for sustenance and growth.
- Offer members emotional security, respect, safety, and a place for appropriate sexual expression.
- Provide a haven for privacy and rest.
- Assist, protect, and advocate for members who are vulnerable or who have special needs.
- Provide support for members' meaningful connection and contribution to community life.
- Facilitate the transmission of cultural heritage.
- Provide a socially and legally recognized identity.
- Create an environment in which children can be nurtured and socialized.

The responsibilities reflected here emphasize not only the functional roles of individual family members and their needs and identities but also the connection to their communities and the overall societal environment. This compilation is not prescriptive; that is, it does not specify *how* children should be nurtured but allows for individual and cultural interpretation. Rather than seeking individual or family dysfunction, the exploration of these tasks in various areas of family life tends to highlight strengths as well as areas for improvement. In that respect, this set of functions serves as a useful guide for assessing the degree to which societal expectations are met in any particular family.

While families serve a critical function in society, we must recognize they have needs which may not be met, thus requiring the involvement of a helping professional. Needs may be categorized in many different ways, but one strategy is to consider the family as a system with a practical purpose. Building on the work of Kilpatrick and Cleveland (1993), Kilpatrick (2009) presents a strategy for classifying family needs by levels, including:

Level 1—basic survival. Needs are in the areas of food, shelter, medical care.
Level 2—structure and organization. Needs center on setting limits and safety concerns.
Level 3—space. Needs relate to issues of privacy, access, and boundaries.
Level 4—richness and quality. Needs exist in areas of inner conflict, intimacy, and self-actualization (p. 4).

Taking into consideration the type and level of need, the social work assessment and intervention process can be focused within a conceptual framework that encompasses both context of those persons with whom you work (i.e., family/community, couple/dyad, and individual) and orientation of the intervention you choose with the family. Will the intervention have a focus on behavioral/interactional change, experiential aspects of clients' thoughts and feelings, or historical, emphasizing family-of-origin issues (Kilpatrick, 2009, p. 10)? Incorporating these complexities of the family system into your social work intervention can help to gain insight into the way in which the family functions, their needs, and the most effective strategies for helping them to reach their desired goals.

Family as a System

Because social work focuses on interactions among people and between people and the environment, the profession enthusiastically adopted systems theory in the 1970s. **Systems theory** posits that a system involves a series of components that are highly organized and dependent upon each other in an orderly way. In the profession, this conceptualization is applied to the multiple levels of practice (individual, group, family, organization, and community). The systems perspective remains influential in social work theory in spite of a growing number of critiques that suggest it is too rigid and tends to place the social worker outside of the work. Nevertheless, systems theory holds continuing authority in many views of structural arrangements, particularly the family. Three elements of systems theory are particularly important:

- Change in one component

- Subsystems and boundaries

- Family norms

Change in One Component   Possibly the most powerful idea in the systems theory for social workers is that change in one part of the system will affect all other components. Social workers, therefore, seek to learn about aspects of a client's environment, including family functioning. For example, knowing that a child's father has just been sent to prison is useful information when exploring reasons for that child having angry outbursts or sullen withdrawals in school. In another example, a mother being physically abused by her intimate partner is likely to have repercussions on her daughter's fragile health. Social workers often find such connections intuitive and useful. Still, you will want to guard against seeing such a situation as automatically producing one type of response. In systems theory, one set of actions can predict multiple sets of reactions. For example, the child whose father goes to prison may respond by being more attentive to her or his mother or by working harder in school. As another example, consider the family in which the primary wage earner becomes unemployed. This event can create a ripple effect within the family. While one of the adolescent children may seek employment to help the family, another may demonstrate anger that her or his perceived needs will not be met.

Subsystems and Boundaries   Another dimension of systems theory that is relevant to social work is the concept of subsystems. **Subsystems**, or components of a system that also have interacting parts, provide a mechanism for organizing relationships and planning ways to engage with them. In systems theory, the individual is a subsystem of the family, the family is a subsystem of the community and the community is a subsystem of the culture. A family may also be a subsystem of more than one larger system, or of differing systems, so you must be thoughtful about making unqualified judgments regarding the place that an individual or group occupies within the system. For example, in a blended family, there may be members who consider themselves part of the family and community from the previous marriage. Social workers have the opportunity to support and facilitate such healthy transition relationships for client systems. Exhibit 6.2 depicts the components of a family system.

In thinking about the concept of family, social workers often distinguish between the subsystem of the parents and the subsystem of the children. The relationship between these subsystems is reflected in the types of **boundaries**, or limits that separate the systems that the family constructs. If the children are included in all decisions the family makes, the boundaries are **permeable**, meaning that information and interchange goes easily across them. However, permeable boundaries can be taken to the extreme of **diffuse boundaries**, which means that boundaries are too loose and that parents should assume more decision-making authorities to maintain appropriate boundaries between them and their children.

Appropriate boundaries between the subsystems of parents and children may vary considerably depending on culture, the times, background, and/or the boundaries within which the parents themselves were reared. While you may consider a boundary to be inappropriate if a mother who uses her pre-teen daughter as her primary confidante, particularly related to her relationship with the child's father, other families may view this as acceptable. Can you cite examples in your own family in which these boundaries are similar or different from those of your friends or other extended family members? Further, consider the way in which your experience with your own family might relate to families with whom you will be working.

Family Norms   Most families establish **family norms**, or rules of conduct, that are related to boundaries and subsystems. These can be similar to the norms described in Chapter 7, in application to groups, but have an additional complication in

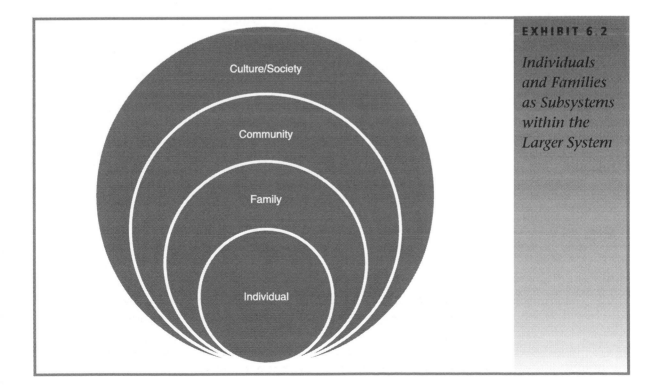

EXHIBIT 6.2

Individuals and Families as Subsystems within the Larger System

Culture/Society

Community

Family

Individual

families in that they are often held as sacrosanct and not negotiable. Family norms may also never be articulated, despite the fact that all members of the family are clear about them. This implicit aspect of norms can make them difficult to address and challenge, so much so that family members may not even realize that there are rules; nevertheless, everyone understands the allowable behavior of family members. For example, all the members of a family may understand that no one will enter into a dispute with Dad at the dinner table, or that everyone will attend religious services, or that all of the children will go out of state to college. Some of these rules apply to everyday boundaries and may simply make some mundane things a lot easier (for example, if a door is closed, one is not to enter without knocking). Other such rules may signal secrets that are taboo or too difficult to talk about or that perpetuate unjust or oppressive situations, such as all the female children know not to find themselves in the same room alone with Grandpa, or no one asks Mom how she got a bruise on her face.

In applying a systems analysis to family norms, a social worker may recognize that interrupting the family's patterns of behavior or relationship by breaking a rule is, in fact, feasible and desirable. For example, a social worker may point out that one of the children spoke out one day about the bruises on Mom's face. This disclosure led to the mother talking about the violence which began the process of empowerment and decreased family violence. Yet, this process might also result in violence directed at the child who raised the issue in the first place; therefore, the social worker would not likely encourage a child to take such an action without support and protection. Family norms in such cases are often very powerful and need to be carefully evaluated before any family member is put at risk.

Implications of Family Systems Theory for Generalist Practice

Along with the notion of function, two specific dimensions reflected in family systems theories have influenced generalist social work practice with families. The first is family structure, and the second is intergenerational patterns. In many respects these represent a classic approach to Western ideology of the family and have influenced the responses of many social service policies and agencies. You may discover that these ideas have shaped your thinking also. As a basis for considering family structure and intergenerational patterns, consider the following practice principles for social work assessment and intervention with families that have evolved from a systemic perspective (Logan, Rasheed, & Rasheed, 2008, pp. 184–185):

1. Family is considered within a "context" that is comprised of multiple systems.
2. Rooted in the basic systemic foundation, the family "is more than the sum of its individual parts," all of which serves to function as a unique system.
3. A change within one part of the family system creates change in the entire system.

4. Viewing the family as a unique system provides the practitioner with the opportunity to focus on the issues that present challenges to the entire system.
5. A systemic-focused assessment and intervention provides a view of the family as a complex system.
6. Behaviors are viewed as a product of the multi-faceted system and not the result of one individual or action.
7. A systems perspective that promotes a strengths-based perspective frames the family assessment within the context of the environment in which they live.
8. Family members, particularly those in minority groups, are viewed within the context of their family system as well as the larger societal system.
9. Family function may be impacted by the legal, social, and economic biases and discrimination that affect families with members who are considered to be in cultural, racial, ethnic, religious, or sexual minority groups.

Family Structure   The relationship among the generations of a family, especially between the subsystems of children and parents, is the **family structure**. One of the early scholars in defining models for family interventions, Minuchin (1974) posits that the boundaries between children and parents should be clear and that parents should be in charge of major decisions as they also carry out appropriate caretaking functions. The difficulties that families experience are usually thought to be a result of blurring the boundaries between these two subsystems.

Some family boundaries may be so diffuse that the result is **enmeshment**, which suggests that family members are too close, have few distinctions in role or authority, and enjoy little autonomy or independence (Nichols, 2011). As an example, consider the family in which the adult children share virtually every life experience with their mother, seldom make decisions without consulting one another and her, and are emotionally dependent on one another. In an enmeshed family, a routine life experience, such as a job change or change in dating relationship by one member of the family creates strong reactions by others in the family.

On the other hand, if the boundaries between family members are too rigid, the family is thought to be **disengaged**. In this situation, the subsystems are so separated that there is little sense of family identity, and parents are apt to relinquish much of their caretaking role as they pursue their own interests and the children pursue their own. In a disengaged family, a significant job change or a divorce in the family may be barely acknowledged by other family members.

When you read about or work with young people who seem to have little or no effective connections with their families, and have experienced school or legal problems, you may wonder about the boundaries (or lack of) established by the parent(s) and the parents' perception of their own roles. In some respects, you are using this theory when you cite the misplacement or laxness of parental boundaries when children go astray. The work in such cases is to restore clear and appropriate boundaries between children and parents that support parent caretaking and authority. This particular position is reflected in court decisions regarding the delinquency of adolescents and a judge's requirement, for example, that parents receive training on managing their children or that they supervise a child's curfew restriction.

Intergenerational Patterns The identification of **intergenerational patterns** has likewise played an important role in social work practice with families. The term refers to the assertion that families transmit their patterns of relationship from one generation to the next (Papero, 2009). For example, if your adult client, Jane, is so closely connected to her mother that she experiences great anxiety when they are separated and therefore cannot work outside the home, Jane is likely to establish that kind of relationship with her daughter as well. The anxiety generated by any effort to be separate from her mother is contagious and makes it difficult for Jane to think clearly because her feelings are so intense. In this way she becomes dysfunctional, tends to be dominated by feelings, and passes that pattern on to the next generation. This emphasis on feeling in a family sometimes results in constant emotional uproar, frequent violence, major feuds, difficulties with the law, and generalized struggle in accomplishing the basic family functions considered earlier.

Although the whole of intergenerational theory is complex, it has influenced attitudes about families. The use of the phrases "welfare families" and "incestuous families" reflects the use of intergenerational theory. These terms reflect the assumption that problematic, emotional patterns and the resulting behavioral consequences (such as violence, inability to focus on work, and substance abuse) appear to be transmitted from generation to generation. Social work then involves breaking such thinking and cycles, bolstering the strengths, and supporting an appropriate level of autonomy in individual family members. These principles are reflected in many social and educational programs that are designed to break patterns of economic dependence, addictions, lack of educational focus, and build up healthy bonds between family members.

Professionals incorporate a systems orientation when considering crime, addiction, or school violence among youth, and assume the causes to be related to the families involved, regardless of evidence. Sometimes children behave illegally or violently even when they have caring, hardworking parents who are doing the best they can. Therefore, the applicability of family systems theory, like any other theory, can be questioned, and does not fit in all situations. The desire to find a rational cause for human behavior sometimes sets in motion the use of methods that have been used before. People are also often eager to blame a child's problems on the family's behavioral shortcomings or background while not addressing the broader (also systems) aspects of poverty, disenfranchisement, challenging school situation (e.g., bullying), or racism.

Systems perspectives can be helpful and support logical approaches in assessment, especially of complex arrangements, like the family. These perspectives have a general cultural appeal, and they do not constitute a magic, one-stop answer for conceptualizing about, assessing, or working with families. While systems views have strong currency in today's analyses of social issues, other ways of thinking about them may be relevant. You must be careful not to blame or scapegoat any particular person simply because it seems logical or fits with a possible systemic interpretation of a family situation. As you continue with your social work education and develop your approach to practicing social work with families, you will be exposed to a wide array of philosophical and theoretical frameworks. Your professional obligation is to consider all the available options and determine the approach(es) that is(are) most appropriate for the families you serve.

THE CONTEMPORARY CONTEXT FOR FAMILY SOCIAL WORK

The real-world, real live family of today is only rarely the idealized outdated television version with a stay-at-home mother, fully employed father, and two bright, talented (and usually white) children. Social workers often work with families who were once seen as "other"; that is, they did not fit dominant fantasies of family life. The following sections briefly identify several types of family constellations that have not traditionally been considered mainstream but are a vital part of the contemporary family landscape—grandparents rearing grandchildren; gay, lesbian, bisexual, transgender, and questioning parents; single-parent families; families of multiple racial and ethnic heritage; families that include persons with disabilities; blended families, international families, and families with multiple problems—and suggest some specific implications for working with them.

Beyond these types of families, other current configurations appearing in the literature (and in practice) include adoptive families, foster families, step families, dual wage earning families, multi-generational families, LGBTQ families, and, of course, many combinations of these types. In the coming generations, the social work profession will need to expect and remain open to ever-evolving forms of the family. This kind of sociocultural change provides social workers with the opportunity to contribute to a sustained and meaningful impact that will benefit clients across all levels of society. Exhibit 6.3 provides additional insights into this growing and changing population with whom social workers are working.

Grandparents Rearing Grandchildren

In many cultures of the past, grandparents had a significant and ongoing role in the nurturing and socialization of children. Some cultural groups today, typically those who are less mobile or those of strong ethnic identification(s), have maintained those patterns, as consistent with their cultural and instrumental needs. The extended family is an age-old pattern of organization that has been obscured by the mainstream societal changes of the industrialized and "informationalized" 20th and 21st centuries. These changes reflect a break with the traditional cultural patterns of their parents and their parents' parents.

EXHIBIT 6.3

LGBTQ Couples as Families

According to the 2010 Census, there were 131,729 same-sex married couple households and 514,735 same-sex unmarried partner households in the United States (U. S. Census Bureau, 2011a). This information does not include LGBTQ couples who do not live in the same households. As of 2013, nine states and Washington, D.C. have passed legislation allowing same-sex marriage, but 31 states continue to constitutionally prohibit same-sex marriage.

Social workers need to stay abreast of public policies that impact LGBTQ couples:

The Defense of Marriage Act (DOMA) "amends the Federal judicial code to provide that no state, territory, or possession of the United States or Indian tribe shall be required to give effect to any marriage between persons of the same sex under the laws of any other such jurisdiction or to any right or claim arising from such relationship" (Library of Congress, 1996). DOMA establishes that the Federal definition of marriage includes legal unions only between one man and one woman. It also defines "spouse" as only a person of the opposite sex (Library of Congress, 1996).

As of the beginning of 2013, there are 37 state laws and/or constitutional provisions, which limited marriages to relationships between a man and a woman. There are nine states, along with the District of Columbia, which issued marriage licenses to same-sex couples. In California, a federal appeals court found that the state constitution's restriction on same-sex marriage was invalid, but has postponed enforcement pending appeal.

Five states allow civil unions, providing state-level spousal rights to same-sex couples. Same-sex marriage has recently replaced civil unions in three other states, Connecticut, Vermont, and New Hampshire. In six states, domestic partnerships are provided by states, which grant nearly all or some state-level spousal rights to unmarried couples (National Conference of State Legislatures (NCSL), 2012).

Upcoming Supreme Court Cases—Two same-sex marriage cases were heard by the U.S. Supreme Court in 2013. The Supreme Court reviewed California's ban on same-sex marriage along with New York's previous ruling eliminating a benefit of DOMA which addresses same-sex couples' access to the same benefits as heterosexual couples.

Sources: Library of Congress, 1996; National Conference of State Legislatures, 2012; U.S. Census Bureau, 2011b; Stempel, 2013.

In contemporary society, there has been a reemergence of grandparents assuming primary (rather than supportive) parenting roles (Hayslip & Kaminski, 2005), many as a result of family violence, drug addiction, and/or incarceration of their adult children. In fact, over 1.8 million children currently reside with a grandparent and have no parent living in the home (Kreider & Ellis, 2008). The growing pressures on many child protection agencies have contributed to the increase in parenting grandparents because child protection workers often see biological relatives as preferable to, and more available than, unrelated foster parents. Many grandparents are healthy, active, and potentially able to take on the responsibility of raising their children's children.

Much more is involved, however, than simply being "able." Many grandparents have reached a point in their lives when they can pursue their own interests and dreams that have been put on hold while they worked and reared families or they may still be engaged in the workforce on a full or part-time basis. Others may find it exhausting to keep up with young children, who have come from dysfunctional situations and have multiple needs, demands, and activities. Some grandparents take on the unexpected role joyfully and fully, and others are enormously burdened, and sometimes guilt ridden because of their own children's inabilities to parent. They may also be struggling with aging, illness, and their own continued need for employment. In any event, many parenting grandparents, even if eager to care for grandchildren, are likely to want and need significant support from social agencies.

Social workers working with grandparents need to use the skills and perspectives of generalist practice. Generalist practitioners are trained to recognize the need for and offer several types of support as they sensitize themselves to the complexities of the emotional and instrumental stresses that grandparents experience. Grandparents may need financial support (Fuller-Thomson & Minkler, 2005), empowerment, self-esteem building communication skills, resource navigation, advocacy, policy development (Cox, 2008), and frequent reassurance that they can and do offer their grandchildren a secure and stable home (Bullock, 2005). In some situations, birth parents recovering from substance abuse or other challenges have visiting privileges or partial child caretaking responsibilities on a preliminary or trial basis. In such cases, there may be considerable tension between grandparents and parents. These and other conflicts may indicate the need for additional ongoing assistance from social workers. Despite the multiple challenges present in such situations, the focus of assessment and intervention should remain on the child(ren) and can include the school system and other involved community resources (Shakya, Usita, Eisenberg, Weston, & Liles, 2012).

Lesbian, Gay, Bisexual, Transgender, and Questioning Couples and Families

Social workers work with LGBTQ clients in a variety of ways, as individuals, as couples, as families, and as groups. Social workers can play an important role when working with same sex couples as families, at all levels of social work practice. One of the most difficult and distinctive issues that LGBTQ couples face, compared to heterosexual couples, is a lack of marriage equality. At the micro level, sexual and gender minority therapy (SGMT), previously referred to as "gay affirmative therapy" or "sexual affirmative therapy", is integrated by practitioners within existing therapy models (Butler, 2009). SGMT is based on the premise that practitioners should first educate themselves, understand and challenge the context of heterosexism, self-reflect, and locate her or his own position and transparency in order to ensure he or she can provide competent services to this population (Butler, 2009; Rostoky & Riggle, 2011).

LGBTQ couples may be considered unique in that they struggle for recognition of their commitment, both culturally and legally. They may have to overtly psychologically support one another (more than a comparable heterosexual couple). They also have unique challenges regarding gender roles; that is, traditional gender roles may not be applicable, and they must negotiate them.

In the context of couples and family interventions, social workers need culturally specific knowledge, skills, and values. Practitioners can convey support and validation of the same-sex relationship in the context of promoting psychological health and wellness by supporting self-determination and human intimacy needs (Rostoky & Riggle, 2011). In working with SGMT couples, social workers should have specific cultural competence in the area of gender roles as they have been found to have a greater impact on the relationship than sexual orientation (Butler, 2009). Using a narrative approach, social workers can work with LGBTQ couples to deconstruct gender role expectations (Butler, 2009).

Social workers can also support same-sex couples by collaborating with the couple to learn strategies for coping with the stress that accompanies discrimination in the form of marriage inequality. Incorporating a strengths-based approach and intervention is also helpful as such an approach can enable the same-sex couples to identify and appreciate the stresses they have faced and overcome together (Rostoky & Riggle, 2011). Taking into consideration the marriage inequality faced by these clients in most states in the U.S., social workers need skills to support and encourage the couple to obtain legal documentation to protect their relationship (Rostoky & Riggle, 2011).

At the group practice level, social workers again have the opportunity to address the discrimination couples face in the form of marriage inequality by providing psychoeducational support and consultation within the community (Rostoky & Riggle, 2011). SGMT promotes connecting LGBTQ couples with wider systems and valuing multiple perspectives (Butler, 2009). Social workers can work with local community organizations, school, religious groups, and other agencies to reduce the prejudice and discrimination directed toward LGBTQ couples and to find support for marriage equality (Rostoky & Riggle, 2011).

In working with organizations and communities, social workers' knowledge and research skills position us well to facilitate marriage equality, thus filling in the gaps in societal awareness of LGBTQ couples (Rostoky & Riggle, 2011). Currently, most of the research that has been conducted with this population has focused on LGBTQ family issues, specifically children or heterosexual family members. Social workers can also engage in political advocacy efforts to show their support for marriage equality among all sexes and genders, as well as educate themselves about local and state laws that affect their clients (Rostoky & Riggle, 2011).

Parenthood may come differently to lesbian, gay, bisexual, transgender, and questioning (LGBTQ) individuals and couples who are married or in civil unions. The person may be the custodial parent of children from an earlier relationship. Some seek artificial insemination and give birth to children, while others become adoptive parents. Also, many informal arrangements for parenting still exist, particularly in some cultural communities.

Recently, states have begun to support the adoption or foster care placement of children in state's custody with LGBTQ parents. The ever-increasing pool of children needing a home and the increasing number and type of adoptions being granted has led, just as with grandparents, to less traditional, more creative efforts in placement planning (Barth, 2008). These types of arrangements have been a highly controversial strategy for dealing with child custody arrangements within the child welfare community because it strikes at the core of the idealized version of the family. Concerns about identity issues for the child, parental adequacy, and the overall mental health of the parents and children have been raised and researched. More than two decades of research on the impact of parental sexual orientation on children's well-being, however, demonstrates no detrimental effect on children's emotional, psychosocial, or behavioral well-being as a result of being reared by LGBTQ parents (Pawelski et al., 2006).

As prospective parents, lesbian, gay, bisexual, transgender, and questioning persons (LGBTQ) still face obstacles in their pursuit of foster care or adoption and experience institutionalized stigma even at the hands of social workers. Lacking any evidence to the contrary, social workers in both the practice and policy components of the profession should support the efforts of all persons seeking parenthood who are deemed eligible through the assessment process (i.e., able to provide a loving, supportive, and stable environment for a child). Social workers can de-emphasize the search for dysfunction and pathology as they give expression to the strengths and resilience of the parents (see Van Den Bergh & Crisp, 2004). Social workers can also recognize the effects of lingering cultural discrimination directed against the LGBTQ community and draw upon a critical awareness of their own biases regarding the strengths and viability of such people as parents.

Issues that the social work practitioner can be aware of to work effectively with gay, lesbian, bisexual, transgender, and questioning parents revolve around sensitivity to the strengths and challenges they face. The strengths perspective may be particularly helpful in working with LGBTQ parents. Helping the parents identify not only their own strengths, but the strengths of the system in which they live, can be an informative, and even transformative, strategy for your work together.

Competent practitioners must also engage in reflection regarding their own attitudes, values, and expectations about LGBTQ parents (Hull & Mather, 2006, p. 218). You must be aware of the challenges they face that are similar to those encountered by all parents (e.g., parental roles, disciplines, etc.) and those that are different (e.g., lack of legal protections, community supports, etc.). Consider the family in which one of the partners of a lesbian couple gave birth to a child conceived through a sperm donation. Given that gay marriage is currently legal in only fourteen states and Washington DC, the two mothers cannot be married in their state. Should the couple end the relationship, the non-biological mother may

have no legal rights to share custody of the child. These family arrangements are strengthened by strong social worker emotional and logistical support.

Single Parent Families

While competence in the language that you use is critical in all social work encounters, speaking with and about client systems in a linguistically competent manner is particularly important when working with families. Social workers are ethically bound to develop awareness of appropriate language (oral and written) to be used with the diverse communities with whom they work (NASW, 2012–2014a). Single-parent families are a group that is often referred to in ways that are not strengths-based or empowering and can be, in fact, derogatory. A term like "single parent," without the corresponding "double parent," implies that one is normal and does not require description, while the other is "other." As you notice the implications of such terms as "broken family" or "split family," consider how deprecating labels influence initial perceptions and may affect the work that follows. It is easy to lose sight of the strengths and commitment of women or men managing families on their own if the work is prefaced with a sense of deficiency or deviance.

Society and, at times, some social service providers, have historically viewed single parenthood as a blight that necessarily leads to insecure, delinquent, and otherwise unhappy and dysfunctional households. At any point, over one quarter of children live in a household headed by a single parent; particularly if the household head is female, these families report lower levels of income than two-parent families or single male-headed households (Kreider & Ellis, 2011). A review of research on single fathers yields findings that indicate that single fathers, who tend to come to single parenting later than single mothers and often do not experience the financial disadvantages that single mothers do, and may approach parenting differently than their female counterparts, are shown to be caring and effective parents to their children (Biblarz & Stacey, 2010).

While research has shown that children reared in single parent families can face more economic, educational, and well-being challenges than their counterparts in two-parent families, having two parents in the home does not ensure the absence of negative outcomes (Musick & Meier, 2009). Growing up in a family in which conflict is handled in a healthy manner also has a strong correlation to the children's later well-being (Musick & Meier, 2009). On the other hand, there is some support in both the scholarly literature and practice community for recognizing the unique challenges of single parenting, and the way that such parenting impacts both the social work relationship and the parenting functions. Four major issues that frequently arise from divorce or separation are: (1) a lack of resources to cope with stress, finances, or other responsibility; (2) unresolved family-of-origin (family in which you grew up) issues often brought on by the single parent's need for assistance from her or his parents, at least temporarily; (3) unresolved divorce or relationship issues, such as anger, grief, or loneliness; and (4) an overburdened older child. Known as a **parentified child**, an older child who is not yet an adult may be pressed into providing excessive household chores or care for another family member. This, in turn, creates concern that the parentified child's own physical and emotional health and well-being is compromised due to the developmentally inappropriate life experiences (Earley & Cushway, 2002).

With the focus of engagement, assessment, and intervention being on the family itself, social work practice with single parent families can use the family's strengths to create and stabilize coping skills. Drawing from the work of several family scholars, skills for social work practice with single parent families include (Atwood & Genovese, 2006; Jung, 1996):

- *Joining*, similar to engagement, reflects the social worker's effort to show clients that they are cared about and that the social worker understands them and their struggles.

- *Empowering clients* supports their activities and capacity to address their own issues; and values their uniqueness and skills. In particular, the social worker can aid the parent in clarifying and reinforcing the parental role, while serving in a nurturing and supportive role to the other family members.

- While maintaining a focus on the family, the social worker can *aid the individual members in identifying strengths and resources within the family unit* (e.g., the parent is fully employed with a flexible work schedule to allow for involvement at the children's school), the extended family (e.g., grandparents are committed to helping with childcare), and the community (e.g., the community has active, well-organized after school programming).

● *Involving significant family members* emphasizes collaboration, reduction of stress, and pooled resources.

● *Allocating agency resources* focuses on agency planning, outreach, and networking.

● *Highlighting small changes* emphasizes the strategy of making small shifts that ease overextended schedules and increase energy; such changes can also highlight success and autonomy.

● *Articulating self-efficacy* emphasizes competence, accomplishments, and empowerment for greater control over management of family issues, ideally for all family members.

These knowledge and values are consistent with a strengths-based, empowering approach that recognizes both internal and external factors and supports single parents in their ongoing efforts to provide security and nurturance for their children. Further, these knowledge and values can enhance the social worker's efforts to engage and assess the family system by conveying a sense of care and concern for the individual members, identifying and building on the strengths of each member and the unit as a whole, and emphasizing the self-efficacy of the family system.

Families of Multiple Racial and Ethnic Heritages

The growth of families with multiple racial and ethnic backgrounds is notable in most U.S. cities and towns. Interracial and interethnic marriages, civil unions, and partnerships have been gaining in acceptability since the 1960s and are certainly increasing in real numbers (Amato, Booth, Johnson, & Rogers, 2007; U.S. Census Bureau, 2011a; 2012a). Shifting immigration patterns in the United States, globalization, and the breakdown of ethnic barriers all appear to have an effect on the incidence of racially and ethnically mixed families.

Perhaps few other developments in the everyday life of our communities offer a greater opportunity to view people differently than the growth of families with multiple racial and ethnic heritages. While some members of society will persist in grieving for the purity of "race" (a social and cultural construction by most accounts), others value the contributions of other cultures and challenge the notions associated with racial privilege. It is important, however, not to minimize the negative power of the persistent oppression met by many ethnically diverse families in our culture. In this arena, professionals also need to be educated in this multicultural society regarding our national and cultural history. Gaining awareness of our own cultural and ethnic heritages and values and oppressions we have encountered is a critical step of becoming a culturally competent practitioner (Kohli, Huber, & Faul, 2010). Flexibility, shared goals, and a willingness to explore and address challenges are critical in working across culturally and ethnically diverse communities. As with all other aspects of individual and family assessment and intervention, race and ethnicity should not solely define the family (Logan et al., 2008). The social work assessment should encompass the many and varied facets of the family's life and the environment.

There are a number of models that can guide your intervention with a multi-racial family. Two are highlighted briefly here. "**Posture of cultural reciprocity**," proposes an approach for working with diverse families (Kalyanpur & Harry, 1999). This process requires that social workers recognize the cultural aspect of their own personal values as well as those of the social work profession. As depicted in Exhibit 6.4, this posture occurs in four steps. These principles imply a constructionist under-standing of cultural difference and value the distinctions without the value of the social worker's orientation. These principles also provide a useful framework for working with people of other cultures and are consistent with social justice and human rights as they enhance inclusion and reflect a basic assumption of cultural strengths.

Cultural attunement embodies self-awareness as a strategy to inform practice. The concept of cultural attunement requires the social worker to go beyond just raising her or his own self-awareness of their own and others' cultural heritages. Being culturally attuned to multiracial families means that you must gain new information about "racial legacies" and the way in which they influence racial self-identity (Jackson & Samuels, 2011, p. 239). Social work skills that can be particularly helpful in assessing and planning with a culturally attuned frame include: asking open-ended questions regarding racial identity; connecting multiracial families to resources for multiracial families; being vigilant about racial/ethnic-focused issues; and developing practice wisdom regarding the unique experiences of the multi-racial family (Jackson & Samuels, 2011).

Families Including Persons with Disabilities

Social workers work with families in which one or more members has a physical, cognitive, and/or mental health disability. Grounded in systems theory and currently used in developing health care-supported patient care programs, **family-centered care** is a ". . . philosophy of care that permeates all interactions between families and healthcare providers. This philosophy places a high value on the contributions made by the family members in relation to their healthcare needs" (Bowden & Greenberg, 2010, p. 5). Family-centered care has evolved into a concept of services as "family-driven." This shift expands the notion of family-centered services and is based on the assumption that the family determines what it needs (Seligman & Darling, 2007, p. 14). This strengths-based approach is a promising method for working with families in many arenas, especially with those who have

© Rob Hainer

disabilities. When children have comprehensive and severe health challenges, such as neuro-developmental delays, interdisciplinary teaming (group of professionals from different disciplines who work together toward the client's goals) is an appropriate response. Social workers can play an important role on such teams that address many needs of children with disabilities that go beyond medical and educational requirements across the life span. With our commitment to a strengths-based, person-in-environment perspective, we have the opportunity to empower families by recognizing that they are the experts on their lives (Tomasello, Manning, & Dulmus, 2010). As an example of a role for a family-centered social worker, many agencies employ social workers to help higher-functioning adults with developmental disabilities learn job skills. Thus, it is very valuable for social workers to be able to work with adults who have developmental, mental or physical health, or cognitive disabilities and their families and support networks.

As one example, when a child with profound disabilities is born, family members usually have to reorganize their everyday lives as well as their long-term dreams. One parent may have to stop working to facilitate the services and treatments the child needs. The time requirements for involvement with school teams, health care teams, and interprofessional teams, as well as the individual services of a speech therapist, audiologist, occupational therapist, pediatrician, psychologist—the list is sometimes quite long—can turn a family upside down. Siblings are affected, family interactions are affected, and parents often struggle with the

STEP	EXAMPLE
"Step 1: Identify the cultural values that are embedded in the professional interpretation of a student's [or client's] difficulties or in the recommendation for service."	This would lead to asking why, for example, a culturally different client's behavior is bothersome to you. (Is she late for appointments? Does she interrupt you? Does she respond to you indirectly? How do you interpret her behavior?)
"Step 2: Find out whether the family being served recognizes and values these assumptions and, if not, how their view differs from that of the professional."	For example, you may discover that your client has a different sense of time from yours and that punctuality has little meaning for her. Here you would want to explore how she approaches time, what it means to her, and whether she recognizes your approach to it.
"Step 3: Acknowledge and give explicit respect to any cultural differences identified, and fully explain the cultural basis of the professional assumptions."	This requires you to enter into a dialogue regarding your assumptions and beliefs and how they are different from those of your client. For example, you might recognize and appreciate the less frantic approach to time and deadlines while you explain the need in your agency to abide by a schedule.
"Step 4: Through discussion and collaboration, set about determining the most effective way of adapting professional interpretations or recommendations to the value system of this family."	Work out a solution that respects the nature of the family's values. You might settle on a more flexible appointment time at the end of the day, or agree on a time range, or make outreach visits if that is possible.

Source: Kalyanpur & Harry, 1999, pp. 118–119

EXHIBIT 6.4

Working with Diverse Families: Racial and Ethnic Diversity

emotional ramifications as well as the physical consequences of exhaustion. Social workers can offer support, time management ideas, and help with expanding parents' ability to identify resources. In addition, many families struggle with gaining access to services they are entitled to receive and therefore may need social work advocacy to negotiate a complex system.

When working with families that have members with disabilities, social workers may be challenged by societal views of individuals and families that are not strengths-based, but emphasize the individual's deficits. In response, social workers and disability scholars have proposed the following set of beliefs as a foundation for working with such families (Mackelprang & Salsgiver, 2009):

- Persons with disabilities are capable, have potential, and are important members of society.

- Devaluation and a lack of resources, not individual pathology, are the primary obstacles facing persons with disabilities.

- Disability, like race and gender, is a social construct, and intervention with people with disabilities must be political in nature. There is a Disability culture and history that professionals should be aware of in order to facilitate the empowerment of persons with disabilities. There is a joy and vitality to be found in disability. Persons with disabilities have the right to self-determination and the right to guide professionals' involvement in their lives (pp. xvi–xvii).

Grounded in a strengths-based perspective, each of the previous statements is critical not only for the social worker to adopt, but equally as important for the person with a disability, the family, and the community in which the individual and family live. Social workers have the opportunity and ethical responsibility to empower clients systems to embrace such a belief system. Individuals and organizations within our society may not always embrace a strengths-based perspective. In such situations, the social worker may need to become an advocate for the person with a disability.

Disabilities should be viewed within a diversity (or social) model in which societal attitudes, structures, policies, and institutions are seen as responsible for imposing limitations on persons with disabilities. Person-first language is currently in use but "disability identity language" may more appropriately frame the disability as a characteristic of diversity. These ideas imply a social work presence based on advocacy, structural principles, and the skills to work for social justice.

There is a growing need for social workers who want to work with families in which a member has a disability. Social workers may engage in significant relationships that focus on empowering the family and the person with a disability to engage in self-determination, self-advocacy, and independence negotiating life transitions with family members, organizations, and themselves (Beaulaurier & Taylor, 2007). Working within a framework of promoting client needs and self-determination, the person with a disability and their family members can be empowered to: (1) expand their range of options and choices; (2) prepare them to be more effective in dealings with professionals, bureaucrats, and agencies that often do not understand nor appreciate their heightened need for self-determination; and (3) mobilize and help groups of people with disabilities to consider policy and program alternatives that can improve their situation (Bueaulaurier & Taylor, 2007, p. 65). It is important, however, to clarify the role the family members, including siblings, would like to have in the assessment and intervention process. While family-centered care is built on the notion of the professional partnering with the family members, not all families may want a high-level role (Lotze, Bellin, & Oswald, 2010).

Lastly, consider that the social worker often has a dual responsibility when working with the family that includes a member with a disability (Hull & Mather, 2006). In working with both the person with the disability and the other family members, the social worker may have to balance differing or conflicting needs and goals. For example, a child may resent the attention given to her or his sibling related to the sibling's disability, not understand the reasons the family cannot participate in a certain activity, or be embarrassed to have a sibling with a disability. The social worker's role may be to focus a part of the intervention on the needs of the sibling who does not experience a disability.

Blended Families

Families become blended in different ways. For U.S. Census purposes, a blended family is considered one in which a parent remarries and the children who reside in the home do not share a biological parent (Kreider & Ellis, 2011). Other blended families include domestic partner relationships, civil unions, and nonrelated families who co-reside in the same household. Approximately 16 percent of children are part of a blended family (Kreider & Ellis, 2008).

While most blended families come together without specifically seeking the services of a helping professional, working with families prior to or as they blend into the new family constellation can be a valuable experience for the family. An intervention with the "pre-blended" family can incorporate four stages of work: discovery, education, parental unification, and family unification (Gonzales, 2009). These stages of work entail working through the process of getting to know one another, helping the family anticipate upcoming events through education about child and family development, and aiding the parents and then family member to unify can serve to prevent challenges in the future.

Social workers working with a blended family should be aware of the family history and be sensitive to the dynamics that may occur when two families merge into one. Incorporating information about family blending is a critical component of the assessment and intervention process. Family members may not be aware of or able to articulate challenges they are experiencing regarding the "merger" of the two family units. The social worker who strives to stay attuned to the issues that can occur when families consolidate can identify the reason the family is struggling.

When working with a blended family, the social worker begins by identifying the strengths of the individual members and the family as a unit. Within the assessment process, the social worker can help the family to identify and discuss the roles of individual members along with boundaries between members and the multiple families coming together. It is important to remember that each member of the newly created family unit must be viewed within the context of both of the systems (i.e., original family and new family) in which they exist. There are, however, unique aspects of working with the blended family. Competence in working with blended families requires knowledge of family development and transitions, involvement of noncustodial parents, extended families, and helping families negotiate new and different family roles, boundaries, relationships, and traditions. The social worker can also engage with and assess blended families by helping them identify their expectations for the forming of this new family. Adapted from Shalay and Brownlee (2007, p. 24), the following questions for clients may be helpful to this process:

1. What do you perceive others think it means to be a family?
2. How do you think your views of what it means to be family have been shaped by what other people think it means to be a family?
3. How might ideas about families on TV have influenced how you expected things would be as a family?
4. If you were a nuclear family what might be different in how you relate to each other?
5. What do you think expectations about a perfect family encourage you to believe about each other?
6. How might expectations about what a family should be have influenced what you expect from each other?

International Families

Social work practice with families in contemporary society requires global competency. If you are practicing social work with families in the U.S., you will likely encounter families who have arrived in the U.S. as immigrants or refugees. If you are a social work practitioner outside the U.S., you must have extensive knowledge of international issues. While working with families in the U.S. or abroad may require a different knowledge base regarding immigration, legal and governmental issues, cultures, and customs, there is a practice skill set that is common to work to all international family social work practice.

To gain competencies in thinking and working internationally, you can learn as much as possible about the family or families with whom you will be working. As with social work practice with individuals who have relocated to the U.S., it is important to be prepared for working with families who are new to their country. Before you meet the client(s), expose yourself to information about their culture, heritage, relocation history and experience, language, customs and traditions, spiritual practices, and community. While reading about your client's country of origin and culture can be helpful, seek out others who can provide you with personal or professional experiences and guidance. Remember also that the client family can ultimately be your best source of information and insight. Allow yourself to learn from them, especially about them as a unique family. While it is important with all client systems to explore their views about working with a helping professional, it is particularly important to understand the perceptions and beliefs about receiving help from the family and those who share their cultural beliefs and traditions.

While there may be common characteristics to groups of people who share a country of origin, culture, or traditions, each person within the family and the family itself should be viewed as individual. You may find as many similarities between a family from the U.S. and a family from Ghana as between two families from Ghana. Learning about the lives of the families that you work with will be an ongoing process that can unfold as you build rapport and trust with the members.

While much of social work practice knowledge and skills you learn is applicable to all families, there are certain competencies that are unique to working with a family that has relocated from their country of origin. First, it is critical to understand the cultural norms of your client family related to the definition of family. Be certain that you have a clear understanding of who is considered to be a member of the family, the relationships of family members to one another, the meaning of those relationships, and any hierarchical traditions that may exist within the family unit. For example, is "family" considered to be the nuclear unit or the larger, extended family? Are persons who are not biologically or legally linked considered to be part of the family? What rules and tasks guide the family members in their daily lives and in making major life decisions such as marriage, parenting, residential arrangements, education, careers, religion/spirituality, and financial priorities?

Regardless of the family's origins, you can use a strengths-based approach in completing your assessment and intervention planning. Using the International Family Strengths Model, DeFrain and Asay (2007a, p. 452) suggest that family strengths can be assessed on the basis of: (1) appreciation and affection; (2) positive communication; (3) commitment to the family; (4) enjoyable time together; (5) sense of spiritual well-being; and (6) ability to manage stress and crisis effectively. While these attributes can be applied to families of any ethnic, cultural, or heritage background, they are particularly helpful when considered within the cultural context of the family with whom you are working, as these family dynamics can have different meanings when viewed within the cultural background of the client system.

Building on family strengths can serve as a particularly helpful strategy as families work to adjust to their new country and environment. Parents, for example, may struggle with their children adopting the customs, language, and dress of their culture or older adults may find it challenging to live in a world that is unfamiliar to them. Using the family's strengths can empower the family members to find their place within their new home while maintaining their connections to their heritage. You can help to make the global connections between the world from which the family has come and the one they have entered by pointing out and affirming a family's ability to enjoy being together despite the challenges of adjusting to life in their new country and home.

While the preceding discussion has focused on family situations and circumstances, the real world of social work practice means that individuals and families may present multiple concerns and dilemmas. Known as **multi-barrier families** or families with multiple problems, the challenges may encompass economic, health, behavioral, social, and psychological issues (Hull & Mather, 2006). For example, a grandparent who is rearing her adolescent grandson may also be faced with a custody battle with his biological parent; a couple with a child born with a disability may, at the same time, be grappling with the grandmother's cognitive impairment; a blended family may be coping with employment layoffs and foreclosure proceedings on their home.

Even with multi-barrier families, the social worker's role is to approach each family as a unique system with strengths and individualized needs. Listening to each member of a family unit enables the social worker to gain insight into the perceptions, relationship dynamics, and possibilities held by the individuals within the collective family system. To be a social worker who is competent in working with families, one needs to develop a repertoire of practice behaviors that encompass family-focused knowledge, skills, and values. Prioritizing problem-solving into short and longer-term goals and promoting a supportive and nurturing environment can be a focus for the social worker. The social worker may find that attention must be given to both the internal and external challenges that are confronting the family (Janzen, Harris, Jordan, & Franklin, 2006). For example, an internal challenge faced by a family may be the substance abuse by one of the members; an external challenge may be the family's inability to qualify for subsidized housing or financial assistance. While different practice behaviors may be required to address internal versus external issues, the social worker may find that she or he serves a number of roles (e.g., broker, advocate, counselor, or educator). The remainder of this chapter will highlight a sampling of those practice behaviors needed to work effectively with families.

CONTEMPORARY TRENDS AND SKILLS FOR ENGAGEMENT AND ASSESSMENT WITH FAMILIES

Theoretical perspectives are a reflection of the sociocultural context and climate. As with social work practice with individuals, there is a wide array of theoretical approaches for assessing and intervening with families. "Most practitioners, educators, and researchers tend to practice, teach, and do research based on multiple and interrelated theories" (Logan et al., 2008, p. 177). While three different theoretical perspectives are presented here, you will, through your career, develop the approach that is most consistent with your philosophical and practice perspectives. Each of the frameworks presented here has some components in common and others born in reaction to each other. All correspond in some ways to traditional perspectives and offer an evolving focus. Regardless of the theoretical approach(es) that you use, social work practice with families should encompass goals that are situational-focused, structured, realistic, concrete, and achievable (Logan et al., 2008).

Narrative Theory in Family Engagement and Assessment

Narrative theory is based on a postmodern, constructionist perspective that enables client systems to make sense of their lives through "stories" or the client system's perception of an individual or a situation. Family interpretations of ongoing events either tend to support the ongoing narrative or refute it. Those stories that are included in the interpretation serve to organize subsequent experience. When there are exceptions, they are often dismissed or forgotten as not representative of the real family. The language used to interpret and describe various family stories is significant and fits into the context of the ongoing experience. Exhibit 6.5 outlines the tenets on which a narrative family assessment and intervention is based. The following examples explore the main ideas of an approach based in narrative theory.

As opposed to seeking a diagnosis, conducting an assessment from the narrative perspective involves helping the family members to tell their stories which then be used to inform the intervention. In essence, this results in the intertwining of the assessment and intervention processes as the stories bring to light the family's strengths and provide direction for the deconstruction of the old stories (Williams, 2009).

The issue that brings the family to the social worker is daughter Liza, who is six years old, the last child, and the only girl in a single parent, male-headed household from an economically affluent neighborhood. Liza has been cast in the role of the family "misfit." Her behavior has been described as oppositional and she is clumsy, speaks disrespectfully, and is disruptive. Any incident involving Liza (e.g., forgetting her pencil or knocking over her milk) becomes just another piece of evidence

---

1) Language (i.e., word choices) helps to define the meaning given to life events and provides the context for change.
2) Related to language, stories help families give meaning to their life events.
3) The stories created by each family member shape subsequent life events and family members may recall life events differently. It is important to acknowledge all stories as multiple realities can and do exist simultaneously.
4) Social context, specifically culture, influences values, social roles, gender relationships, and concepts of justice which, in turn serve to define the stories family members create.
5) Stories are not necessarily based in facts, but may be the result of family members' perceptions of life experiences, causing strengths and positive alternatives to be lost.
6) Stories can be used to encompass possibilities for growth and healing.
7) Families are not the problem, families experience problems.

EXHIBIT 6.5

Tenets of a Narrative Family Intervention

Source: Adapted from Van Hook, 2008, p. 169–170.

for the family. Her family members often shake their heads at her behavior, and wonder what to do with someone who is (in their view) oppositional, clumsy, disrespectful, and disruptive. When Liza's first-grade teacher reports to her father that Liza is exceptionally well liked by both her peers and teachers and that she is bright and fun, Liza's father is incredulous. He suspects the teacher has her confused with another student. Or Liza must be faking at school. The *real* Liza, as everyone knows, is oppositional, clumsy, disrespectful, and disruptive, as misfits are inclined to be.

The story that Liza tells about herself is different from that which is perceived by her family. Each story carries the power of the context to perpetuate and expand it, which in turn will influence the way Liza, as the major character, plays her role. If her family persists in maintaining the original perception of Liza, she is likely to respond over time by becoming increasingly rude, failing, or developing truly disruptive conduct. On the other hand, if her family at any point re-authors their perceptions by recognizing the exceptions to their ideas about Liza, her story may unfold quite differently. Everyone has multiple stories, but some have more power, relevance, and a wider audience than others. Liza's alternative story (told by her teacher) has the potential to influence her future in positive and significant ways.

Thickening the Story   Although the narrative framework has many components, the exploration in this chapter will include the most relevant concepts for generalist social work practice with families. You may recall, **thickening the story** refers to the social worker's effort to expand "thin" (Morgan, 2000) or **problem-saturated stories** that are one-dimensional perspectives on a truth that client systems created about themselves that may or may not be based in fact (Kelley, 2009). This attempt to create a more complex story may achieve the larger goal of instilling hope that the client system can make positive changes. Liza's first story is a good example of a thin account. She has no redeeming virtues and is simply perceived as an oppositional, clumsy, disrespectful, disruptive misfit. A thickened version of Liza's story reveals her likable personality, talents, and ability to connect with people in spite of, or in addition to, any behaviors that are oppositional, clumsy, disrespectful, and disruptive.

The Smith family provides another example of the value of reframing the story. The Smiths are a family that has experienced considerable challenges in child rearing. One child has been taken into the custody of the state, and now child protection workers are investigating to determine if another child should be removed for reasons of safety. The mother disparagingly claims, in defeat and sarcastic resignation, that her family is "just one of those families." Child welfare workers may likewise view the family as "just one of those families" because various children have been in custody for three generations (as was Ms. Smith).

Rather than assessing the narrowly-defined dysfunction of this family, the narrative social worker would search for the exceptions to this story to enrich it and make it more complex. Rather than maintaining a focus on problem-solving, the narrative social worker collaborates with her or his clients to "enhance their awareness of how cultural forces have lulled them into accepting problem-saturated ways of living (Williams, 2009, p. 205). For example, the social worker can ask about Ms. Smith's ability to keep a family together for ten years in the face of poverty, or to overcome a major childhood health challenge, or survive homelessness. The effort is to expand the narrow failure story so that the family can see itself as having potential, which in turn can support re-authoring the story to reflect the way the family would like it to be. The family then can shape its future to fit the new story. Such an approach can be contrasted with the one taken in the section "Intergenerational Patterns," earlier in the chapter. Through a different lens, it offers the potential for hope through development of the family's resilience and positive attributes.

A family's ability to be resilient can be a powerful aspect of the social work intervention. In keeping with the practice approaches presented here, the social worker should perceive that resilience begins with the beliefs, socially constructed judgment and evaluations that each family member creates (Benard & Truebridge, 2013, p. 205). Some families need help identifying their areas of resiliency. Social workers can help families by incorporating the following resiliency-oriented beliefs about families (Benard & Truebridge, 2013:

- All people have the capacity for resilience.

- Most individuals do survive and even thrive despite exposure to severe risk.

- One person can make a difference in the life of another person.

- Coming from a risk environment does not determine individual or family outcomes.

- Challenging life experience and events can be opportunities for growth, development, and change.

- There is nothing wrong with you that what is right with you cannot fix.

- Bad behavior does not equate with being a bad person.

- As a practitioner, it is how you do what you do that counts.

- To help others you need to help yourself.

- Resilience begins with what one believes (p. 207).

Resiliency can be viewed as the family's ability to "absorb the shock of problems and discover strategies to solve them while finding ways to meet the needs of family members and the family unit" (Van Hook, 2008, p. 11). The social worker's role is to focus on empowering the family to operationalize their capacities for resilience, specifically by:

- Incorporating a sense of hopefulness and purpose into the meaning that families assign to situations (e.g., pointing out ways in which the family can show support for the family member who has just completed a substance abuse treatment program);

- Supporting organizational structures that provide effective leadership and a balance of flexibility and stability;

- Promoting clear, empathic and supportive communication patterns;

- Emphasizing existing positive family members' relationships;

- Promoting family members' problem-solving abilities; and

- To the extent possible, improving and enhancing the social support system and community and economic resources accessible to the family (Van Hook, 2008, p. 23).

Externalizing Problems   The notion that problems are external to the family is known as **externalization**. The problem is not the client system itself but is the result of an issue that is separate from the client, thus the focus of the intervention is also then external to the client system (Kelley, 2009). Narrative social workers try to identify and help the family to name the issue that is creating difficulties. By objectifying or personifying it, family members can develop a relationship with the challenge, rather than be consumed by it, and ultimately they may control the challenge.

For example, if a family feels overwhelmed by the demands of a child with disabilities, the resulting "worry" may be externalized. The social worker can then ask about the feelings of being overwhelmed, help family members label their feelings, and support all those times when the family takes control over "the Worry." By separating the issue (or "worry") from the family's identity, the worker and the family can explore ways to defeat their concerns or at least keep them at bay.

Unearthing the Broader Context   One of narrative theory's most relevant contributions to generalist social work practice is a consistent emphasis on the political context of the family. The social worker will be highly sensitized to the danger of reinforcing the oppressive dimensions of a dominant pattern (such as racism) in society. For example, consider six-year-old Damion who appears to be having a fearful reaction about going to school. The reaction seems to be extreme for the situation. The social worker will not want to externalize this prematurely as "the school creeps" or "school scares" if in fact Damion is being taunted and bullied because of his ethnicity. Using narrative theory, the social worker seeks to identify any political factors that impact the situation with the client and family, and through a partnership with the client and family, address those factors that are negatively shaping and impacting the challenge at hand. In the case of Damion, it will be important for the social worker to elicit the client and his family's perception of the climate at his school that may promote an environment in which Damion does not feel safe. The social worker can work with Damion and his family to develop an externalized version of the story and incorporate the school into the situation.

Solution-Focused Family Work

Like many contemporary family models, **solution-focused practice** deemphasizes history and underlying pathology by defining the family broadly and not limiting its conception to traditional forms or even requiring all the members to be present in the meeting with the practitioner (Nichols, 2011). With a focus on brief interventions that narrowly define the arena of specific problems within the context of particular environmental variables, solution-focused family work has been used with a range of life situations and groups.

Approaching family social work from a solution-focused perspective is grounded in a set of assumptions, that include (Koop, 2009):

1) The family is the expert.
2) Problems and solutions are not connected.
3) Make unsolvable problems solvable.
4) Change is constant and inevitable.
5) Only a small change is needed to make an impact.
6) Keep it brief.
7) Keep the focus on the future.
8) Focus on the family members' perceptions (pp. 147–148).

Solution-focused social workers usually emphasize a cognitive approach, support a collaborative stance with clients, and tend to reject notions that problems serve any unconscious or ulterior motive. Accordingly, solution-focused social workers believe that individuals want to change, and they believe that any attribution of resistant behavior to families is more about the interpretation of the practitioner than it is about families. The founder of solution-focused therapy, Steve deShazer (1984), early in his work declared resistance "dead" and in turn redefined clients' balking at practitioner directives as their way of educating the social worker about what is needed to help them. As depicted in Exhibit 6.6, solution-focused family interventions are based on seven tenets.

Solution-focused social workers emphasize the future, in which solutions can be used within the specification of clear, concrete, and achievable goals. The simplicity of this approach, along with its time-limited, specific emphasis, has enabled solution-focused work to become an important contemporary model as it generally aspires to short-term, specific, and direct results, thus avoiding costly protracted professional relationships. Long used in community-based programs, solution-focused interventions have shown promise in addressing family and relationship problems, particularly related youth and families (Kim, 2008b; Nowicki & Arbuckle, 2009). The strengths and client perspectives of the approach have made

EXHIBIT 6.6

*Tenets of
Solution–
Focused Family
Interventions*

1) Emphasis is placed on future and ways in which the future will differ from the past.
2) Solutions may be unrelated to the problem's history; therefore, understanding the past may not be helpful to developing future-focused solutions.
3) Negativity and pessimism can overshadow family members' abilities to see positive alternatives to their ways of relating to one another.
4) People do want to change. Resistance can be a trigger for the social worker to identify different strategies for intervening.
5) Because family members can be influenced by the social worker, the practitioner's focus should remain on empowering the family to understand their problems as well as the possibilities for solutions.
6) Family members can be encouraged by the social worker to change the language they use to discuss their issues from problem-oriented to solution-oriented.
7) Family members determine those areas of focus that are important to them and, as a result, are most appropriate to formulate goals for the intervention.

Source: Adapted from Van Hook, 2008, p. 148–149.

solution-focused practice popular for use in schools settings along with the other components, including portability, adaptability, brief time frame, and the opportunity for small changes to matter and applicability for cultural competence (Kelly, Kim, & Franklin, 2008, p. 8).

In solution-focused work with more than one person, the emphasis is on the relationship, not the individual, and finding and maintaining a common goal (De Jong & Berg, 2013). Solution-oriented social workers need specific skills for working with families as families often have one or more of the following pre-conceived notions about their needs (Sebold, 2011):

- Families see problems stemming from the relationships among the members.

- Families often see one person as needing to change.

- Families often view problems as belonging to the child(ren) (p. 211).

With a focus on developing solutions and de-emphasizing ongoing descriptions of the problem, the solution-oriented social worker enables the family to "know" their problems *and* the solutions and work together as a group to build on one another's ideas and competencies to develop solutions (Sebold, 2011). This strengths-based, future-oriented perspective approach is established from the outset of the social worker-family relationship and continues on throughout the intervention.

The assessment process begins with a series of questions to elicit the perceptions of each of the members. The social worker then uses the family's perceptions to co-construct a plan for intervention (De Jong, 2009). With family groups, questions can focus on the relationships among the members, even the ones who are not present. Questions included in a solution-focused intervention are:

- *Joining:* Initially chatting informally with the family members can serve to build rapport (Koop, 2009).

- *Normalizing:* When possible, helping the family to view the current situation as normal behavior can help them to envision a future in which they do not feel in crisis (Koop, 2009).

- *Goal-formulation:* Questions aimed at goal-formulation prompt the client system to consider the way in which their situation will be different if the problem that brought them to the social worker is no longer a problem (e.g., "What will be different if the problem is resolved?"). The "miracle" questions are aimed at helping the client system envision life if a miracle occurred that eliminated the presenting problem. In the engagement and assessment phases of work, goal-formulation can serve several purposes: enable the members of the family to communicate their individual goals to one another; focus on the members of the family on a common direction for work; and enable the members to identify and build on individual strengths in pursuit of the agreed-upon goals.

- *Exception-finding:* Critical to the assessment process, questions developed to identify exceptions help family systems recall experiences in which they were successful and provide opportunities for identifying strengths and resources available to incorporate into the intervention.

- *Scaling:* Scaling questions invite the family system to quantify the past and the future within the context of the problem or crisis they are experiencing. The clients are asked to use a scale of 1–10 to rate their belief that a solution will be found for the issue that brought them to see you. During the engagement and assessment phases of the intervention process, scaling questions help to establish the point from which the work will begin and identify current and past successes and areas for future growth.

- *Coping:* Aimed at issues of coping, coping questions strive to connect the family members to coping strategies that have or could be used. Such questions can engage and invest the client system in the helping process as well as assess past successes and failures, which can then be used to develop the plan of work for the current intervention.

- *Circular questions:* A family member's perception about what other people (including other family members) think of her or him can influence actions;

therefore, asking questions about what the family members believe others think of them can serve to can deeper understanding of the clients' concerns (e.g., "What do you believe your wife thinks about you when she feels you are not listening to her?") (Koop, 2009).

Like narrative proponents, solution-focused social workers concentrate on identifying and bolstering the attempts made by families that have been successful. The goal for solution-focused social work with families is to increase the discussion about solutions and decrease the focus on problems. Client families are asked to remember when their efforts worked, even when they seem like very small incidents occurring rarely. Solution-focused social workers direct their attention to those exceptions and explore the contextual factors that made the exception possible. In this way social workers give families the message that they have the strengths to cope, that they have in fact done it successfully before, and that they can do it again (De Jong & Berg, 2013).

Solution-focused practitioners have been credited with promoting a greater emphasis on strengths through their "exception" question that encourages the identification of, and focus on, those strengths that are relevant to the current situation (Weick, Kreider, & Chamberlain, 2009). The "exception" questions allow the client to place strengths within the context of times and situations in which the client was successful.

Environmental Focus   With an ecological orientation, solution-based social work looks to the community as a resource and always seeks to understand the problem in terms of the relationship to the surrounding context.

For example, when the social worker views the client within the context of the environment in which the client has grown up and currently lives, the social worker may assume that a woman who abandoned her children is embedded within a culture in which she experiences gender discrimination, poverty, unmet mental health challenges, or perhaps racism. Such assumptions can then help to explain the client's current inability to care for her children. The intervention may then be focused on education, mobilizing resources, and developing a support system to support the client.

All of these considerations are part of the assessment, and a single diagnostic or strictly pathology-oriented label as a unitary explanation is viewed as inadequate. The overall understanding of the family's difficulty is rooted in their everyday, unique living experience to which they bring their own personalities and idiosyncrasies within the larger context, which also shapes that experience.

Constructionist and Social Justice Approaches to Family Social Work

The contemporary approaches described in this chapter are complementary and share some constructionist notions that are represented in the client-defined meanings of family, the lack of rigid ideas of so-called normal family development, and the collaborative partnerships built with client systems. Constructionist theories incorporate a critical perspective that reaffirms the long-standing tradition in social workers working with families. Critical social construction suggests the development of a perspective that focuses more directly on constructionist ideas and social justice principles. Narrative and solution-focused concepts also relate to social justice issues in their concerns for the contextual locations that clients experience on a daily basis. In that respect, the approaches are more alike than they are different, and each has a positive contribution to make to contemporary social work practice.

Approaching social work interventions with families from a constructionist perspective enables the social worker to help families to tell their stories and acknowledge the meaning of those stories. During the assessment and planning phase of the intervention, the social worker can use the family's stories to identify individual and family strengths and those times when the family was able to respond in a way that is different from their current situation, and explore the implications of those stories on their functioning and potential change (i.e., exceptions) (Holland & Kilpatrick, 2009). Principles of constructionist thought that are helpful in working with families include (Holland & Kilpatrick, 2009):

- Family stories transmit meaning, create coherent sequences, shape identity, organize values and explain choices, and involve alternative interpretations.

- To function as a family, the family members must have shared meanings.

- The meanings of stories cannot be changed by outside individuals (p. 26).

Critical Constructionist Emphasis   Critical constructionist social workers explore the meaning of family within a broad contextual framework. Constructionist descriptions of the family present the social worker with multiple perspectives that can be viewed critically; as such models have implications for social work practice. For example, if you assume a theoretical model in which the family is considered basic for human survival and absolutely sacrosanct, you might view the family, as many did for centuries, as immune from external interference in its internal dynamics, including family violence. If, however, you assume that the meanings (or realities) that client systems make of their lives occur within a larger community, you and the client system can view their situation within the context of the groups to which they belong (e.g., racial, ethnic, socioeconomic, religious, etc.) (De Jong & Berg, 2013). Using the critical constructionist emphasis, the solution-focused intervention then becomes a collaboration between the social worker and the family in which the members of the family explore their realities (i.e., problems, miracles, successes, strengths, and solutions) while incorporating the influence of the multiple groups within which they live their lives (e.g., neighborhood, extended family, schools, job settings, and church) (p. 369).

Social Justice Emphasis   Constructionist theorists call for consideration of both external and internal dimensions of social justice. From an external perspective, social workers will direct their attention to ensuring that all families are granted the rights and privileges of society, not just the idealized families of dominant groups. Diverse families, by ethnicity, sexual orientation, socioeconomic class, or any other difference, should be guaranteed the same access to the benefits of the culture. When they are not, the social worker is called upon to intervene in whatever ways are applicable, including legal advocacy, legislative advocacy, public education, or other forms of social action.

The internal focus requires the social worker to look within the family itself for reflection of justice for all family members. Clearly the social worker needs to challenge overt and specific oppressive behaviors, such as intimate partner violence and child abuse, but also is encouraged to look at family structure, gender dynamics, and roles. Critical constructionist social workers will address concerns that are external and have clear internal ramifications through such activities as education and advocacy for more just family practices and policies. Such a perspective reflects the continuous and energetic efforts of the social work profession to develop practice models that both confront and support various dimensions of contemporary society.

Generalist Practice Skills Guidelines for Family Engagement and Assessment

While models may differ somewhat, the following list includes family-oriented engagement and assessment practice behaviors that are common to most models:

- Ensure the family is as physically comfortable as possible.

- Work to facilitate a respectful tone throughout the meeting.

- Transmit positive regard or warmth, support, and respect in verbal and nonverbal practice behaviors.

- Engage with and hear from each member of the family, inviting each to share their perception of the family's strengths, areas for concern, and reason for seeking services.

- Ask each family member her or his perception of the family's purpose and goals in coming to see you.

- Agree upon the expectations and focus of the work.

- Recognize your own biases around family forms and norms.

- Observe the family's communication and interrelationship patterns.

- Inquire about, observe and discuss the role and emotional function that each family member fulfills within the family, particularly within the context of intergenerational relationships.

The engagement and assessment process should be completed expediently and thoroughly. A key aspect of the engagement process is for the social worker to "join" with the client system. The concept of joining entails the development of an alliance between the members of the family and the social worker in order to build trust and rapport (Van Hook, 2008). Having established a trusting relationship is critical for moving forward into the assessment and planning phases of the social work intervention. Family systems are complex and encompass multiple perspectives, thereby creating the need for an assessment process that is unrushed and includes the perspective of each member of the family system (Logan et al., 2008). Family assessment processes are typically determined by the agency. Your agency may or may not use a standardized assessment process. If you are in a setting where you will use a formalized assessment protocol with standardized measures, your role is to clarify for the family the purpose and logistics related to the measures to allay potential anxiety and frustration. Specifically, you can clarify the timing, place, and format of the standardized assessment measure, the meaning of the score or outcomes, and the way in which the information will be shared and used (Corcoran, 2009). While your agency protocol may provide a structure for completing the assessment, content areas that a social worker can address include (Van Hook, 2008, p. 67):

- What are the sources of distress in the lives of family members?

- How do family members view these issues?

- What aspects within the family and their extended world contribute to these sources of distress?

- Are there additional factors that contribute to this distress?

- What are the resources for coping and support possessed by family members, the family as an organization, and their external world?

- How can family members use these resources?

- How can these resources be enhanced?

- What barriers are preventing family members from using these resources?

Family assessment encompasses an array of assessment tools and strategies. The following discussion highlights the use of mapping with families.

Mapping: A Family Assessment and Planning Tool Competency in assessing and planning for family-focused interventions includes gaining skills in mapping the family situation. Chapter 4 presented genograms and ecomaps as helpful mapping tools for assessment and planning with individuals. Typically included as part of the assessment phase of the social work intervention, these two tools are also widely used with families and offer many uses for social workers. Genograms and ecomaps are compatible with most models of social work practice with families. For example, everyone in the family may be asked to participate in the genogram, or each member could make her or his own map of the family relational patterns (see Exhibit 6.7). Whether it is a genogram or another mapping activity, the process of collecting the information can provide the social worker with insights into the way in which the family members define membership in the family, perceive the current situation, or

EXHIBIT 6.7

Graphic Representation of Different Views of the Family within the Family

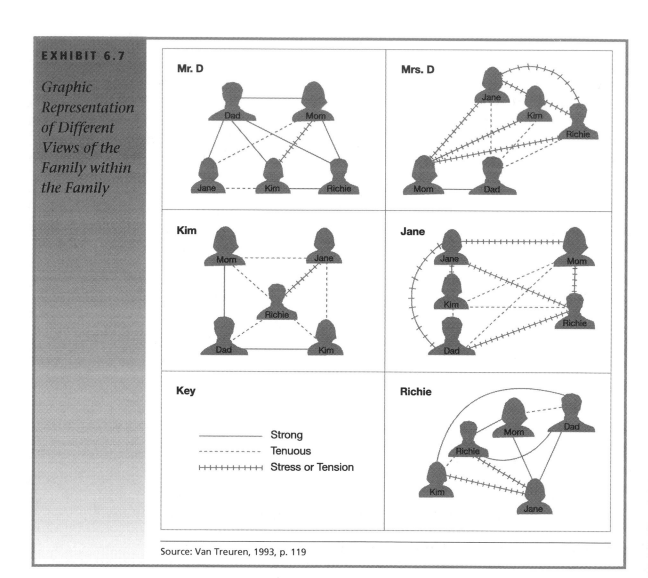

Source: Van Treuren, 1993, p. 119

interact with one another within the interview and on paper (Minuchin, Colapinto, & Minuchin, 2007).

Genograms are particularly illuminating assessment tools for family social work because they provide a visual depiction of family structure and patterns. Beyond collecting factual information about multiple generations of family members, the genogram can serve as an "information net" in which data and insights can be all captured within the larger, intergenerational content. For example, the presenting problem may be viewed within the larger context of the family's issues, the immediate household may be seen within the context of the extended family and community, the current family situation is a component of the history of similar patterns, nonthreatening questions are perceived in light of painful inquiries, and facts are compared to judgments regarding family patterns and function (McGoldrick, 2009, p. 411).

The Cultural Genogram is a tool with multiple purposes—building rapport, family assessment, and supporting culturally competent practice (McCullough-Chavis & Waites, 2008). Building on the foundation of the genogram, a cultural genogram incorporates the perspective of the client system in areas of culture that impact the family's experience. Culture can encompass, but is not limited to, race, ethnicity, sexual orientation, social and political influences and oppression, socioeconomic status, and religious and spiritual influences. McCullough-Chavis and Waites (2008) outline the role of the social worker in completing a cultural genogram, which is to identify intergenerational patterns and focus on strengths within the context of the larger socio-cultural-political world in which they live.

First developed by Hardy and Laszloffy (1995) for use with family therapy students, the cultural genogram has been adapted for use with social work practice (Warde, 2012). Students are encouraged to complete their own cultural genogram to gain insight into the way in which the client family might experience this aspect of assessment. Modifying the original list of items to consider when developing a cultural genogram, Warde (2012) presents a list of eleven questions one should consider when completing a cultural genogram. For a list of the questions, see Quick Guide 18.

Ecomaps and other variations that demonstrate relationship patterns can liberate some families from what seems like endless talking, and they usually enjoy developing them and examining the final product. Ecomaps can be used in a visual service evaluation when the goal has been to expand community connections in general or specifically (for example, engaging in more recreational pursuits) or to alter them (such as improving the relationship between the school and the family of a child with disabilities).

Other mapping techniques can prove helpful in social work practice with families. Maps are limited only by imagination. They represent relationship complexities, the passing of time, interacting components, interpersonal interactions and patterns, levels of intimacy, specific types of connection (such as material or emotional support), and spirituality (Hodge, 2005b). One such map is the

---

QUICK GUIDE 18   QUESTIONS TO CONSIDER WHEN
                 COMPLETING A CULTURAL GENOGRAM

1. If other than Native American, under what conditions did your family come to the United States?
2. What were/are your family's experience with oppression?
3. What significance do race, skin color, and hair play in your family?
4. What role does religion play in your family?
5. How are gender roles defined in your family?
6. How does your family view people who are lesbian, gay, bisexual, transgender, or questioning?
7. What prejudice or stereotypes does your family have about your own racial group and other racial groups?
8. What principles shaped your family's values about education, work, and interaction with people outside the family?
9. What are the pride/shame issues of your family?
10. What impact do you think these pride/shame issues will have on your work with clients who are culturally similar and dissimilar?
11. As you reflect on the answers you gave to the previous questions, how do you think the values you got from your family of origin will influence your ability to work with clients whose values and beliefs are discernible different from yours?

Source: Warde, 2012, pp. 574–575.

culturagram which represents a family's experiences in relocating to a new culture (Congress, 2004; 2009). With the increase and diversity of families coming to this country as immigrant or refugees, social workers need culturally competent assessment skills. Targeted specifically for use in engaging and assessing families who have immigrated to this country, the culturagram enables the social worker and family to empower from the perspective of their culture. The resulting map illustrates content from inquiries relating to the specifics of the family's experience from immigration to their celebration of values in a new land. A culturagram and the areas for discussion are shown in Exhibit 6.8. This visual and interactive tool helps social workers and families understand the family's internal experiences, recognize differences between and within families, see ways in which the family has been successful, and, importantly, pinpoint areas for planning for potential intervention (Congress, 2009).

More literal maps of physical arrangements such as floor plans in housing situations illustrate concretely the challenges in daily living or the disparities in economic circumstances. For example, a map that shows five children's cots in a tiny bedroom or an older adult on the couch demonstrates a person-in-environment reality that may be difficult to comprehend fully through verbal means. A legend on the map that describes the meaning of all symbols and a general heading to orient the reader can be used and are especially helpful in any creative or unconventional mapping.

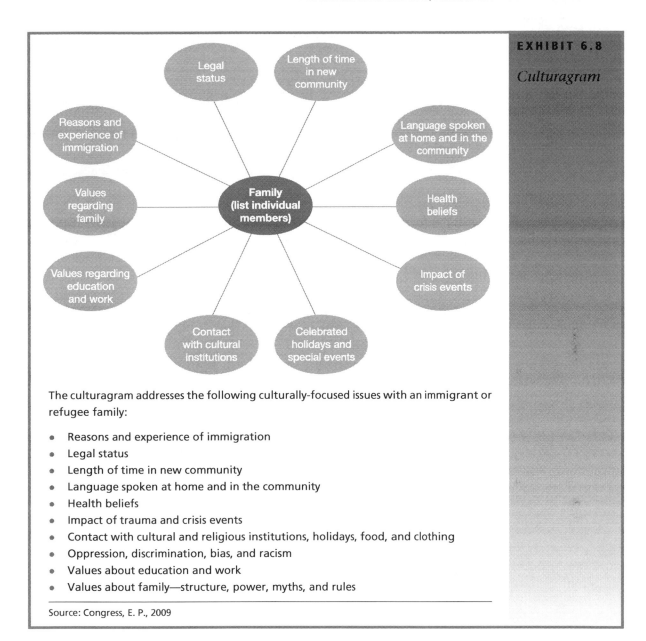

EXHIBIT 6.8

Culturagram

The culturagram addresses the following culturally-focused issues with an immigrant or refugee family:

- Reasons and experience of immigration
- Legal status
- Length of time in new community
- Language spoken at home and in the community
- Health beliefs
- Impact of trauma and crisis events
- Contact with cultural and religious institutions, holidays, food, and clothing
- Oppression, discrimination, bias, and racism
- Values about education and work
- Values about family—structure, power, myths, and rules

Source: Congress, E. P., 2009

STRAIGHT TALK ABOUT FAMILY SOCIAL WORK PRACTICE

Families are a powerful ingredient in our lives. Families have inspired fierce loyalties, lethal conflicts, abject miseries, and quiet pleasures throughout all of history, and continue to do so today. Whether you view your own family as supportive or toxic or any of the many positions in between, reaching a peace with your own feelings can enhance your work with other families. It is difficult to assess and intervene with the family situations of others if they trigger feelings because of their similarity to (or even difference from) your own. Your family concerns need not be perfect or even fully resolved, but your feelings and concerns about your own family situation should not intrude into or influence your work in ways you do not recognize. Such an issue falls into the arena of supervision, and you will benefit from sharing with your supervisor any current struggles you are engaged in with your own family, especially if you are also seeing families in your practice.

Granting that you enjoy a sense of comfort with your own family, you will likely experience situations, at some point, in which your feelings are difficult to manage. Egregious abuse exists in some families, and although contemporary theoretical perspectives can help you temper your responses and recognize that individuals and families do the best they can, the litany of injuries or aggressions emerging in court reports or police accounts or living room conversations can bring on powerful emotions in the most seasoned and balanced social worker. Fortunately, you can use supervision, agency supports, peer connections, and personal strategies for coping with such feelings and reactions. Many social workers see professional therapists to process their family challenges and avoid the influence of their family challenges on their professional practice. The most encouraging dimension for social workers today is that the majority of families inspire the greatest admiration for their resiliency, spirit, resourcefulness, and agency in the midst of potentially demoralizing circumstances.

Another important "straight talk" item for social work practice with families is the need for accurate and comprehensive documentation. Family-related interventions to be documented can include child welfare situations (i.e., child protective services and foster care), adoptions, and early interventions. Such cases may require the creation of documents for: intake assessments, progress/case notes, reports, case referrals or transfers, treatment planning, and evaluations (McGowan & Walsh, 2012). Social workers working with families of adults may have to initiate documents plans for members with addictions, disabilities, health concerns, relationship challenges, or residential placements. In all situations, the social worker's focus should focus on including only relevant information framed in a strengths-framework that represents the perspectives of all members of the family.

As in social work practice with individuals, documentation is a critical, although more complex, practice behavior. You will recall from Chapter 4 the components of basic documentation (see Chapters 4 and 5 for a review of the guidelines). While these guidelines are applicable to documenting the family interventions, there can be the "group" aspect of family work to consider. Recording the assessment and intervention phases of social work practice with families requires the social worker to encompass all members of the family who are participating in the intervention process and to give voice and perspective to the individual contributions. Quick Guides 19 and 20 provide guidelines for family-focused information to be included in the documentation of a family assessment.

QUICK GUIDE 19   DOCUMENTATION OF A FAMILY ASSESSMENT

Demographic Data (include all members participating in the assessment and intervention):
- Names, birthdates, and relationship to others
- Contact Information

Family Information:
- Presenting need(s) or concern(s)
- Living situation (e.g., members of household and level of stability)
- Family (composition—parents, siblings, spouse/significant other(s), children, and others (extended family and friends); level of support; and family history of mental illness, as applicable). A genogram and/or ecomap may be a helpful tool to gather and depict this information
- Timeline—significant life events experienced by the family
- Cultural environment (traditions and cultural view of help-seeking)
- Religion/spirituality (statement of beliefs and levels of activity and satisfaction)
- Family strengths and significant events, including trauma (e.g., physical, sexual, and/or emotional abuse or neglect and experience with perpetrator(s))

Individual Family Member Information:
- Individual family member information, including:
  - Educational history of each member (highest level achieved, performance, goals, and challenges)
  - Substance use/abuse (history of addictive behaviors—alcohol, drugs, gambling, sexual, or other)
  - Emotional/behavioral functioning and treatment history
  - Risk factors
  - Physical health
  - Legal status or concerns
  - Financial/employment circumstances (employment status, satisfaction, financial stability, areas of concern or change)

Social and Environmental Information:
- Strengths and concerns regarding physical environment, as applicable (e.g., structure, neighborhood, and community)
- Safety issues (e.g., risks or concerns of and for individual members and social worker)

Summary
- Current providers (including psychiatrist, primary care physician, therapist, caseworker, etc.)
- Community resources being used (including support groups, religious, spiritual, other)
- Family goal(s) for intervention
- Summary of social worker's observations and impressions

Adapted from St. Anthony's Medical Center, St. Louis, Missouri; Missouri Department of Social Services

QUICK GUIDE 20   DOCUMENTATION OF A FAMILY INTERVENTION PLAN

Intervention Plan:

- Preliminary assessment
- Preliminary plan for intervention and plan for change (to be developed at first visit), including:
  - What will each family member do differently?
  - How does each family member view themselves accomplishing changes?
  - What support and services are needed to accomplish plan for change?
  - Who will provide support and services?
  - Who will arrange for support and services?
- Interventions and plans for emergency/safety needs
- Other interventions needed
- Needs (include date, identified need, status (active, inactive, deferred, or referred), and reason for deferral or referral)
- Strengths
- Facilitating factors for intervention
- Limitations
- Barriers to intervention
- Other care providers/referrals and purpose (including plan for service coordination)
- Plan for involvement of individual family members, extended family members, significant others and friends
- Review and termination criteria/plan
- Planned frequency and duration of intervention

Adapted from St. Anthony's Medical Center, St. Louis, Missouri; Missouri Department of Social Services

CONCLUSION

The contemporary family both supports and challenges the social worker. Social workers engage with the family and its struggles in our culture, thus a goal for the profession is to develop additional and relevant models for working with them that recognize their strengths, agency, and resilience. Education and advocacy for shifts in the structural and political arrangements that exist for families are also required.

As a culture we still value the family, and social workers can be a part of the solution for creating environments in which families of all kinds are validated and supported.

As the structure and meaning of family itself continue to change, you can stay alert for your own capacities to honor those notions of others. As a form of "group," the family has particular resonance and serves as a grounding point for understanding human collectives. With that dimension in mind, this exploration will move to the intervention, termination, and evaluation of your work with families.

MAIN POINTS

- Historical antecedents for involvement with families, including family function and systems theories, shape the way in which social workers engage with and assess families.

- An idealized, or fantasy, notion of the American family still exists today, but social workers recognize and work with many forms, including grandparents raising grandchildren; lesbian, gay, bisexual, transgender, and questioning families; single parent families; families of multiple racial and ethnic heritages; families with members who have disabilities; blended families; and families with multiple challenges.

- Several contemporary theoretical perspectives have emerged that are consistent with critical social construction, the strengths perspective, and social justice orientations, including narrative, solution-focused, and constructionist approaches.

- Your practice setting will guide much of your work with families, but the skills and practice behaviors that you have learned for engaging and assessing individuals and groups will be applicable in working with families. Additionally, family-oriented skills and practice behaviors are needed, including engaging the whole family, reframing, and recognizing your own biases around family forms.

- Mapping tools can be helpful in assessing and evaluating the work with families; they can also help to empower families to change.

EXERCISES

a. Case Exercises

1. Go to www.routledgesw.com/cases and review the case file for Roberto Salazar. As the undocumented nephew of Hector and Celia Sanchez, Roberto has consistently earned an income but has also experienced several health challenges. He is currently living with the Sanchez family due to an injury that prevents him from working. He has a number of skills but his current injury and inability to work has him feeling defeated.

   You are the social worker charged with monitoring the status of Hector and Celia's Section 8 housing voucher. While Hector and Celia generally manage their rent payments, they are having difficulty meeting the schedule due to extra expenditures in support of Roberto. Your agency is responsible for controlling expenses and complying with federal regulations. Your supervisor is especially concerned with this aspect of the program.

   On your visit to the Sanchez home, Hector assures you that—even though he knows the landlord can evict him and his family for violating regulations regarding occupancy—Roberto is family and, of course, he and Celia will house and feed him. He remembers his own loneliness when he came to the U.S. and that his uncle helped him. He has no doubt that he can assist Roberto by providing temporary housing and support. Hector explains to you that it is important for immigrants to stick together and support one another, especially family. You are feeling some pressure from the agency to report and help resolve the issue of Roberto's unacceptable presence in the Sanchez home. You are concerned that your supervisor will look unfavorably on you if you allow Roberto to continue to live in the house.

Respond to the following questions:

a. How might a family focus differ from an individual focus in this situation?
b. How will you respond to Hector? Your supervisor?
c. How might diversity be a factor in this situation? Compare your responses to those of your peers. Identify and team up with another student whose approach seems similar to yours and develop a unified approach. Brainstorm in class regarding different or creative ways to approach this situation.

2. Go to www.routledgesw.com/cases and review the case for Carla Washburn. Create a genogram of her family. Explore the connections between Carla and her family members and the ways in which those connections impact her relationships within the family. Address the following:
* What are the strengths of the family?
* What are the issues that have impacted the family?
* How have those issues impacted the various relationships?
* Who has the "power" in the family?
* If you were a social worker working with this family, what issues would take precedent?

3. Go to www.routledgesw.com/cases and review the case for Brickville, focusing on Virginia Stone and her family. Review the genogram for the Stone family and provide an analysis of the information provided by responding to the following questions:
* What additional information do you believe would be helpful to include in the genogram?
* What are the patterns that emerge as you review the Stone's genogram? In addition to general patterns, comment on potential themes in the areas of:
  ○ boundaries between family members
  ○ intergenerational family relationships and patterns
  ○ single parenting
  ○ grandparents rearing grandchildren
* What strengths and areas of challenge are evident?
* What information does the genogram yield that suggests priorities for planning a social work intervention?

4. Go to www.routledgesw.com/cases and review the case for Hudson City. The Patel family is one of the many families displaced by Hurricane Diane. The members of the Patel family include:

- Hemant and Sheetal—both in their late 40s, they immigrated to the U.S. from India fifteen years earlier. They own and operate a family restaurant in the neighborhood in which they live.
- Rakesh, Kamal, and Aarti are the Patel's three children. The oldest son, Rakesh, is 18 years old and is his first semester at the University of the Northeast. He lives at home and commutes to school each day. The younger son, Kamal, is a 16-year-old high school junior. The Patel's daughter, Aarti, is age twelve and in the 7th grade.
- Bharat and Asha are Mr. Patel's parents who came to live with their son and his family when the restaurant opened ten years ago. They are both in their early 70s and help part-time in the restaurant. Bharat had cardiac bypass surgery last year. Asha suffers from hypertension. Their medications were left behind when the family fled from the storm.

All the members of the Patel family work full- or part-time in the restaurant. The Patel's restaurant and home sustained major damage as a result of the storm. While the restaurant and house can be rehabilitated, the damage is extensive and repairs will likely take weeks to months to complete. The Patels are staying with friends in a nearby community, but feel they should make other arrangements and not impose on their friends' hospitality. As the entire family income is derived from the restaurant, money is a concern.

You are working as a volunteer mental health disaster responder and have been asked to work with the Patel family who has come to the Emergency Center. Develop a list of culturally competent engagement strategies that will be helpful as you begin your work with the Patel family, including additional information and resources you will need.

5. Go to www.routledgesw.com/cases and click on the case for Hudson City. Returning to the information provided regarding the Patel family, click on Assess the Situation and respond to My Assess Tasks. Using the Patel family information provided here, complete Tasks #1 and #2 as it relates to the Patels. Once you have assessed the issues and needs facing the Patels, develop a preliminary plan for your social work intervention.

b. Other exercises

6. You are a social worker on an interprofessional team that works with children who are on the autism spectrum, and their families. Three-year-old Jenny is referred to your team. She is the light of her father's life—she is lively, energetic, and bright-eyed. In the last year, she has become quiet, preferring to play by herself, and is less interested in the special outings her father loves to share with her.

After a series of anxious appointments with the pediatrician, Jenny was referred to a specialist in developmental pediatrics. Many observations and checklists later, Jenny was diagnosed on the autism spectrum. Her parents, Catherine and Jason, were devastated. Her two-year-old brother, Sammy, was oblivious.

Over a period of a month, Catherine began to adjust to the diagnosis. She connected with a supportive group of parents coping with children on the autism spectrum and read all she could about autism. She also spent considerable time with Jenny, playing and coaxing her to interact with her. Jason, however, was notably uninterested in Catherine's activities. He began to refuse to go to Jenny's doctor's appointments. During one argumentative dinner with Catherine, he stated that he did not believe the diagnosis; he thought Jenny was fine, just going through a stage, and accused Catherine of "selling out" her own daughter. Jenny was referred to the interprofessional team by her pediatrician. Catherine engaged in the process enthusiastically, if painfully. Jason attended the assessment and seemed sullen, participating very little. The team concurs that the family would benefit from your "support work" around the diagnosis. Catherine and Jason agree to meet with you and you speculate this might be your only chance to engage Jason.

Respond to the following questions to compare with your peers.

a. What is your assessment of this family? Identify a theoretical perspective that is most applicable to working with this family. How does your choice of perspective influence your approach to this family? Be specific.

b. Generate a list of three questions or issues that you think are important to address in your first meeting.

c. How might you attempt to engage the family, especially Jason? As you compare responses with your peers, what different (from your own) perspective was most useful to you?

7. In working with Jenny's family (from exercise #6), develop a solution-focused plan for the assessment and planning phase of the social work intervention.

8. In an alternative approach to working with Jenny's family (from exercise #6), develop a plan for assessment and planning using a narrative approach.

9. This chapter includes a range of contemporary family structures about which social workers must have knowledge and skills for competent practice. Select one of the groups discussed in this chapter and conduct a search of the evidence-based practice approaches currently utilized with the group you have selected. Prepare a brief summary of the knowledge and skills needed for effective practice.

10. Create a genogram of your family. Explore the connections within the family and the ways in which those connections impact the relationships within the family, particularly your relationships. Address the following:

- What are the strengths of your family?
- What are the issues that your family has faced?
- How have those issues impacted the various relationships?
- Who has the "power" in the family?
- If you were the social worker working with your family, what issues would take precedence?
- What have you learned about your family from this exercise?

CHAPTER 7

Social Work Practice with Families: Intervention, Termination, and Evaluation

. . . to think larger than one to think larger than two or three or four this is me this is my partner these are my children if we say, these are my people who do we mean? how to declare our bond how to keep each of us warm we are in danger how to face it and not crack

Melanie Kaye/Kantrowitz, in *We Speak in Code: Poems & Other Writings*, 1980

Key Questions for Chapter 7

(1) What competencies do I need to intervene with families? (EPAS 2.1.10(c))

(2) What are the social work practice behaviors that enable me to effectively intervene with families? (EPAS 2.1.10(c))

(3) How can I engage in research-informed practice and practice-informed research to guide the processes of intervention, termination, and evaluation with families? (EPAS 1.2.6)

(4) What potential ethical dilemmas might I expect to occur in intervening with families? (EPAS 2.1.2)

SOCIAL WORKERS HAVE A LONGSTANDING HISTORY OF intervening in family situations and crises. Dating to the era of Mary Richmond and the Charity Organization Society, and Jane Addams and the Settlement House movements, social workers have focused on intervening with families (Logan et al., 2008). Just as social work practitioners respond to societal changes, social workers have also adapted their practice behaviors, including knowledge, skills, and values, to the developmental stages of families or changing structure of families. While families are generally self-sufficient in meeting their ongoing financial, emotional, and caregiving needs, when they seek help outside the family, they require a response that is developed for that family's needs (Briar-Lawson & Naccarato, 2008). Family social work interventions require competencies to address the complexities of the contemporary family, which may include culture, racial and ethnic diversity, financial and legal challenges, and intergenerational relationships and dynamics. Different from family therapy, family social work is an approach based on generalist social work skills for intervening with families who are at-risk. Family social work assumes the intervention is family-centered, and support can be provided in the family's home or in the social worker's office and in times of crisis (Collins, Jordan, & Coleman, 2013). Family social work practice interventions may be focused on: (1) reinforcing family strengths to prepare families for long-term change, such as a member arriving, leaving, needing care, or dying; (2) creating concrete changes in family functioning to sustain effective and satisfying daily routines independent of formal helpers; (3) providing additional support to family therapy so families will maintain effective family functioning; and/or (4) addressing crises in a timely way to enable the family to focus on longer-term concerns (Collins et al., 2013, p. 3).

Essential to an effective intervention with families is the engagement and assessment. Building on the assessment that focused on the family's strengths and self-determined needs, the intervention process is an opportunity to collaborate with the family to facilitate growth and change. With its emphasis on brief, efficacious interventions, managed care has influenced contemporary social work practice with families by focusing the social worker clearly, systematically, and succinctly on identifying and assessing the problem or concerns, developing and implementing an intervention plan, and terminating the working relationship (Jordan & Franklin, 2009, p. 429). This chapter will highlight theoretical frameworks and practice behaviors that will be helpful to your work with families.

THEORETICAL APPROACHES TO INTERVENING WITH FAMILIES

Just as with individuals, family intervention is a planned change process in which the social worker and client system work together to implement the steps to reach the goals established in the assessment and planning process. Research has shown that families, even those who endure intense or chronic stressors, can be resilient (Benard & Truebridge, 2013). An array of theories provide the underpinning for the approaches to social work practice with families, including systems, ecosystems, family life cycle, cultural and social diversity, strengths-based, and empowerment (Logan et al., 2008, p. 184). The social worker who is well grounded in theoretical approaches to working with families can select the approach(es) and techniques that is (are) best suited for the social worker's practice philosophy and the needs of the client family. In an effort to best serve the families with whom they work, practitioners "often use a combination of family techniques from different models rather than adhering to one particular approach. Integrationism [i.e., blending models and techniques], technical eclecticism [i.e., using different techniques] and the use of common factors are the preferred ways that most practitioners work" (Franklin, Jordan, & Hopson, 2009, p. 434). As a practitioner, you will recognize that theories each have their own limitations within the context of the client family's ethnic, cultural, familial traditions and your scope of practice as a social worker (Hull & Mather, 2006). However, the family intervention models that are most effective typically share certain elements, including education, opportunities to practice and model new behaviors and skills, and multi-faceted intervention plans (Franklin et al., 2009). As an ethical and culturally competent social work practitioner, you are responsible for gaining the training necessary to use the evidence on available family models and select the approach or combination of approaches that you believe will be most effective for your client.

Created by Hull and Mather (2006), Exhibit 7.1 depicts a framework for approaching family intervention from a multidimensional perspective. As shown in the exhibit, the process of developing a family intervention begins with viewing the family within their environment, followed by the selection of a relevant theoretical approach, and concludes with the creation of an intervention that uses appropriate techniques.

Following is a discussion that explores theoretical approaches using the strengths and empowerment, narrative, and solution-focused perspectives. In keeping with the overall approach of this book, the theoretical perspectives presented here will be within the postmodern grouping of frameworks. We will also learn to help families grow and change by aiding them in identifying and building on strengths, reconstructing their life experiences, and developing new realities. While postmodern approaches may be perceived as counter to traditional systemic approaches, Logan and colleagues (2008) posit that, in fact, the two complement one another. Systems theories help the practitioner to view and frame the family within the context of environment, while the postmodern constructs help the social worker to view the family within the context of the meaning of the presenting issues that are provided by the family.

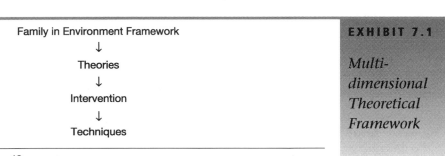

Family in Environment Framework

↓

Theories

↓

Intervention

↓

Techniques

Source: Hull & Mather, 2006, p. 18.

EXHIBIT 7.1

Multi-dimensional Theoretical Framework

Strengths and Empowerment Perspectives and Family Interventions

Having completed an assessment and planning process that identifies the strengths on which the family can build, an intervention in the strengths tradition strives to not only enhance the family's assets but to empower the family to develop new coping and resiliency strategies. From a strengths perspective, challenges are viewed as opportunities and possibilities (Saleebey, 2009). An intervention that is grounded in a strengths perspective is one that builds capacity and assets and focuses on solutions (Briar-Lawson & Naccarato, 2008). For example, an intervention that emanates from the traditional deficit-based perspective will view the family as the source of its problems (e.g., poor parenting, dysfunctional relationships, or irreparable problems). A strengths-based intervention identifies the family's assets (e.g., the parents are committed to placing value on being together as a family, productive children and the family has remained together in the face of adversity) and capacities and focus on solutions (e.g., the family is willing to work on the challenges that brought them to a social worker).

In the face of the complexities of "being" a family in contemporary society, practitioners may find it easier (although not ultimately as effective) when developing an intervention plan with a family to focus on the problems that exist within the family, rather than the strengths. Instead, a strengths-based approach could entail asking the family members, for instance, to describe the characteristics of a healthy, strong family can lead to a discussion of those characteristics that are shared by their own family. To address the family's current situation, it is important to understand their views on strengths and to incorporate those strengths into the planning and intervention process (DeFrain & Asay, 2007b). Engaging in an exploration of those actions that the family members can take individually and collectively can result in an altered perspective for the family (and possibly the social worker, as well). Gaining insight into interacting in a new way with one another and the environment can be an empowering experience for the family. For example, consider the case of the resident of Brickville (www.routledgesw.com/cases), Virginia Stone. You will recall from previous chapters that Virginia is currently struggling with an array of challenges. In your initial intervention, you might lean toward focusing on the most compelling issues (e.g., her caregiving stressors or home ownership dilemma). It may be most helpful to Virginia in the longer term to help her to identify the strengths and previous coping experiences that her family and she possess so that she may use those strengths-based experiences to help resolve her current challenges. For example, strengths possessed by this family include their commitment to caring for older members and remaining connected to one another even through times of adversity.

Grounded in a commitment to build on family strengths, empowerment-oriented practice is applicable to family interventions. Having been developed over the past several decades from a concept to practice principles and methods, empowerment-oriented interventions extend beyond intentions to encapsulate the principles and actions (Parsons, 2008, p. 124). As an example, Dunst and Trivette (2009), leading scholars in this arena, have continued to evolve their family-systems model for early childhood and family support interventions. Moving away from the traditional paradigm that emphasizes an intervention based on the family's deficits and the professional's expertise to a contemporary capacity-building paradigm, Dunst and Trivette propose that effective interventions should focus on models that embrace: promotion, empowerment, strengths- and resource-based, family-centered models (2009, p. 128). This paradigm shift is intended to aid professionals in developing help-giving practices that encompass a capacity-oriented approach to identifying and mobilizing family concerns and priorities, abilities and interests, and supports and resources. In a capacity-building approach, the role of the social worker in providing services becomes more relational and participatory. In addition to demonstrating traditional social work practice skills of active listening, compassion, empathy, and respect, the capacity-focused family social worker emphasizes possibilities for change in an individualized, flexible, and responsive manner (Dunst & Trivette, 2009, p. 131).

Empowerment-driven interventions are helpful for families in crisis—especially those who have experienced a history of crises—and can aid the social worker in helping the family to create solutions. An empowerment perspective is grounded in a belief that the stressors and trauma a family experiences can be buffered by a repertoire of protective factors available in their environment (i.e., family provides caring relationships, maintains high expectations, and provides opportunities for meaningful participation and contributions) (Benard & Truebridge, 2013, p. 208). In essence, the family has the skills and resources they need to be resilient and adaptive in the face of adversity. For example, consider the family who has survived 2012 Hurricane Sandy but lost their home. They are a large family and could not find temporary housing in one place. As a result, they are staying in different locations with family, friends, and in shelters, but they come together once each week to work on rehabilitating their home so they can return there as a family.

An empowerment approach uses seven principles of practice (Wise, 2005):

1. Build on strengths and resources and diminish oppressive factors. As noted, the strengths that can be used in addressing the challenges must first be identified by the family. The family then reviews existing and potential resources and begins the mobilization of those resources. Oppressive factors or barriers to growth and change may lessen as family members perceive options and possibilities. Strategizing with the family about ways in which oppressive factors may be confronted, ameliorated, or minimized can also be empowering. A first step, however, may be for family members to confront the oppressive factors that may be perceived to exist *within* the family. For example, family members may be stifling anger at another member for having a disability. Until this anger is acknowledged and addressed by all the members of the family, the family is

unlikely to be able to collaboratively work toward resolution of the presenting concern.

2. Multicultural respect. Multicultural respect requires attention to a wide and growing array of phenomena, including race, ethnicity, gender, age, socio-economic status, religious/spiritual beliefs, sexual orientation, differing ability, language, and developmental phases. The family may be impacted in multiple areas. The role of the social worker in helping the family create an intervention is to ensure awareness by all members of their perceptions and experiences in these areas and to help the family understand the meaning and impact and confront stereotypes and barriers.

3. Recognize needs at the personal, interpersonal, and community levels of empowerment. While family members may well know their needs, they may have difficulty, particularly during a crisis, articulating those needs to one another or the social worker. A successful empowerment-focused intervention relies on the social worker to help the family create solutions that are directly linked to their needs. Regular individual and group check-ins are necessary to ensure that the needs have been accurately and thoroughly voiced and are being addressed through the plan that was developed for the intervention. Needs can and do change and the social worker can help the family to recognize that and maintain a realistic approach to success.

4. With sufficient resources, family can empower themselves. Upon identifying and mobilizing resources, the social worker can facilitate an intervention plan. The social worker does not have the ability to complete the work for the family, but can, however, monitor and interpret the family's use (or lack) of the resources. For example, consider the value of helping the family to mobilize resources from their religious or spiritual community. The religious community can not only provide tangible and emotional resources but can serve as a recognized authority that may be of help in engaging and intervening with family members (Nakhaima & Dicks, 2012).

5. Support is needed from each other, from other families, and from the community. Receiving support from within the family and outside the family can be the impetus that the family needs to reach their agreed-upon goals and solutions. The role of the social worker may be to normalize experiences, provide information, and connect the family to one another and others.

6. Establish and maintain a "power WITH" relationship. Power can serve as an asset and a barrier in a family intervention. Being able to convey to the family that each member has power, the family as a group has power and the social worker has power. Most importantly, the social worker and the family can share the power for the purposes of achieving the goals of the intervention. Interventions should not rely on the "power" of any one or sub-group of the family-social worker partnership, but should be shared in a collaboration using the strengths, assets, and resources that each person or sub-group brings to the intervention.

7. Use cooperative roles that support and assist family members. Implementing the intervention requires that each member assume a variety of roles over the course of the relationship. Such roles may include co-consultant, co-collaborator, guide, co-teachers/learner, co-investigator, and co-creator (pp. 86–88). The social worker is in a position to point out ways in which the family members can use their knowledge and skills.

To bring these principles to life, consider the situation being faced by the Murray family, a three generation family living together in the same house. William (55-year-old high school teacher) and Genevieve (50-year-old occupational therapist) Murray have been married 27 years and have three children: Elle, a 25-year-old unemployed licensed practical nurse, recently returned to her parents' home with her two children ages four and two years, following a divorce; Stephen, 21, who lives at home, works part time, and attends a local university; and Samantha, 16, who is a high school junior. Recently, William's mother, Edna, 76, moved into the house. Edna suffers from Alzheimer's disease. After a recent car accident, the family determined that Edna is no longer able to safely live in an independent living situation. The arrival of Elle and her children and Edna resulted in Stephen and Samantha having to give up their bedrooms. A makeshift bedroom has been created for Samantha in the basement and Stephen is sleeping on the fold-out couch in the den. While the crisis that brings the Murray family to your agency is Samantha's arrest for driving under the influence of alcohol and subsequent suspension from school, it is immediately evident that this is a family in crisis in several additional areas: Edna's illness and increasing need for care; Elle's adjustment to divorce and single parenting; displacement and lack of privacy for Stephen and Samantha; and the stress experienced by William and Genevieve in supporting the family.

Using strengths and empowerment-based perspectives, how can this family be supported through this challenging period in their lives? Consider first the strengths and resources that exist. William and Genevieve have a longstanding marriage; are employed; have opened their home to their daughter, her children, and William's mother; and are willing to provide care for these family members. The family can be a resource for itself but an important intervention strategy is to ascertain from each member their perception of the issues and any factors that are barriers to resolution of the issues. Samantha, for example, may view her grandmother as the problem as Edna has taken her room and her parents' time and resources, leading her to argue constantly with her parents and spend as much time away from home as possible. Helping the family to articulate their needs within a multiculturally respectful way is the next step in creating alternatives with the family. For example, providing information about Alzheimer's disease and its course may help the family better understand and accept Edna's behavior and needs. The social worker can enlist ideas from each family member about ways in which each member of the family can help and offer suggestions for accessing resources outside the family.

Collaborating with the family to access and mobilize resources can not only provide a model for them, but can also engage them in an alternative to their present incapacitation. Examples of collaborating with the family include: (1) Capitalizing on Elle's professional expertise as an LPN, the family can apply for a family caregiver program in which a family member can be paid to care for an older adult. Elle can contribute financially to the family, care for her grandmother and her children, and work toward rebuilding her life; (2) Co-investigating with Stephen options for using his experience as a camp counselor to apply for a live-in resident assistant position on campus can provide him with space and privacy; and (3) Samantha's substance use and arrest has effectively gotten her parents' attention. Guiding the family to consider the various responses and treatment options can enable them to make choices together and learn from one another to co-create a new way of being a family. While the social worker may guide the family members toward resources, the members would be encouraged to handle as many of the logistics of accessing resources as possible.

Narrative Theory and Family Interventions

Like strengths and empowerment approaches, a narrative approach to intervening with families also incorporates strengths, viewing the family as experts on the family unit, and collaboration between the social worker and the family. In hearing each family member's perceptions of the family and the problems that brought them to a social worker, the family members are able to identify the meaning of the problem and discover alternatives to those meanings that will aid in changing the family's interactions (Kelley, 2009). As with the narrative approach to working with individuals, the problems confronting the family are assessed through respectful listening. The family's perceptions are reflected upon and deconstructed, then reconstructed by challenging perceived truths. Through the reconstruction process, the family members in collaboration with the social worker are able to create an intervention plan that enables them to arrive at unique outcomes that have meaning and viability within their family unit.

Implementation of an intervention based in a narrative approach is the culmination of the family-social worker partnership. The family members' story is the intervention and the social worker's role is to empower the clients by supporting their strengths (Williams, 2009). A narrative approach provides the social worker with a variety of strategies that will optimize the family's strengths to expand their perceptions of themselves and their problems and to create a new vision. Within the discovery process, the family has the opportunity to envision a future in which the current problem persists or a future in which the problem is altered or resolved through their actions. Through collaboration with the social worker, clients can become empowered when their stories are heard, resulting in their capacity to construct new stories (and realities) (Williams, 2009).

Narrative social workers find that a variety of strategies and questioning formats are needed to empower clients for change. Interventive techniques can include (Van Hook, 2008, pp. 171–175; Williams, 2009):

- *Listen to the family tell their problem.* Ask each member of the family to share their concerns.

- *Normalizing.* Listening and empathizing can serve to build rapport and trust between the family and the social worker.

- *Externalizing the problem.* Collaborating with the family, the social worker can help them to separate the problem from the person. The partners can then create a team to tackle the problem and not the individuals who make up the family.

- *Mapping the problem.* Looking for patterns from the past can help the family to understand the ways in which the problems were sustained and the impact the problems had on the family. Mapping strategies can include:

  - Searching for a unique outcome—identify a time when the family experienced success in overcoming a problem.

  - Asking spectator questions can help the family to consider how others view them.

  - Reauthorizing the family's story to emphasize a perception in which the members have been able to overcome adversity (also known as significance questions).

  - Identifying cultural messages that support problem-saturated stories— examining the influences of previously received cultural messages on the family's ability to interact with one another and handle life's responsibilities.

- *Findings news of a difference:* Identifying evidence of change, no matter how small, can serve to move the family beyond the problem behaviors and into a new, alternative reality.

- *Asking therapeutic questions:* Ongoing use of questions can help the family to continue to tell their stories and re-author new stories that create a healthier environment. Therapeutically-focused questions can focus on: opening space for possibilities; effecting change to amplify news of differences; preference questions (i.e., comparing past and the present times); developing stories that highlight differences; meaning questions (e.g., getting to the perceived meaning of family's actions); and future-oriented questions.

- *Scaffolding:* Once the old story begins to be deconstructed, a new story (i.e., competence) can be constructed to replace it. Scaffold questions can help to move the story focus from exceptions (e.g., a time when the family was happy) to significance questions (e.g., acknowledging the importance of an action/event) to spectator questions (e.g., understanding ways in which family members experience one another).

- *Collapsing time and raising dilemmas:* The past and present are both important aspects of evolving the problem to a newly constructed story.

- *Enhancing change:* The social worker may challenge the story to motivate clients to examine the old stories to abandon their reliance on their problem-saturated stories.

- *Predicting setbacks:* The social worker can help the family to accept that they will find it difficult to give up on problem-saturated stories and setbacks are likely to occur. Anticipating such an eventuality may disempower the potential impact of the setback.

- *Creating ways to reinforce the new narrative:* Encourage the family members to identify strategies and individuals who can help them maintain the changes they have made.

Returning to the Murray family, their crisis could be framed using a narrative approach by the social worker listening to each member sharing her or his views on the family situation. As you can imagine, the perceptions of the Murray family are likely to be quite different. William and Genevieve may share that they are doing their best to provide for all the members of the family and they feel betrayed by Samantha's arrest and suspension. Samantha (as noted earlier) may believe her parents have given more to other members than they have to her, particularly her grandmother. Struggling with cognitive impairment, Edna may feel she is a burden to her family but feels powerless to change the situation. Elle likely also feels guilt for being unable to support her children and herself. Stephen is feeling pressured to hold down a job and maintain his grades and scholarship so that he can move out of the house. After reaching a consensus with the family regarding the desired outcomes, the social worker can begin deconstructing and reconstructing the family's perceptions by externalizing the problems (i.e., focusing the family on the issue not the person). Should the family, for example, choose to focus on stabilizing the situation with Edna's care; the social worker can externalize Edna's behaviors as, because of her cognitive impairment, outside her control. The social worker can then help the family to envision life if they take no steps to improve their perceptions of Edna's situation, as well as a life in which they understand the disease, coping strategies, care and respite options, and ways to enjoy their remaining time with Edna. While reconstruction of Alzheimer's disease will not alter the course of the disease, the quality of life for Edna and her family can be enhanced through challenging the family's problem-saturated perceptions. Edna will be viewed as an honored member of the family, not as a burden.

Solution-Focused Family Interventions

Like other postmodern family-focused interventions, solution-focused family interventions also emphasize strengths and empowerment, client self-determination, and collaborating with the social workers to construct new realities by mobilizing assets and resources. Solution-focused interventions differ from other similar approaches in the use of questions that move the client from crisis to solution through the co-construction of solutions (De Jong, 2009).

With an emphasis on the relationships and competence, family solution-focused interventions incorporate explorations of the skills, strengths, and competence of clients in a way that is consistent with many postmodern approaches. This focal point solidifies the social worker-client relationship as a collaboration focused on building on past successes (De Jong & Berg, 2013). The assumptions inherent in their ecological orientation encourage social workers to look to environmental, structural supports for solutions rather than to internal dynamics for pathology. Social workers do not disregard individual responsibility for behaviors such as those that occur in child abuse or intimate partner violence, but these are viewed as evidence that the client's external resources and skills are underused (Christiensen, Todahl, & Barrett, 1999). For instance, family violence may occur when a family is experiencing stress but not using resources outside the family (e.g., social, health, mental health, or financial aid services).

Employing a solution-focused intervention with a family shifts the family's focus to a hopeful, future-oriented view of their lives. The social worker's role in creating interventions is to help the family mobilize their energies to achieve solutions using existing and expanded coping strategies in a concrete behavioral approach (Van Hook, 2008).

You will recall from Chapter 6 that a series of questions are used for assessment and planning of the solution-focused intervention. Upon completion of the assessment, the social worker and the client develop a plan for change. While the intervention process should be driven by the family's identified needs, goals, strengths, and resources, the social worker can be most helpful by developing a repertoire of skills and strategies for aiding the family in achieving their mutually-determined goals. Exhibit 7.2 provides a listing of six tasks for implementing a solution-focused approach.

Once the plan is underway, the social worker checks in with the members to monitor progress through the use of "what's better?" questions. The scaling questions that were included in the assessment phase can become a part of the intervention as well. Scaling questions can help review previously discussed solutions and exceptions, and highlight changes as they are made (Nichols, 2011). For example, the social worker might ask the family members to rate on a scale from 1–10 their feeling when they first came to see the social worker and compare those responses with a current rating of feelings on the same issue. Scaling questions can also be used to anticipate future feelings and behaviors (e.g., "on a scale from 1–10, how confident are you that you will be able to sustain this change?").

EXHIBIT 7.2

Solution-Focused Social Worker Tasks

- Create interactional balance and balance between the problem and solution-focused talk. Maintaining a balance between the persons who speak ensure that each voice is heard. Maintaining a balance of content ensures that the emphasis remains on talk about solutions and not about the problems.
- Use effective listening and summarizing skills. Staying tuned in to the contributions of each member of the family and regularly summarizing their input can serve to re-energize the clients and the social worker to remain focused on the goal and ensure that each person's voice is heard.
- Introduce transitions that will move the conversation from problem-focused to solution-focused. To maintain the focus on solutions, the social worker can regularly infuse affirmations or the family's strengths that include statements that convey support of the family's efforts.
- In order to sustain the solution emphasis, the social worker must focus on the family's current realities, that is, key in on those issues being raised by the members. When issues are not addressed, there is a tendency for them to be forgotten. The social worker can help to redefine such issues from problem- to solution-focused, but can promote the process by staying focused.
- Facilitate a "family reflecting pool." As issues are discussed and solutions initiated, each member must be given the opportunity to reflect on the issue or change from her or his vantage point, as each is likely to view the item or issue in a different way from other family members.
- Establish goals and negotiate tasks. While "homework" is typical in any social work intervention, solution-focused tasks should emphasize the desired goals and solutions versus the problems. For example, one of the stated goals is for the family to spend more time together. Before the next session, the family agrees to have dinner together three times in the upcoming week.

Source: Sebold, 2011, p. 217–228.

Additional techniques for completing the solution-focused intervention include (Koop, 2009, pp. 158–160; Nichols, 2011, pp. 257–258; Van Hook, 2008, p. 155):

1) Providing compliments as often as possible to emphasize strategies that have been successful. Compliments may be direct, indirect, or self-compliments.
2) Focusing attention on the family members' relationships with one another. Questions such as "how is your relationship different when you actively listen to her when she speaks?" can help the family keep the emphasis on changing their relationships.
3) Maintaining throughout the intervention phase a positive, future-oriented emphasis on "who, what, when, where, and how" questions.

4) Once the family has identified goals, the social worker can shift the focus to a range of behavioral tasks in which the family can engage, including:

- Doing more of what works—if the family has identified interactions and strategies that work, encourage them to continue them.

- Doing something different—introducing the idea of trying new strategies for relating to one another.

- Going slowly—promoting a slow, incremental approach to change.

- Doing the opposite—encouraging the family members to engage in behaviors and interactions that are the opposite of what they are currently doing.

- Predicting tasks—helping the family to predict outcomes and to identify patterns that occur when changes are experienced.

5) Finding, amplifying, and measuring progress for the purposes of monitoring signs of positive movement toward desired goals.

6) Taking a break within an interaction for the social worker and family to provide feedback to one another.

7) Recapping at the end of the meeting the work of the family and social worker and providing suggestions for future work (i.e., "homework" that focuses on observing successes, engaging in new tasks, and predicting desired changes). Tasks can be considered as three different types: formula—general tasks in which family is asked what they might do differently; perception tasks—observational tasks in which family is asked to pay attention to and note differences in one another's behaviors; and behavioral tasks—tasks in which the family members take action and interact different with one another.

Returning to the Murray family once again, consider the intervention from a solution-focused perspective. During the assessment and planning phase, you questioned each of the members regarding their perceptions of life in their household. Imagine William and Genevieve responses to goal-formulation or miracle questions. They will likely talk about having imagined a near "empty-nest" household with their two older children living on their own and Samantha about to head off to college. Samantha's response to the "miracle" question might likely involve her grandmother not living with the family, and not facing legal issues related to her driving under the influence arrest and her school suspension. Elle's "miracle" might include employment, a supportive partner and a home of her own. Stephen's miracle may involve a room of his own. Edna may wish desperately to have her memories back and to return to independent living.

Given the diverse array of goals that may be expressed by each of the family members, helping the Murray family to connect with a realistic set of goals can be a complicated process. Highlighting the fact that the family members care for one another is a technique to foster goodwill and find common ground on which the members can agree (De Jong & Berg, 2013). In working with the Murray family, reminding them of their commitment to and concern for one another can be a regular part of the intervention. Imagine that the agreed-upon goal is to find a solution to the overcrowded housing situation. Your role can be to work with the family to develop concrete and achievable short- and long-term solutions, check in regularly with "what's better?" questions, and terminate when you and the family agree the goals have been achieved.

Regardless of the theoretical approach that guides your work with families, your social work values and ethical standards will provide the foundation on which you will develop interventions to be implemented with families. Ensuring that the client system's rights to self-determination, strengths, and diversity are honored is the priority for your intervention with families. Emphasizing the practice behaviors to enable you to become a practitioner who is competent in social work practice with families will be the focus of the following discussion.

CONTEMPORARY TRENDS AND SKILLS FOR INTERVENING WITH FAMILIES

Families are made up of individuals and are a form of a group, and as such, benefit from the careful use of the same practice behaviors that assist individuals and groups. Interventions with families are likely to vary according to your practice setting and the constraints of your agency. For example, if you are a member of the intake unit of a child protection team, your intervention with the family will probably differ from that of your colleague who works in a community mental health agency. The theoretical lens through which you are working will also influence the way you intervene with family members. You may develop an intervention plan from a systems focus, or you may work with family members to build an intervention plan based on their perceptions of ways in which you might help them, as in solution-focused work. In any situation, your intervention begins with an invitation to the family to tell their story, but your agency mission and purpose and your own theoretical biases will influence the intervention approaches you take.

Social work practice interventions with families shares many similarities with individually-focused interventions. Recall from Chapter 5 the discussion on roles that a social worker may have in working with individuals (e.g., case manager, counselor, broker, mediator, educator, client advocate, and collaborator). Each of these roles is applicable for working with families, particularly in light of the setting in which the intervention may occur. Family-focused social work practice can take place in an array of settings, including child welfare, mental health centers, schools, health care facilities, and community centers, to name a few. For example, the social worker working with a family in a child welfare setting may engage in all the roles, but emphasize case management and brokering activities in particular in order to reunify the family. In a health care setting, the social worker's emphasis may be on education and advocacy, for instance. With families, it is important to consider all the members' perspectives, recognize relationship dynamics, and the strengths-based intervention goals that will optimize family functioning.

As with individuals, the setting in which you practice and your philosophical and theoretical approach will frame your intervention activities. While model-specific practice behaviors have been previously discussed, there are additional general practice skills and behaviors that transcend the continuum of practice approaches. Building on the practice behaviors used in the phases of family engagement and assessment, the following strengths-based skills for intervening with families can be helpful (Benard & Truebridge, 2013, p. 207; Hull & Mather, 2006):

- Identify the issues and concerns, use active listening to enable the family members to tell their story and reflect on the information shared by the family members.

- Acknowledge the pain.

- Look for and point out strengths.

- Ask questions about survival, support, periods of time that were positive for the family, interests, dreams, goals, and pride.

- Resilience begins with what one believes and all people have the capacity for resilience.

- Link strengths to the goals and dreams of the family members (both group and individual).

- Find opportunities for family/members to contribute to the intervention by helping to educate other members and serve as helping agents in achieving the family's agreed-upon goals.

- Use brainstorming for solutions as a strategy to help the family to view the situation and themselves differently.

Written homework assignments, task designation, and teaching are strategies that can be helpful but must be appropriate to the family's situation, their investment in the process, and the framework you are using in the intervention. Return for a moment to the Murray family scenario and consider the above list of practice behaviors. In order to mobilize the family into working toward crisis resolution, you can begin by asking each member her or his priorities for change. From those verbalizations, it is likely that you will glean the pain, strengths, goals, and dreams felt by each of the family members. You may choose to engage the family in brainstorming strategies they can use to address each of the areas of concern or assign the family homework in which they develop ideas to address their concerns. An alternative may be to ask the family to have a family conference and bring their ideas back to the next meeting with you. Once the family has developed a range of ways to address their prioritized concerns, consider asking them to have a discussion in which specific tasks are identified and assigned. Be sure to explore the possibility that the expertise for problem-solving resides within the family, and individual members can function as teachers and guides.

In addition to the individual and group skills, reframing, perspectival (or circular) questions, family group conferencing, motivational interviewing, and reenactments (i.e., role-playing or rehearsals) are a sampling of practice strategies with good applicability in many family situations when carefully used. Discussed in Chapter 6 as an assessment tool, mapping is an important strategy for the intervention process.

Reframing

An approach used frequently in social work with families, as well as with individuals, groups, communities, and organizations, reframing is a practice skill in which the social worker conceives of and describes a situation in different terms. Reframing can be particularly helpful in family conversations in which one member makes an incendiary statement to or about another family member. Carefully used, reframing can assist both recipient and "sender" to view the situation with diminished uproar so they can begin to listen to each other. Focused on strengths and positive alternatives, reframing must be appropriately timed to have a meaningful impact. To have a significant effect, the social worker must stay carefully attuned to the family members' dialogues to identify opportunities for reframing as they occur (Minuchin et al., 2007).

For example, a 15-year-old boy, who sees his mother as an autocratic barrier to his enjoyment because she will not allow him to go out with friends who drive, says, "There is no person on the face of this earth who is more controlling and overprotective than my mother. She is just like Hitler! She keeps me locked up in the prisoners' camp." You may suggest that the teen's mother cares so much for him that she fears he will be hurt in an automobile accident or in some other way if he goes out with his friends.

When reframing, offer a plausible alternative that does not resonate as a "gimmick" or "Pollyanna" type of effort to diffuse strong feelings and, accordingly, will be heard by the parties involved. The danger lies in interpreting the thoughts or feelings of another person without directly having been told. In the example of the 15-year-old teen and his mother, it is quite likely that the mother is not intentionally attempting to torture her son (and that she truly worries about his going out with friends who drive), but he may or may not be able to "hear" her sentiment as reframed. Another reframing effort may be more effective depending on the nature of the relationship and the people involved. Developing the content of such interpretations requires judgment and skill; use reframing cautiously and only when conditions are relatively straightforward. How can reframing be used with the Murray family? Instead of focusing on the upheaval created by Edna, Elle, and Elle's children moving into the house, consider emphasizing the strong family commitments and caring environment for members who are in need of support.

Perspectival Questions

Mentioned in connection with individuals in Chapter 5 and with group work practice in Chapter 9, perspectival questions can be effective in family social work. By seeking the perspective of another family member, you can help clarify the feelings and meanings of one member's view of another. If the family is experiencing stress because the eldest son is leaving home, you may ask the teenage daughter, "What do you think your mother will do to prepare for Johnny's leaving?" Or you may ask the mother, "What will your daughter miss most about Johnny?"

The responses to these questions can communicate ideas and feelings that no one in the family has openly or previously recognized. Such assistance in communication is relevant when family members assume they know all they need to about the responses of other family members as a result of long-term, "stuck" patterns of argument or difference or "saving face." As in the use of reframing, perspectival questioning is a strategy to be used carefully and only when you are confident that you can respond appropriately to any statement. The daughter in the example just provided may respond with, "Mother will sew name tags in Johnny's underwear so he won't lose it in the dorm laundry room at college," or, alternatively, she may say, "Mother will no doubt start to drink again." The same element of the unexpected that can create new ways of thinking for families can also throw a curve ball to the unwary or unprepared social worker. Utilizing perspectival questions with the Murray family could potentially yield some illuminating insights. Consider, for example, the new perspective that could be uncovered if Edna were asked about the support that her son and daughter-in-law have provided to Elle and her children and her response was that they should not have invited Elle and the children to move into their house. By expecting the unexpected and maintaining flexibility, the social worker who incorporates perspectival questions can be prepared for the unexpected.

Family Group Conferencing

A strategy for use with families, family group conferencing (FGC) is an empowerment-focused intervention aimed at creating or strengthening a network of support for families as they are experiencing a crisis or transition. Originally developed for work with families in which children were at risk for abuse or neglect, this practice strategy has been extended to other family situations, including families with older adults experiencing life changes. The family conference usually occurs after the assessment and planning phases, and is a collaborative effort that includes the social worker, family (including extended family members), and members of the family's community who are existing or potential resources for the family (Wise, 2005).

The conference goal is to gather the family members with people who are connected to the family in order to make decisions regarding a response to the presenting challenge. The social worker confers individually with each potential participant prior to the conference to ensure that each agrees to be actively involved in decision-making and action steps. During the conference itself, the desired outcomes are determined by the participants themselves with the strengths and areas for concern being included in the planning process (Brody & Gadling-Cole, 2008). Once a plan is established, the group adjourns and may reconvene after a period of time has elapsed in which the plan is implemented to discuss progress or re-negotiate the plan.

Return for a moment to the Murray family. If you were to convene a family group conference with the Murrays, consider those people you would invite to participate and the reasons for inviting them. What goals could provide the Murrays

© Lisa F. Young

with a much-needed support network and stronger coping skills? What are their strengths? How might they benefit individually and collectively from participation in a family group conference? What is your role? While answers to these questions can only be speculative, you can use this exercise to begin to see yourself in the role of a family social work practitioner.

Motivational Interviewing

As you recall from Chapter 5, motivational interviewing (MI) has been introduced for use with a variety of clients, including those who are non-voluntary, experiencing substance abuse, or intimate partner violence (Wahab, 2005). Using the four processes of MI—engaging, focusing, evoking, and planning—to promote behavior change, MI is well-suited for situations in which the client is uncertain about making the change and/or the time is limited (Miller & Rollnick, 2013; Wagner, 2008).

Motivational interviewing is also applicable to social work with families. Miller and Rollnick (2013) offer strategies for engaging in motivational interviewing with families with adolescents. Using the FRAMES strategy, the social worker can conduct a Family Check-up that provides the social worker and the family with the opportunity to: (1) Feedback of personal status relative to norms; (2) Responsibility for personal change; (3) Advise to change; (4) Menu of change options from which to choose pursuing change; (5) Empathetic counselor style; and (6) support for Self-efficacy (Miller & Rollnick, 2013, p. 375). For example, using the FRAMES strategy in the case of the Murray family, the social worker may opt to check in with the family in this way:

1) Provide Feedback to each family member regarding their proposed solutions for the identified stressors currently affecting the family.
2) Designating Responsibility for change—clarifying roles and tasks for each member of the family.
3) Advise on change—provide information to the family on the impact of ongoing stress, Alzheimer's Disease, and potential positive outcomes of collaborative change.
4) Provide a Menu of options—presenting family members with an array of options for behavior change, healthier interactions, services and resources that may be mobilized within and outside the family.
5) Empathetically validate the concerns expressed by each member of the family and acknowledge the pain they are experiencing.
6) Self-efficacy of each member of the family can be regularly reinforced by the social worker identifying their strengths, commitment to change, and progress toward goals.

Motivational interviewing provides another opportunity for you, as a social worker, to engage in a collaborative partnership with the client family aimed at providing the opportunity for change. If you opted to employ motivational interviewing with the Murray family, you may begin by providing feedback to them based on the perceptions they have shared with you regarding the various crises that are occurring within the family. They also have the opportunity to provide feedback to one another. After they share feedback, you can empathetically reiterate that the family members are the experts about themselves and it is within their power and responsibility to implement change. Try to create an environment in which the family is empowered to view themselves as having the capacity to respond to their crises with resilience and efficacy.

Incorporating motivational interviewing into social work practice with families can present challenges that do not exist in working with individuals. Given that families come to social workers often times as the result of their inability to resolve relationship challenges on their own, ensuring that each member of the family is given adequate opportunity to speak is critical (Miller & Rollnick, 2013). Moreover, the social worker will need to be able to balance the interactions to maintain a focus on positive change as opposed to continuing with arguments and negative discussions.

Re-enactments

Participation in experiential activities can provide insights and alternatives for families who are experiencing a crisis or challenging transition. Re-enactments can be completed in the form of role-plays or rehearsals. Re-enacting a particularly challenging interaction and then role-playing or rehearsing the scenario with alternative behaviors can be a powerful experience mechanism for the family to "try on" new ways of being a family. Viewed as a safe way to share feelings and rehearse, role-plays can be carried out in a variety of settings, for a range of situations, and revisited as the work progresses. Considerations for developing role-played activities include (Hull & Mather, 2006, pp. 163–164): (1) purpose and parameters are to be discussed; (2) members can play themselves or other members of the family; (3) members should play the roles accurately and consider their feelings as they move through the rehearsal; (4) the role-play can be stopped so members can discuss, reflect, and change interactions; and (5) the role-play should be de-briefed so alternatives can be explored. A re-enactment may be an ideal activity in which to engage the Murray family. Once the family has identified the issues that are their highest priority to address, you can help them to "rehearse" the change strategies they have brainstormed through an experiential exercise. With the multiple generations, issues, and priorities that exist within the Murray family, role-playing can enable the family to work on individual issues and relationships. Imagine a role-play in which William and Genevieve share with Samantha their feelings and concerns regarding her life choices and decision-making. More importantly, envision a dialogue between Samantha and her parents that any one of the three may stop so they may regroup, change course, or ask for input. Such a rehearsal may enable the family to change a pattern of interactions.

Mapping as an Intervention

Mapping strategies have been examined primarily within the context of assessment. They do, however, have a place within the intervention process itself. If not completed during the assessment phase, mapping can be incorporated into the intervention phase of work. During the intervention, mapping in the form of genograms, in particular, can be used for clarifying family patterns, framing and detoxifying family issues, and in developing the intervention plans (McGoldrick, Gerson, & Petry, 2008). Using the patterns and unhealthy family issues to identify and facilitate change plans can be a liberating experience for the family members. McGoldrick and colleagues (2008) promote the use of genograms as an intervention strategy to empower clients through changing existing relationships. Being able to see the historical patterns of loss, relationships (healthy and unhealthy), physical and mental health issues, substance use, responses to stress and crisis, and cultural traditions can illuminate for family members the options they have for change. Creating and discussing genograms can also provide family members with an opportunity to engage in intergenerational dialogue. Consider the Murray family, for instance. Providing the family members with the opportunity to engage in dialogue with Edna could provide them with family history they had not previously known, as well as have meaningful time with her before her memory fades away. The social worker can incorporate the genogram into the development of the intervention by asking members to identify strengths on which they can build in relationships and cultural traditions (McCullough-Chavis & Waites, 2008). They may also be able to identify patterns of substance abuse, trouble with the law, or other troubling patterns that can inform their present collective and individual work.

Ecomaps and culturagrams are also assessment tools that can be integrated into the intervention. Using the baseline ecomap or culturagram as a strategy for monitoring change throughout the intervention provides the family and the social worker with a visual depiction of their work. Updating the ecomap or culturagram can indicate those areas in which progress is being made or not made and the barriers preventing success. New maps can be constructed as a means for ritualizing a successful outcome. As with social work practice with individuals, documentation is a critical component of the family intervention. Along with maintaining copies of the assessment tools mentioned here, the client portfolio also includes a record of the intervention plans. Exhibit 7.3 offers a sample family intervention plan template followed by Exhibit 7.4 which provides an example of documenting the intervention plan developed with the Murray family.

As with social work practice with individuals, intervening with families requires the social worker to complete comprehensive and ongoing assessments from which flexible, individualized interventions can be created through collaboration with the family. The outcome of a successful intervention is, of course, the termination.

EXHIBIT 7.3
 	
Documentation of a Family Intervention Plan
	Intervention Plan:
	• Preliminary assessment
	• Preliminary plan for intervention and plan for change (to be developed at first visit), including:
	    ○ What will each family member do differently?
	    ○ How does each family member view themselves accomplishing changes?
	    ○ What support and services are needed to accomplish plan for change?
	    ○ Who will provide support and services?
	    ○ Who will arrange for support and services?
	• Interventions and plans for emergency/safety needs
	• Other interventions needed
	• Needs (include date, identified need, status (active, inactive, deferred, or referred), and reason for deferral or referral)
	• Strengths
	• Facilitating factors for intervention
	• Limitations
	• Barriers to intervention
	• Other care providers/referrals and purpose (including plan for service coordination)
	• Plan for involvement of individual family members, extended family members, significant others and friends
	• Review and termination criteria/plan
	• Planned frequency and duration of intervention
	
	Adapted from St. Anthony's Medical Center, St. Louis, Missouri; Missouri Department of Social Services

Our focus will now turn to the phases of terminating and evaluating the family intervention.

ENDING WORK WITH FAMILY CONSTELLATIONS

The general principles for ending interventions apply to all levels of practice. The need for culturally sensitive practice plays out in individuals, families, groups, communities, and organizations. Just as you need to explore the meaning of endings with individual client systems, so you do for families. This discussion assumes a cumulative recognition of the important areas for ending the work, based on considerations covered so far, and a flexible application with greater emphasis on some principles than others according to context. With those principles in mind, we will look to some specific variations as additional perspectives.

In voluntary family work, endings tend to occur when the family is satisfied that they achieved their goals. In some circumstances , the restrictions of third party

Preliminary assessment

The Murray family has requested services from this agency in order to address a number of crises and concerns within the family. The family is comprised of William (55 year-old-high school teacher) and **Genevieve** (50-year-old occupational therapist) and their three children (Elle, 25-year-old divorced single mother of two- and four-year-old children; Stephen, 21-year-old college student; and Samantha, 16-year-old high school junior). Along with William's mother, 76-year-old Edna, all members of the family reside in the couple's now overcrowded home. Stephen and Samantha were displaced when their bedrooms were given to Edna and Elle and her children. Edna can no longer live alone as she is experiencing cognitive impairment (Alzheimer's Disease) and presents safety concerns at her assisted living facility.

The Murrays sought help because Samantha was recently arrested for driving under the influence and suspended from school. During the assessment, information was shared that suggests the family has been in crisis for some time and that Samantha's arrest and suspension brought the crisis to a climax.

Preliminary plan for intervention and plan for change

During the first meeting of the entire family, each member was asked to share their concerns, needs, and potential solutions. After each person had the opportunity to artic-ulate each of these areas, the social worker provided a summary of the information shared and the family and social worker collaborated on a list of highest priorities, resulting in the development of the following preliminary plan:

1) **Genevieve** agreed to contact the local chapter of the Alzheimer's Association to obtain information on services available for families experiencing dementia. She will specifically inquire about programs that provide information on Alzheimer's Disease and provide financial support for care needs (e.g., Family Caregiver Support Program).
2) All family members agreed they would attend a presentation for families on Alzheimer's Disease provided by the Alzheimer's Association.
3) Stephen will contact the campus housing department at his university to learn about applying for a resident advisor position which would provide him with tuition support and free housing in exchange for living in one of the student residential facilities. In the interim until this goal may be achieved, William and **Genevieve** agreed to work with Stephen to fix up the basement so he may have a private bedroom and workspace.
4) **Genevieve**, William, and Samantha all agreed to attend the court-mandated program for teens arrested for driving under the influence.
5) Samantha agreed to attend the alternative school program for the duration of her school suspension.

Interventions and plans for emergency/safety needs

Two areas of immediate concern were raised regarding safety needs. The following plan was agreed on by all members of the family:

EXHIBIT 7.4

Documenting the Intervention Plan: The Murray Family

EXHIBIT 7.4

Continued

1) Elle agreed to provide care for Edna during the day when Genevieve and William are at work. This will allow Genevieve and William to feel that Edna is safe and that Elle can provide supervision and transportation. Elle will be able to remain home with her children and not bear the expense of child care.

2) Samantha agreed not to drink and to be subject to a periodic alcohol test if her parents are uncertain about her sobriety.

Other interventions needed
Longer term interventions may include:

1) Exploration of all family members' feelings regarding the overcrowded living conditions in the home and stressful relationships.

2) Discussion regarding Stephen's living arrangement if he is not accepted for a resident advisor position.

3) Decision-making regarding Edna's ongoing care needs as her disease progresses.

4) Discussions regarding Elle's permanent employment and living situations.

5) Reestablishing a trusting and healthy relationship between Samantha and her parents.

Needs
At this preliminary phase, the primary needs are:

1) Establish a safe environment for Edna
2) Address Samantha's legal, school, and alcohol issues
3) Provide adequate space for Stephen and Samantha
4) Address family stressors

Strengths
The Murray family has several important strengths on which to build, including:

1) Couples' longstanding marital and employment history and commitment to each other and their family members as evidenced by their willingness to open their home to Edna and Elle and her children.

2) Families' commitment to care for one another

Facilitating factors for intervention
The family members' willingness to care for one another is the primary facilitating factor in the intervention. Additionally, having resources within the family (e.g., Elle's availability to care for Edna and her expertise in nursing care) and outside the family (e.g., Alzheimer's Association, alternative school, etc.) enable the family to access needed services.

Limitations and barriers to intervention

1) While plans are underway to address the overcrowding situation, change may take some time. Coping with the stress of the cramped living quarters in the interim presents a particular challenge.

2) While she agreed to attend the alternative school and participate in family therapy, Samantha appears angry and may become reluctant to actively engage, particularly if the home continues to be a stressful environment and she feels she has no attention or privacy.

3) The family may be feeling so overwhelmed by the multitude of issues and tasks to be addressed, they may be unable to mobilize for change.

Other care providers/referrals

1) School personnel at the alternative school and Samantha's regular school
2) Alzheimer's Association
3) Family Caregiver Support Program
4) Substance Abuse Treatment Program

Plan for involvement of individual family members, extended family members, significant others and friends

At this time, there is no plan to include other family members. However, the family mentioned that William has two sisters that live out of town and may be willing to help with Edna's care at some point.

Review and termination criteria/plan

All agreed that the goals will be achieved when the family members feel they are in better control of their lives, specifically:

1) Overcrowded housing issue has been resolved
2) Plans for Edna's long term care needs are determined
3) Elle's employment and housing plans are clarified
4) Samantha has successfully completed the court-mandated program and has returned to a student in good standing at her school (one-year minimum of no school violations)
5) Stephen has a permanent solution for housing.

Planned frequency and duration of intervention

Initially, the plan is for the social worker and family members to meet weekly. Social workers will be available by telephone in the interim. After the first four meetings, social worker and family will discuss the plan for ongoing contact. As issues resolve, meeting frequency will decrease with the plan being for termination to occur within three months of the intake.

Adapted from St. Anthony's Medical Center, St. Louis, Missouri; Missouri Department of Social Services

insurers or managed care companies may mandate an earlier end point. With a lessened focus on the relationship with the social worker than exists in individual work, the emphasis often falls on examining the ways the family wanted the dynamics of their relationship or their relationship with outside entities to be different, and the extent to which they have been successful. There is also likely to be some focus on translating the gains into future situations that the family can anticipate so that responses can be predicted. For example, if a family is struggling with the decision to allow an adolescent son freedom to develop a unique identity when he has a history of legal altercations, it will be useful for the family to consider ways in which they will manage that issue when he leaves home for college or when the next sibling reaches an age to declare herself or himself a separate person. These positions are all consistent with the principles of review and exploration highlighted earlier in Chapter 5 regarding terminating with individuals.

To take a different perspective, the following discussion will identify brief responses to ending work with families from the perspectives of the previously discussed theoretical frameworks, strengths and empowerment, narrative, and solution-focused approaches. These approaches strive to minimize the difficulty of endings. In general, they propose a naturalized and comfortable process that is flexible and controlled by clients whenever possible.

Endings with Strength and Empowerment

Viewing each family as individuals and as a whole means that the social worker's role is to help each person articulate her or his feelings about the ending (Wise, 2005). The family members, individually and as a group, can benefit from the opportunity to talk about feelings and insights about the strengths that each individual and the group brought to the intervention. These insights can reinforce the successes that have been accomplished through the intervention and enable the members to acknowledge the ending. Keeping in mind the original goal, the social worker and the family can focus on the future and the ways in which the changes can be sustained. If the goals they hoped for were not achieved, but the relationship is terminating nonetheless, the termination phase can focus on lessons learned that the family can use in the future.

Just as assessment and intervention with families is approached from a strengths perspective, so too is the termination process. Building on the strengths that were identified in the assessment process and those identified or created during the intervention phase, the strengths- and empowerment-oriented social worker can focus the termination process on strengths as well. The existing and new strengths can become the basis for the family to sustain the changes they have made. Together, the social worker and the family can review the family's strengths. The social worker can then ask the family to consider the way in which they can apply these strengths to future situations. The family can also use these strengths as coping skills when and if they encounter new challenges. For example, the social worker may ask the family members, "How can ending our work together help you to achieve your goals?" (Wise, 2005, p. 215). To further probe, you can ask the family members to speculate on their motivation and ability to continue working on their goals even after the formal intervention has ended.

Endings in Narrative-Focused Work

In narrative work, as with other approaches, there is an emphasis on normalizing the point at which the family decides to end. Narrative social workers often punctuate the ending of their work by working with families to develop rituals or ceremonies in which the family invites an audience to witness the changes they have made and to rejoice in their achievements. For example, asking siblings, other professional who have worked with the family, and/or extended family to attend a meeting can be a validating experience for all. Public acknowledgment of the family's successes not only celebrates them but also provides a structure for their supportive maintenance when the work is over (Morgan, 2000). This focus, like solution-focused work, represents a departure from traditional views of endings while acknowledging the same concerns about maintaining gains. The reduced emphasis on stages with specific boundaries in both these approaches emphasizes the view of the client as the expert.

Endings in Solution-Focused Work

Solution-focused work emphasizes ending almost from the beginning. As a short-term intervention approach, a solution-focused approach stresses that the clients have abilities to manage their lives competently. Because this approach is built on the premise that change can occur within a brief, time-limited period, using scaling,

© Lisa F. Young

an early question the social worker might ask is, "What [number] do you need to be in order not to come and talk to me anymore?" (De Jong, 2009). This question refers to the number from 1 to 10 that reflects the degree of well-being that the client reports. In this approach, a family's concerns about needing further work are honored, and they determine the number and content of further sessions. There is very little emphasis on the relationship between the social worker and the family because "not coming to talk to me anymore" is seen as the preferred reality and a natural and comfortable conclusion to a problem for which the family is already likely to have a solution (which the social worker helped to bring to light). In this sense, then, ending is seen as success almost by definition. Termination, as well as follow-up (e.g., checking in with the family, inviting the family for a return of follow-up session, or making referrals) when possible, is as equally important to the planned change process as any other stage of the relationship. Bringing the intervention to a close can serve as an opportunity for the social worker and family to engage in: (1) a recital (i.e., a review of the work); (2) creating an awareness of the changes made; (3) consolidating gains (i.e., celebrating changes and successes), and; (4) providing feedback to the social worker; and (5) preparing the family for handling challenges that arise in the future (Collins et al., 2013, p. 447).

Evaluation of Social Work Practice with Families

Evaluation of family interventions helps you as a practitioner and your organization determine if the intervention has been complete and is effective. By gathering information from the family members at the beginning of the working relationship, evaluation of family interventions also enables the social worker to identify the changes that were or were not made and the reasons that the intervention was or was not effective (Hull & Mather, 2006). If you have used evidence-based practice approaches, your evaluation can provide insights and contributions to your practice or your agency's practice. While evaluation of your interventions with individuals can yield similar information about your practice, it is critical that you do not assume that evaluation of family interventions can be completed with the same evaluative strategies. Just as families are unique, so, too, are evaluations of family interventions.

The ongoing evaluation of the family work that social workers do is often largely contingent on, and defined by, the context and goals of the original contact. For example, if you are working with a family in child or adult protective services, the first goal may be the continued safety of a child(ren) or adult. There may be other goals, such as the parents' improved skills in managing a family, meeting the health needs of a particular child or adult, or providing appropriate care for an older adult. Goals in cases such as these, in most cases, are documented in written goals, and you will want to pay attention to them along the way.

QUICK GUIDE 21   STRENGTHS-BASED MEASURES FOR FAMILIES

Caregiver Well-Being Scale

I. ACTIVITIES

Below are listed a number of activities that each of us do or someone does for us. Thinking over the past three months, indicate to what extent you think each activity has been met by circling the appropriate number on the scale provided below. You do not have to be the one doing the activity. You are being asked to rate the extent to which each activity has been taken care of in a timely way.

1. Rarely	2. Occasionally	3. Sometimes	4. Frequently	5. Usually

	1	2	3	4	5
1. Buying food	1	2	3	4	5
2. Taking care of personal daily activities (meals, hygiene, laundry)	1	2	3	4	5
3. Attending to medical needs	1	2	3	4	5
4. Keeping up with home maintenance activities (lawn, cleaning, house repairs, etc.)	1	2	3	4	5
5. Participating in events at church and/or in the community	1	2	3	4	5
6. Taking time to have fun with friends and/or family	1	2	3	4	5
7. Treating or reward yourself	1	2	3	4	5
8. Making plans for your financial future	1	2	3	4	5

II. NEEDS

Below are listed a number of needs we all have. For each need listed, think about your life over the past three months. During this period of time, indicate to what extent you think each need has been met by circling the appropriate number on the scale provided below.

1. Rarely	2. Occasionally	3. Sometimes	4. Frequently	5. Usually

	1	2	3	4	5
1. Eating a well-balanced diet	1	2	3	4	5
2. Getting enough sleep	1	2	3	4	5
3. Receiving appropriate health care	1	2	3	4	5
4. Having adequate shelter	1	2	3	4	5
5. Expressing love	1	2	3	4	5
6. Expressing anger	1	2	3	4	5
7. Feeling good about yourself	1	2	3	4	5
8. Feeling secure about your financial future	1	2	3	4	5

Source: Berg-Weger, Rubio, & Tebb, 2000; Tebb, Berg-Weger, & Rubio, 2013.

QUICK GUIDE 22   STRENGTHS-BASED MEASURES FOR FAMILIES

Family Support Scale

Name _____   Date _____

Listed below are people and groups that oftentimes are helpful to members of a family raising a young child. This questionnaire asks you to indicate how helpful each source is to your family.

Please circle the response that best describes how helpful the sources have been to your family during the past three to six months. If a source of help has not been available to your family during this period of time, circle the NA (Not Available)

How helpful has each of the following been to you in terms of raising your child(ren):	Not Available	Not at All Helpful	Sometimes Helpful	Generally Helpful	Very Helpful	Extremely Helpful
1. My parents	NA	1	2	3	4	5
2. My spouse or partner's parents	NA	1	2	3	4	5
3. My relatives/kin	NA	1	2	3	4	5
4. My spouse or partner's relatives/kin	NA	1	2	3	4	5
5. Spouse or partner	NA	1	2	3	4	5
6. My friends	NA	1	2	3	4	5
7. My spouse or partner's friends	NA	1	2	3	4	5
8. My own children	NA	1	2	3	4	5
9. Other parents	NA	1	2	3	4	5
10. Co-workers	NA	1	2	3	4	5
11. Parent groups	NA	1	2	3	4	5
12. Social groups/clubs	NA	1	2	3	4	5
13. Church members/minister	NA	1	2	3	4	5
14. My family or child's physician	NA	1	2	3	4	5
15. Early childhood intervention program	NA	1	2	3	4	5
16. School/day-care center	NA	1	2	3	4	5
17. Professional helpers (social workers, therapists, teachers, etc.)	NA	1	2	3	4	5
18. Professional agencies (public health, social services, mental health, etc.)	NA	1	2	3	4	5
19. _____	NA	1	2	3	4	5
20. _____	NA	1	2	3	4	5

Source: Dunst, Trivette, & Deal, 2003, pp. 155–157

QUICK GUIDE 23   FAMILY STRENGTHS PROFILE

Recording Form

Family Name _____   Interviewer _____

INSTRUCTIONS
The Family Strengths Profile provides a way of recording family behaviors and noting the particular strengths and resources that the behaviors reflect. Space is provided down the left-hand column of the recording form for listing behavior exemplars. For each behavior listed, the interviewer simply checks which particular qualities are characterized by the family behavior. (Space is also provided to record other qualities not listed.) The interviewer also notes whether the behavior is viewed as a way of mobilizing intrafamily or extrafamily resources, or both. A completed matrix provides a graphic display of a family's unique functioning style.

FAMILY MEMBER	DATE OF BIRTH	AGE	RELATIONSHIP

FAMILY BEHAVIOR	Commitment	Appreciation	Time	Sense of Purpose	Congruence	Communication	Role Expectations	Coping Strategies	Problem Solving	Positivism	Flexibility	Balance			TYPE OF RESOURCE	
															Intrafamily	Extrafamily

NOTES

Source: Dunst, Trivette, & Deal, 1988

The focus of evaluation in interventions with families is on the family unit itself and not the individuals within the family. An examination of the family's ability to have new behaviors and coping skills, realistic attitudes, and new information and learning will promote enhanced well-being for the entire family (Wise, 2005). For example, if you and the family are able to determine that the power has shifted among and between the family members, this could be indicative of an effective family intervention. Alternatively, if family members agree that the stress level within the family has decreased or the quality of the interactions has improved, the intervention may be deemed successful. Family evaluations may focus not only on the outcomes of the goals established during the assessment and planning phases, but on the relationship with the social worker and the agency as well.

Strengths-Based Measures for Families   While family-focused intervention evaluations differ from evaluations of individual and group practice, the evaluative strategies described in Chapter 5 provide a basis for developing plans for evaluation of family interventions. Ensuring that the focus of the evaluation is on the family and not the individual members, strategies such as single-system design, goal attainment scaling, and case studies, can be effective evaluation tools. There has been an array of evaluation measurements developed specifically for use with families. While the scope of this book cannot address the vast number of selections that are available for evaluating your practice with families, the following discussion will briefly explore the selection of measures that are grounded in a strengths-based perspective.

Several instruments are designed to be used in strengths-based practice with families and also to allow social workers to document their service effectiveness (see Early & Newsome (2005) for detailed review of strengths-based family assessment tools). Standardized evaluative tools are increasingly important as agencies and practitioners are held accountable for measurement of outcomes by funders, boards of directors, client advocacy organizations, and the social work profession itself. Additionally, practice evaluation instruments can be helpful in maintaining the social worker's focus. Some of the instruments emerged as strengths-based emphases were introduced in the 1980s and remain useful tools today because they measure family perceptions and assets. With the advent of evidence-based practice, there has been an increase in the number of standardized family assessment measures from which to select. Early and Newsome (2005, pp. 390–391) offer the following guidelines to consider when incorporating standardized assessment measures in family-focused social work practice:

1)   Select a measure with an emphasis on consistency and appropriateness to client situation.
2)   Explain to the family the reasons for using the measure (e.g., method for efficiently collecting information to guide the helping process) and specific details about the measure's contents and outcomes.

3) Allow for adequate and appropriate time, space, and materials for completing the measures.
4) Ensure that the family members understand the items and feel comfortable asking questions about the assessment tool.
5) When at all possible, score the measure in the family's presence.
6) Share the scores and interpretations with the family.
7) Maintain the assessment in agency records for review/comparison in the future.
8) Repeat the measure over time to assess progress toward goals.

While your agency or you will select the evaluative strategy that is most appropriate to your setting and the families that you serve, the following is a small representative list of reliable strengths and empowerment-focused evaluation measures that may be of help to you as you consider evaluating your practice:

- *The Caregiver Well-Being Scale* (Berg-Weger, Rubio, & Tebb, 2000; Tebb, 1995; Tebb, Berg-Weger, & Rubio, 2013): A strengths-based clinical measure to help family caregivers or adults or children identify the strengths and areas for change in their caregiving experience, this scale is applicable for families caring for adults and/or children. See Quick Guide 21 for an example.

- *The Parent Empowerment Survey* (see Dunst, Trivette, & Deal, 2003): Designed to measure parental perceptions of control over life event (see Herbert, Gagnon, Rennick, & O'Loughlin, 2009 for review of empowerment measures).

- *The Family Support Scale* (see Dunst et al., 2003): Measures what is helpful to families. See Quick Guide 22 for an example.

- *The Family Strengths Profile* (see Dunst et al., 2003): Designed to chronicle family functioning, the profile provides a qualitative format for identifying and assessing family strengths and type of resources needed. Quick Guide 23 provides an example of this measure.

- *The Family Resource Scale* (see Dunst et al., 2003): Measures the adequacy of resources in households with young children and emphasizes success in meeting needs, while it also identifies needs.

- *The Family Functioning Style Scale* (see Dunst et al., 2003): Measures family values, coping strategies, family commitments, and resource mobilization. Families indicate to what degree various statements are "like my family."

- *The Family Empowerment Scale* (Koren, DeChillo, & Friesen, 1992): Measures family empowerment on three levels: family, service system, and community/political.

- *The Behavioral and Emotional Rating Scale: A Strengths-Based Approach to Assessment* (Epstein, 2004) and *The Behavioral and Emotional Rating Scale (2nd edition): Youth Rating Scale* (Epstein, Mooney, Ryser, & Pierce, 2004). These scales focus on children and youth to determine the presence of behavioral or emotional conditions.

In strengths- and empowerment-oriented measurements, self-reporting is seen as an asset. Self-reporting can complement the traditionally-oriented, scientific measurements in which the goal of objectivity is thought to conflict with the biases in self-reporting. Bias is inherent within self-reported information because it is difficult to be objective about yourself. The strengths perspective supports the expertise of individuals and families about their own lives, experience, and aspiration, thereby making self-report a natural and theoretically consistent method of data collection.

The empirical procedures in these scales are just a few among many. Some of these are flexible and, with creativity, they can apply to a variety of situations. Using both quantitative and qualitative evaluative strategies can provide the social worker and the profession with a comprehensive picture of the intervention. Remember, as well, that all evaluative measures have their limitations. Drawing conclusions regarding the impact of the change may not be possible because change may have occurred separately from the intervention (Hull & Mather, 2006). Further, just because the goals of the intervention were not achieved does not mean that positive or meaningful change did not occur. In those situations in which the identified goals are not met by the time the intervention must end (i.e., as in the case of managed care, court- or school-mandated treatment, etc.) or the family terminates prior to the agreed upon ending point, the social worker must process this experience. Talking with a supervisor or colleagues regarding the possible reasons for this unanticipated outcome can be an opportunity to professional development for the social worker. While practice evaluations can be immensely helpful tools, they must be viewed within a context for family, yourself as a practitioner, and your agency.

Consider also your own role with the family and reflect on the viability of your relationship with them. Reflection is typically a more introspective and interactive process. Some questions you can ask yourself and members of the family are: Does each member feel valued as if she or he can contribute to the work? Are there issues regarding the family's culture? Does the family still agree with the direction of the work? How is the work changing their experience? Checking in with families, through a dialogic process, to make sure they feel heard, understood, and are invested in the work you are doing with them is just as important as it is with other levels of practice. Be certain as well to remain attuned to non-articulated feedback from the family members. Lack of follow through with assignments or commitments and nonverbal gestures can be indicators of the family members' feelings about the intervention (Hull & Mather, 2006).

STRAIGHT TALK ABOUT FAMILY INTERVENTION, TERMINATION, AND EVALUATION

Practice with families is a complex enterprise, and the intervention, termination and evaluation processes can include the unexpected. Family members have individual relationships, networks, and influences outside the family; therefore, the work that is accomplished within the intervention may be positively or negatively impacted by those external persons or groups. Within the family unit, the members may establish alliances that can also serve to strengthen or, in some cases, undermine the work within the intervention. Staying attuned to the family members' descriptions of their activities, relationships, and pressures may aid you in identifying those potential unanticipated occurrences within the intervention.

Just as with terminations with individuals, ending work with families can be emotional, particularly if the family members have developed a positive relationship with the social worker. The social worker may even be viewed as a member of the family; therefore, termination can elicit powerful reactions from the family. Sensitivity to the family's cultural norms and feelings is important for the social worker (Logan et al., 2008). Terminations can also elicit negative responses from some family members. As a result, members may refuse to participate or may display anger, denial, anxiety, or even regression as the termination approaches (Fortune, 2009). As the social worker, you may find yourself juggling a variety of different responses from family members to the ending of the family intervention. Addressing the individual and collective reactions to termination, while building on the strengths and gains of the intervention, can serve as an intervention in and of itself. Empowering the family to handle a change (perceived possibly as another loss) can be a meaningful new experience for the family as they can provide support to one another, model new behaviors, and mobilize strengths and resources developed during the intervention.

Evaluations of family interventions may also yield unexpected results. As you complete the evaluative process, you may have anticipated a particular outcome. If, for instance, you completed a pre- and post-intervention evaluation and the members of the family provide responses different from those you expected, you may be surprised at the family members' perceptions. Such occurrences can provide you with the opportunity to explore with the family their responses so you may gain insight into differing interpretations.

CONCLUSION

Facilitating an intervention with a family from engagement through termination and evaluation can be an immensely rewarding professional accomplishment for you and transformative for the family. Working with families requires the social worker to develop a repertoire of practice behaviors that include theoretical frameworks and skills to optimize family strengths and create and mobilize needed resources. This chapter further explored the integration of several theoretical approaches into social work practice with families and provided examples of ways in which the approaches can be applied to family situations. While refining your theoretically-driven social work practice is a lifelong process, identifying the theoretical approaches that are most consistent with your professional and personal value systems is an important early commitment for your social work development. Clarifying your own views on families and your role as a family social worker are exploratory exercises that can be helpful along your journey. Along with the provision of theoretical underpinnings for developing family interventions, this chapter highlights strategies for working with families that can be used in a range of settings and with diverse types of families.

MAIN POINTS

- Social workers have an array of theoretical perspectives to choose from to guide interventions with families. The social worker's training, philosophical and value system, and agency orientation will provide the primary influence for selection.

- Approaches for developing intervention plans with family groups highlighted in this chapter include strengths, empowerment, solution-focused, and narrative perspectives. While each shares similarities with the others, each has characteristics that make it unique.

- While specific practice behaviors are included within the various models of family intervention, there are basic skills for working with families that are found in most of the models: collaboration, using strengths, and supporting each member of the family's voice.

- Both your own and your clients' families can impact your work with families. Ongoing evaluation includes not only documenting achievement of externally imposed goals (e.g., school or court system) but also determining if your role and relationship with the family are working.

- Because the social work intervention with a family is multi-faceted and vulnerable to influences in and outside of the relationships, the evaluation process should incorporate candid discussion of those potentially influential factors. When ending work with families, the social worker must consider the potential impact of the professional-client relationships that formed during the work, the theoretical perspectives they used, and the practical, contextual dimensions of the work.

- The evaluation of families may emphasize an empirical process and/or a qualitative one. Standardized tools may be helpful in the evaluation process of family social work. In some family models, the emphasis is on the family's qualitative satisfaction with the outcome of the work; postmodern views of evaluation question the emphasis on standardized, scientific, and quantitative measures and remind social workers of their susceptibility to bias even when their work is supported by numbers.

EXERCISES

a. Case Exercises

1. Go to www.routledgesw.com/cases and review the case files for each member of the Sanchez family. A number of potential ethical issues may emerge as you work with this family. Using the NASW *Code of Ethics* (available at naswdc.org), identify the ethical issues or dilemmas that you anticipate could arise, citing the core social work and ethical principle that is relevant to the issue you identify. Upon completing this task, write a brief reflection on your findings.

2. Go to www.routledgesw.com/cases and review the case file for Carmen Sanchez. Address Critical Thinking Questions 1, 2, and 3. For Critical Thinking Question 2, identify at least one article in each of the two areas that concern Carmen: the impact on families of children with special health needs and outcomes of children in different types of families.

   Gather into groups with four to five of your classmates and develop a potential work plan for the family, based on your answers to the Critical Thinking Questions, focusing on the following areas:
   a. Identify the areas of potential challenge.
   b. Develop strategies for working with the entire Sanchez family around these challenges.
   c. Develop a plan to insure Carmen's maximum participation in the process.

3. Go to www.routledgesw.com/cases and review the case file for Carla Washburn. While Carla does not have the traditional family network, she does have a support system of persons to whom she is connected. Strategize about ways in which you might integrate each of the members of her support network into your intervention with Carla, identifying the strengths and potential contributions each can make. You may wish to review the case files and ecomap to develop your plan.

4. Go to wwww.routledgesw.com/cases and review the case file for Riverton. While the Riverton case initially appears to be a case focused on a neighborhood and a community, each community is comprised of individuals and families. In your role as the social worker who has just moved into the Riverton community, consider the Williams family (not described in the case file). Joyce

is a 48-year-old single mother who lives with her two children, 18-year-old Amanda and 17-year-old Jason, and her mother, 77-year-old Nina. Joyce divorced her husband as a result of his chronic alcoholism and she is worried about Jason as she knows he regularly drinks and uses marijuana. She has come to your agency and asked for your help with Jason. In light of the alcohol and drug issues that are occurring in the neighborhood, she feels powerless to handle the situation alone. She cannot afford to move out of the neighborhood as the house is owned by her mother and Joyce is unable to work full-time due to the caregiving responsibilities she has for her mother. Utilizing your knowledge of the Riverton community, its resources and current culture, develop a strategy for engaging, assessing, and intervening with the Williams family.

5. Go to www.routledgesw.com/cases and review the case for Brickville, focusing on Virginia Stone and her family. Based on the initial assessment and planning process you conducted in collaboration with Virginia (you may want to review Exercise #3 in Chapter 6), you and she have determined that a family conference will help to address the multiple concerns related to her mother's care, home ownership, and response to the proposed plans for neighborhood development. Using the information available to you (including the eco-map and genogram), develop two separate written plans for a family conference from two perspectives (narrative and solution-focused), including your responses to the following questions:

   * Who should be invited to participate in the family conference?
   * What is an optimal location for holding the conference?
   * Based on discussions with Virginia, develop a list of:
     ○ Potential agenda items
     ○ Individual and family strengths and areas of concern
     ○ Available and needed resources (within the family and community)
   * What is your anticipated role?
   * How might the family genogram and/or eco-map be incorporated into the family conference to develop a collaborative intervention, particularly as it relates to providing care for family members.

   Upon completing your plans for a family conference, reflect in writing on three areas:

   * Differences and similarities of a narrative or solution-focused approach;
   * Ways in which strengths-based perspective can be integrated into the conference;
   * Possible motivational interviewing strategies you can employ during the family conference.

6. Go to www.routledgesw.com/cases and review the case for Hudson City and the information provided in Exercise #4 in Chapter 6 regarding the Patel family. Using a strengths-based, solution-focused approach, develop a summary of the four-phase planned change model intervention (engagement, assessment/planning, intervention, termination/evaluation) you would facilitate with the

family. Be sure to include your theoretical approach(es) and incorporate the available resources in Hudson City.

b. Other exercises

7. As a social work practitioner at a community mental health center, you serve primarily individual and family client systems. You have recently been called by the James family and will conduct an intake appointment later on today. You have the following information about the family:
    * The father, self-identified as African American, made the appointment.
    * The family includes the mother, father, their two adolescent sons, and the father's parents who share the home with the family.
    * The elder son, age 18, is of most concern to the family, having expressed suicidal thoughts and recent but increasing withdrawal from school, family, and community life. Last year this son was a well-known school athlete and this year is not active in any school or athletic activities.
    * The parents suspect that alcohol or other drug use is involved in the family.
    * Both sons are reluctant to attend the appointment but will because their father has indicated they will.
    * Mother will "go along."
    In your preparation for this appointment, you consider several dimensions of the work that appear below. In small groups, discuss these and explore the questions. Report back to the entire class.
    a. You are not African American but are of Jewish heritage, which is relatively rare in the town. You know there are cultural differences between this family and you but believe you can bridge those to some extent because you consider yourself different culturally as well. What might you need to consider in your own assumptions?
    b. What model of family social work (of those in the chapter) do you believe will provide the most useful base? Discuss the reasons for this selection.
    c. What specific information will be important to first clarify?
    d. What specific approaches will be most important here with this particular family?
    e. How would you begin your intervention with this family? What practice behaviors will you use? What might you actually say? (Give an example)
7. Using the strengths-based strategies highlighted in this chapter, review the following list and describe in detail the way in which you would incorporate each into an intervention with the James family (from exercise #6):
    * Identify the issues and concerns, use active listening to enable the family members to tell their story and reflect on the information shared by the family members.
    * Acknowledge the pain.
    * Look for and point out strengths.

- Ask questions about survival, support, periods of time that were positive for the family, interests, dreams, goals, and pride.
- Resilience begins with what one believes and all people have the capacity for resilience
- Link strengths to the goals and dreams of the family members (both group and individual).
- Find opportunities for family/members to contribute to the intervention by helping to educate other members and serve as helping agents in achieving the family's agreed-upon goals.
- Use brainstorming for solutions as a strategy to help the family to view the situation and themselves differently.

8. Using a solution-focused approach, identify the techniques reviewed in this chapter that you believe will be appropriate for use with the James family described in exercise #6. Be specific in describing the way in which you would implement the strategies.

9. Termination and evaluation are critical aspects of the social work intervention. Conduct a review of the research to identify evidence-based practice skills and practices that are applicable for facilitating an effective termination and evaluation with a family. Share your findings in class and compare the results of your review.

REFERENCES

350.org. (n.d.). *Homepage*. Retrieved form 350.org

Abramowitz, M. (2005). The largely untold story of welfare reform and the human services. *Social Work, 50*(2), 175–186.

Abramson, J. (2009). Interdisciplinary team practice. In A.R. Roberts, *Social workers' desk reference* (2nd ed.) (pp. 44–50). New York: Oxford Press.

Alissi, A. S. (2009). United States. In A. Gitterman & R. Salmon, *Encyclopedia of social work with groups* (pp. 6–12). New York: Routledge.

American Red Cross. (n.d.). *Red Cross continues to help Joplin recovery from 2011 tornado*. Retrieved from http://www.redcross.org/news/article/Red-Cross-Continues-to-Help-Joplin-Recover-from–2011-Tornado

Amato, P. R., Booth, A., Johnson, D. R., & Rogers, S. J. (2009). *Alone together: How marriage in America is changing*. Cambridge, MA: Harvard University Press.

Anderson, K. M. (2013). Assessing strengths. Identifying acts of resistance to violence and oppression. In D. Saleebey (Ed.), *The strengths perspective in social work practice* (6th ed.) (pp. 182–202). Boston: Pearson.

AP Worldstream. (2010, July 15). *Planned US mosque draws opponents, supporters*. Retrieved from highbeam.com/doc/1A1-D9GV6I8G0.html

Arrington, P. (2008). *Stress at work: How do social workers cope? NASW Membership Workforce Study*. Washington, DC: National Association of Social Workers.

Atwood, J. D. & Genovese, F. (2006). *Therapy with single parents. A social constructionist approach*. New York: The Haworth Press.

Austin, M. J. & Solomon, J. R. (2009). Managing the planning process. In R. J. Patti (Ed.), *The handbook of human services management* (pp. 321–337). Thousand Oaks, CA: Sage Publications.

Banyard, V. L., Plante, E. G., & Moynihan, M. M. (2007). Bystander education: Bringing a broader community perspective to sexual violence prevention. *Journal of Community Psychology, 32*(1), 61–79.

Barker, R. L. (2003). *The social work dictionary* (5th ed.). Washington, DC: NASW Press.

Barsky, A. E. (2010). *Ethics and values in social work*. New York: Oxford University Press.

Barth, R. P. (2008). Adoption. In T. Mizrahi & L. E. Davis, *Encyclopedia of social work* (20th ed.) (pp. 1:33–44). Washington, DC and New York: NASW Press and Oxford University Press.

Beaulaurier, R. L. & Taylor, S. H. (2007). Social work practice with people with disabilities in the era of disability rights. In A. E. Dell Orto & P. W. Power, (Eds.), *The psychological and social impact of illness and disability* (5th ed.) (pp. 53–74). New York: Springer Publishing.

Beck, J. S. (2011). *Cognitive therapy for challenging problems: What to do when the basics don't work*. New York: Guilford Press

Benard, B. & Truebridge, S. (2013). A shift in thinking. Influencing social workers' beliefs about individual and family resilience in an effort to enhance well-being and success for all. In D. Saleebey, *The strengths perspective in social work practice* (6th ed.) (pp. 203–220). Boston: Allyn & Bacon.

Berg, R. D., Landreth, G. L., & Fall, K. A. (2013). *Group counseling concepts and procedures* (5th ed.). New York: Routledge.

Berg-Weger, M., Rubio, D. M., & Tebb, S. (2000). The caregiver well-being scale revisited. *Health and Social Work, 25*(4), 255–263.

Berg-Weger, M. (2013). *Social work and social welfare. An invitation*. New York: Routledge.

Berman-Rossi, T. & Kelly, T.B. (2003). *Group composition, diversity, the skills of the social worker, and group development*. Presented at the Council for Social Work Education, Annual Meeting, Atlanta, February.

Biblarz, T. J. & Stacey, J. (2010). How does the gender of parents matter? *Journal of Marriage and Family 72*, 3–22.

Bittman, M. (2011, May 17). Imaging Detroit. *New York Times*. Retrieved from http://opinionator.blogs.nytimes.com/2011/05/17/imagining-detroit/

Bloom, M., Fischer, J., & Orme, J. G. (2009). *Evaluating practice: Guidelines for the accountable professional* (6th ed.). Boston: Allyn & Bacon.

Blunsdon, B. & Davern, M. (2007). Measuring wellness through interdisciplinary community development: Linking the physical, economic and social environment. *Journal of Community Practice, 15* (1/2), 217–238.

Bobo, K., Kendall, J., & Max, S. (2010). *Organizing for social change*. Santa Ana, CA: The Forum Press.

Boland-Prom, K. W. (2009). Results of a national study of social workers sanctioned by state licensing boards. *Social Work, 54*(4): 351–360.

Bowden, V. R. & Greenberg, C. S. (2010). *Children and their families: The continuum of care*. Philadelphia: Wolters Kluwer/Lippincott Williams and Wilkins.

Bowlby, J. (1969). *Attachment and loss: Vol. 1, Attachment*. New York: Basic Books.

Bowlby, J. (1982). *Attachment and loss: Vol. 1, Attachment* (2nd ed.). New York: Basic Books.

Boyes-Watson, C. (2005). Seeds of change: Using peacemaking circles to build a village for every child. *Child Welfare, 84*(2), 191–208.

Briar-Lawson, K. & Naccarato, T. (2008). Family Services. In T. Mizrahi & L. E. Davis, *Encyclopedia of social work* (20th ed.) (pp. 2:206–212). Washington, DC and New York: NASW Press and Oxford University Press.

Brager, G., Specht, H., & Torczyner, J. (1987). *Community organizing*. New York: Columbia University Press.

Breton, M. (2004). An empowerment perspective. In C. D. Garvin, L. M. Gutierrez, & M. J. Galinsky (Eds.), *Handbook of social work with groups* (pp. 58–75). New York: Guilford Press.

Brill, C. (1998). The new NASW code of ethics can be your ally: Part I. *Focus Newsletter*. Retrieved from http://www.naswma.org/displaycommon.cfm?an=1&subarticlenbr=96

Briskman, L. & Noble, C. (1999). Social work ethics: Embracing diversity? In J. Fook & B. Pease (Eds.), *Transforming social work practice: Postmodern critical perspectives* (pp. 57–69). London: Routledge.

Brody, K. & Gadling-Cole, C. (2008). Family group conferencing with African-American families. In C. Waites (Ed.), *Social work practice with African American families: An intergenerational perspective*. New York: Routledge.

Brookings Institution. (2011, September). *An update to "Simulating the effect of the 'Great Recession' on poverty"*. Retrieved from http://www.brookings.edu/research/reports/2011/09/13-recession-poverty-monea-sawhill

Broussard, C. A., Joseph, A. L., & Thompson, M. (2012). Stressors and coping strategies used by single mothers living in poverty. *Affilia, 27*, 190–204.

Brown, N. W. (2013). *Creative activities for group therapy*. New York: Routledge.

Brown, R., & Hauser, C. (2012, August 10). After a struggle, Mosque opens in Tennessee. *New York Times*. Retrieved from http://www.nytimes.com/2012/08/11/us/islamic-center-of-murfreesboro-opens-in-tennessee.html

Brueggemann, W. G. (2005). *The practice of macro social work*. Belmont, CA: Thomson Higher Education.

Brueggemann, W. G. (2013). History and context for community practice in North America. In M. Weil's (Ed.), *The handbook of community practice* (pp. 27–46). Thousand Oaks, CA: Sage.

Buck, H. G., Overcash, J. & McMillan, S. C. (2009). The geriatric cancer experience at the end of life: Testing an adapted model. *Oncology Nursing Forum, 36*(6), 664–673.

Bullock, K. (2005). Grandfathers and the impact of raising grandchildren. *Journal of Sociology & Social Welfare, 32*(1), 43–59.

Burry, C. (2002). Working with potentially violent clients in their homes: What child welfare professionals need to know. *Clinical Supervisor, 21*(1), 145–153.

Butler, C. (2009). Sexual and gender minority therapy and systemic practice. *Journal of Family Therapy, 31*(4), 338–358.

Butler, A. C., Chapman, J. E., Forman, E. M., & Beck, A. T. (2006). The empirical status of cognitive-behavioral therapy: A review of meta-analyses. *Clinical Psychology Review, 26*, 17–31.

Campbell, D. (2000). *The socially constructed organization.* London: Karnac Books.

Carmeli, A. & Freund, A. (2009). Linking perceived external prestige and intentions to leave the organization: The mediating role of job satisfaction. *Journal of Social Service Research, 35*(3), 236–250.

Carter, L. & Matthieu, M. (2010). *Developing skills in the evidence based practice process.* St. Louis, MO: Washington University in St. Louis. Presented March 12, 2010.

Casa de Esperanza. (n.d.). *History.* Retrieved from https://www.casadeesperanza.org

CHANGE, Inc. Community Action Agency. (n.d.). *County needs assessment.* Retrieved from http://www.virtualcap.org/downloads/VC/US_NA_Examples_CHANGE_Inc_Community_Needs_Survey.pdf

Chaskin, R., Brown, P., Venkatsh, S., & Vidal, A. (2009). *Building community capacity.* New York: Aldine de Gruyter.

Chaskin, R. J. (2010). The Chicago School: A context for youth intervention, research and development. In R. J. Chaskin (Ed.), *Youth gangs and community intervention: Research, practice, and evidence* (pp. 3–23). New York: Columbia University Press.

Chaskin, R. J. (2013). Theories of community. In M. Weil's (Ed.), *The handbook of community practice* (pp. 105–121). Thousand Oaks, CA: Sage.

Chen, E. C., Budianto, L., & Wong, K. (2011). Professional school counselors as social justice advocates for undocumented immigrant students in group work. In A. A. Singh & C. F. Salazar, *Social justice in group work. Practical interventions for change* (pp. 88–94). New York: Routledge.

Christiensen, D. N., Todahl, J., & Barrett, W. C. (1999). *Solution-based casework: An introduction to clinical and case management skills in casework practice.* New York: De Gruyter.

Chu, W. C. K. & Tsui, M. (2008). The nature of practice wisdom in social work revisited. *International Social Work, 51*(1), 47–54.

Chun-Chow, J. & Austin, M. J. (2008). The culturally responsive social service agency: The application of an evolving definition to a case study. *Administration in Social Work, 32*(4), 39–64.

City of Seattle, Washington. (2012). *A community assessment of need for housing and services for homeless individuals and families in the Lake City neighborhood.* Retrieved from http://seattle.gov/realestate/pdfs/Needs_Assessment_data_report.pdf

Clowes, L. (2005). *Crossing cultures in systems of care.* Presented at the 1st New England LEND Conference, Burlington, VT.

Cnaan, R. A. & Rothman, J. (2008). Capacity development and the building of community. In J. Rothman, J. Erlich, & J. Tropman (Eds.), *Strategies of community intervention* (7th ed.) (pp. 243–262). Peosta, Iowa: Eddie Bowers Publishing Co., Inc.

CNN Wire Staff. (2012, August 6). *Missouri mosque destroyed in second fire in a month.* Retrieved from http://articles.cnn.com/2012-08-06/us/us_missouri-mosque-burned_1_vandalism-and-anti-muslim-sentiment-mosque-islamic-center

Colangelo, L. L. (2009, July 12). Queens one of the most diverse places on earth, new figures show. *New York Daily News.* Retrieved from http://articles.nydailynews.com/2009-07-12/local/17929058_1_dominicans-hispanic-queens

Collins, D., Jordan, C., & Coleman, H. (2013). *An introduction to family social work* (4th ed.). Belmont, CA: Brooks/Cole.

Collins, K. S. & Lazzari, M. M. (2009). Co-leadership. In A. Gitterman & R. Salmon, *Encyclopedia of social work with groups* (pp. 299–302). New York: Routledge.

Comer, E. & Meier, A. (2011). Using evidence-based practice and intervention research with treatment groups for populations at risk. In G. L. Greif & P. H. Ephross (Eds.), *Group work with populations*

Conan, N. (2012, February 15). Providing therapy across different cultures. *Talk of the Nation, NPR*. Retrieved from http://www.npr.org/2012/02/15/146936181/providing-therapy-across-different-cultures

Congress, E. P. (2004). Cultural and ethical issues in working with culturally diverse patients and their families: The use of the culturagram to promote cultural competent practice in health care settings. In A. Metteri, T. Krôger, A. Pohjola, & P. Rauhala (Eds.), *Social work visions from around the globe* (pp. 249–262). Binghampton, NY: Haworth.

Congress, E. P. (2009). The culturagram. In A. R. Roberts, *Social workers' desk reference* (2nd ed.) (pp. 969–975). New York: Oxford Press.

Cooney, K. (2006). The institutional and technical structuring of nonprofit ventures: Case study of a U.S. hybrid organization caught between two fields. *Voluntas: International Journal of Voluntary and Nonprofit Organizations, 17*(2), 143–161.

Corcoran, J. (2009). Using standardized tests and instruments in family assessments. In A. R. Roberts, *Social workers' desk reference* (2nd ed.) (pp. 390–394). New York: Oxford Press.

Cordova, T. L. (2011). Community-based research and participatory change: A strategic multi method community impact assessment. *Journal of Community Practice, 19*(1), 29–47.

Council on Social Work Education (CSWE). (2012). *Educational policy and accreditation standards*. Washington, DC: Author.

Cox, C. (2008). Empowerment as an intervention with grandparent caregivers. *Journal of Intergenerational Relationships, 6*(4), 465–477.

Cox, D. & Pawar, M. (2013). *International social work: Issues, strategies, and programs*. Thousand Oaks, CA: Sage Publications, Inc.

Cross, T. L., Bazron, B. J., Dennis, K. W., & Isaacs, M. R. (1989). *Toward a culturally competent system of care*. Washington, DC: Georgetown University Development Center.

Cumming-Bruce, N. (2012, September 12). U.N. says Syrian refugee numbers are surging. *New York Times*. Retrieved from http://www.nytimes.com/2012/09/28/world/middleeast/un-says-syrian-refugee-numbers-are-surging.html?_r=0

Cummings, T. G. & Worley, C. G. (2009). *Organizational development and change*. Mason, OH: South-Western Cengage Learning.

DeFrain, J. & Asay, S. M. (2007a). Epilogue: A strengths-based conceptual framework for understanding families worldwide. In J. DeFrain & S. M. Asay, *Strong families around the world: Strengths-based research and perspectives* (pp. 447–466). New York: Haworth Press, Inc.

DeFrain, J. & Asay, S. M. (2007b). Family strengths and challenges in the USA. In J. DeFrain & S. M. Asay, *Strong families around the world: Strengths-based research and perspectives* (pp. 281–307). New York: Haworth Press, Inc.

De Jong, P. (2008). Interviewing. In T. Mizrahi & L. E. Davis (Eds.), *Encyclopedia of social work* (20th ed.) (pp. 2:539–542). Washington, DC and New York: NASW Press and Oxford University Press.

DeJong, P. (2009). Solution-focused therapy. In A. R. Roberts, *Social workers' desk reference* (2nd ed.) (pp. 253–258). New York: Oxford Press.

De Jong, P. & Cronkright, A. (2011). Learning solution-focused interviewing skills: BSW student voices. *Journal of Teaching in Social work, 31*, 21–37.

De Jong, P., & Berg, I. K. (2013). *Interviewing for solutions* (4th ed.). Belmont, CA: Brooks/Cole.

deShazer, S. (1984). The death of resistance. *Family Process, 23*, 11–21.

Detroit Area Food Council. (n.d.). *Homepage*. Retrieved from http://www.detroitfoodpolicycouncil.net/

Diller, J. (2011). *Cultural diversity: A primer for the human services*. Belmont, CA: Thompson/Brooks/Cole.

Dolgoff, R., Loewenberg, F. M., & Harrington, D. (2012). *Ethical decisions for social work practice*. Itasca, IL: F. E. Peacock.

Dombo, E. A. (2011). Rape: When professional values place vulnerable clients at risk. In J. C. Rothman (Ed.), *From the front lines: Student cases in social work ethics* (3nd ed.) (pp. 185–188). Boston: Allyn & Bacon.

Donaldson, L.P. (2004). Toward validating the therapeutic benefits of empowerment-oriented social action groups. *Social Work with Groups, 27*(2), 159–175.

Dressler, L. (2006). *Consensus through conversation.* San Francisco, CA: Berrett-Koehler Publishers, Inc.

Drumm, K. (2006). The essential power of group work. *Social Work with Groups, 29*(2/3), 17–31.

Dudley, J. R. (2009). *Social work evaluation: Enhancing what we do.* Chicago, IL: Lyceum Books, Inc.

Dunst, C. J., Trivette, C. M., & Deal, A. G. (2003). *Enabling and empowering families: Principles and guidelines for practice.* Newton: MA: Brookline Books.

Dunst, C. J. & Trivette, C. M. (2009). Capacity-building family-systems intervention practices. *Journal of Family Social Work, 12*, 119–143.

Dybicz, P. (2012). The hero(ine) on a journey: A postmodern conceptual framework for social work practice. *Journal of Social Work Education, 48*(2), 267–283.

Earley, L. & Cushway, D. (2002). The parentified child. *Clinical Child Psychology & Psychiatry, 7*(2), 163–178.

Early, T. J. (2001). Measures for practice with families from a strengths perspective. *Families in Society, 82*(3), 225–232.

East, J. F., Manning, S. F., & Parsons, R. J. (2002). Social work empowerment agenda and group work: A workshop. In S. Henry, J. East, & C. Schmitz (Eds.), *Social work with groups: Mining the gold* (pp. 41–53). New York: Haworth Press.

Eaton, Y. M. & Roberts, A. R. (2009). Frontline crisis intervention. Step-by-step practice guidelines with case applications. In A. R. Roberts, *Social workers' desk reference* (2nd ed.) (pp.207–215). New York: Oxford Press.

Edwards, R. L. & Yankey, J. A. (2006). *Effectively managing nonprofit organizations.* Washington DC: NASW Press.

Ephross, P. H. (2011). Social work with groups: Practice principles. In G. L. Grief & P. H. Ephross, *Group work with populations at risk* (3rd ed.) (pp. 3–14). New York: Oxford University Press.

Ephross, P. H. & Greif, G. L. (2009). Group process and group work techniques. In A. R. Roberts, *Social workers' desk reference* (2nd ed.) (pp. 679–685). New York: Oxford Press.

Epstein, M. H. (2004). *Behavioral and emotional rating scale: A strengths-based approach to assessment* (2nd ed.). Austin, TX: PRO-ED.

Epstein, M. H., Mooney, P., Ryser, G., & Pierce, C. D. (2004). Validity and reliability of the Behavioral and Emotional Rating Scale (2nd ed.): Youth Rating Scale. *Research on Social Work Practice, 14*(50), 358–367.

Evans, S. D., Hanlin, C. E., & Prillehensky, I. (2007). Blending Ameliorative and transformative approaches in human service organizations: A case study. *Journal of Community Psychology, 35*(3), 329–346.

Finlayson, M. L. & Cho, C. C. (2011). A profile of support group use and need among middle-aged and older adults with multiple sclerosis. *Journal of Gerontological Social Work, 54*, 475–493.

Fischer, J. & Orme, J. G. (2008). Single-System Designs. In T. Mizrahi & L. E. Davis, *Encyclopedia of social work* (20th ed.) (pp. 4:32–34). Washington, DC and New York: NASW Press and Oxford University Press.

Fleming, J. (2009). Social action. In A. Gitterman & R. Salmon, *Encyclopedia of social work with groups* (pp. 275–277). New York: Routledge.

Freedenthal, S. (2008). Suicide. In T. Mizrahi & L. E. Davis, *Encyclopedia of social work* (20th ed.) (pp. 4:181–186). Washington, DC and New York: NASW Press and Oxford University Press.

Frey, B. A. (1990). A framework for promoting organizational change. *Families in Society, 71*(3), 142–147.

Fortune, A. E. (2009). Terminating with clients. In A. R. Roberts, *Social workers' desk reference* (2nd ed.) (pp. 627–631). New York: Oxford Press.

Franklin, C., Jordan, C., & Hopson, L. M. (2009). Effective couple and family treatment. In A. R. Roberts, *Social workers' desk reference* (2nd ed.) (pp. 433–442). New York: Oxford Press.

Freire, P. (1973). *Education for critical consciousness.* New York: Seabury Press.

Fuller-Thomson, E. & Minkler, M. (2005). American Indian/Alaskan Native grandparents raising grandchildren: Findings from the Census 2000 Supplementary Survey. *Social Work, 50*(2), 131–139.

Furman, R., Rowan, D., & Bender, K. (2009). *An experiential approach to group work*. Chicago, IL: Lyceum Books, Inc.

Gamble, D. N. & Hoff, M. D. (2013). Sustainable community development. In M. Weil (Ed.), *The handbook of community practice* (pp. 215–232). Thousand Oaks, CA: Sage Publications.

Gambrill, E. (2013). *Social work practice: A critical thinkers guide*. New York: Oxford University Press.

Garkovich, L. E. (2011). A historical view of community development. In J. W. Robinson, Jr. & G. P. Green (Eds.), *Introduction to community development* (pp. 11–34). Los Angeles, CA: Sage Publications, Inc.

Garland, J., Jones, H., & Kolodny, R. (1965). A model for stages of development in social work groups. In S. Bernstein (Ed.), *Explorations in group work: Essays in theory and practice* (pp. 12–53). Boston: Boston University School of Social Work.

Garvin, C. D. & Galinsky, M. J. (2008). Groups. In T. Mizrahi & L. E. Davis, *Encyclopedia of social work* (20th ed.) (pp. 2:287–298). Washington, DC and New York: NASW Press and Oxford University Press.

Garvin, C. (2009). Developing goals. In A. R. Roberts, *Social workers' desk reference* (2nd ed.) (pp. 521–526). New York: Oxford Press.

Gilligan, C. (1993). *In a different voice*. Cambridge, MA: Harvard University Press.

Gitterman, A. & Germain, C. B. (2008a). Ecological framework. In T. Mizrahi & L. E. Davis (Eds.), *Encyclopedia of social work* (20th ed.) (pp. 2:97–102). Washington, D.C. and New York: NASW Press and Oxford University Press.

Gitterman, A. & Germain, C. B. (2008b). *The life model of social work practice: Advances in theory and practice* (3rd ed). New York: Columbia Press.

Gonzales, J. (2009). Prefamily counseling: Working with blended families. *Journal of Divorce & Remarriage, 50,* 148–157.

Goode, E., & Kovalski, S. F. (2012, August 12). Wisconsin killer fed and was fueled by hate-driven music. *New York Times*. Retrieved from http://www.nytimes.com/2012/08/07/us/army-veteran-identified-as-suspect-in-wisconsin-shooting.html?pagewanted=allhttp://www.nytimes.com/2012/08/07/us/army-veteran-identified-as-suspect-in-wisconsin-shooting.html?_r=1&pagewanted=all

Goodman, H. (2004). Elderly parents of adults with severe mental illness: Group work interventions. *Journal of Gerontological Social Work, 44*(1/2), 173–188.

Grace, K. S., McClellan, A., & Yankey, J. A. (2009). *The nonprofit board's role in mission, planning and evaluation*. Washington DC: BoardSource.

Grant, D. (2008). Clinical social work. In R. Mizrahi & L. E. Davis (Eds.), *Encyclopedia of Social Work. National Association of Social Workers and Oxford University Press, Inc. Encyclopedia of Social Work: (e-reference edition)*. Oxford University Press. Retrieved April 1, 2010 from http://www.oxford-naswsocialwork.com/entry?entry=t203.e63

Gray, M. (2011). Back to basics: A critique of the strengths perspective in social work. *Families in Society: The Journal of Contemporary Social Services, 92*(1), 5–11.

Gray, M., Coates, J., & Hetherington, T. (2012). *Environmental social work*. New York: Routledge.

Greene, G. J. & Lee, M. Y. (2011). *Solution-oriented social work practice*. New York: Oxford University Press.

Greenpeace. (n.d.). *About us*. Retrieved from http://www.greenpeace.org/usa/en/campaigns/

Gregg, B. (2012). *Human rights as a social construction*. New York: Oxford University Press.

Grieco E. M., Acosta, Y. D., de la Cruz, P., Gambino, C., Gryn, T., Larsen, L. J., Trevelyan, E. N., & Walters, N. P. (2012). *The foreign-born population in the United States: 2010*. Washington, D.C.: U.S. Census Bureau.

Grief, G. L. & Morris-Compton, D. (2011). Group work with urban African American parents in their neighborhood schools. In G. L. Grief & P. H. Ephross, *Group work with populations at risk* (3rd ed.) (pp. 385–398). New York: Oxford University Press.

Grief, G. & Ephross, P. H. (Eds.). (2011). *Group work with populations at risk* (3rd ed.). New York: Oxford University Press.

Hall, J. C. (2011). A narrative approach to group work with men who batter. *Social Work with Groups, 34,* 175–189.

Hamilton, B. E., Martin, J. A., & Ventura, S. J. (2012). Births: Preliminary data for 2011. *National vital statistics reports, 61*(5). Hyattsville, MD: National Center for Health Statistics.

Hanley, J. & Shragge, E. (2009). Organizing for immigrant rights. *Journal of Community Practice, 17*(1), 184–206.

Hardina, D. (2002). *Analytical skills for community organization practice.* New York: Columbia University Press.

Hardina, D., Middleton, J., Montana, S., & Simpson, R. A. (2007). *An empowering approach to managing social service organizations.* New York: Springer Publishing Company.

Hardina, D. (2012). *Interpersonal social work skills for community practice.* New York: Springer.

Hardy, K. V. & Laszloffy, T. A. (1995). The Cultural genogram: Key to training culturally competent family therapists. *Journal of Marital and Family Therapy, 21,* 227–237.

Hartman, A. & Laird, J. (1983). *Family-centered social work practice.* New York: Free Press.

Harrington, D. & Dolgoff, R. (2008). Hierarchies of ethical principles for ethical decision making in social work. *Ethics and Social Welfare, 2*(2), 183–196.

Hartman, A. (1994). *Reflection & controversy: Essays on social work.* Washington, DC: NASW Press.

Harvey, A. R. (2011). Group work with African American youth in the criminal justice system: A culturally competent model. In G. L. Grief & P. H. Ephross, *Group work with populations at risk* (3rd ed.) (pp. 264–282). New York: Oxford University Press.

Hasenfeld, Y. (2000). Social welfare administration and organizational theory. In R. J. Patti (Ed.) *The handbook of social welfare management* (pp. 89–112). Thousand Oaks, CA: Sage Publications.

Haynes, K. S. (1998). The one-hundred year debate: Social reform versus individual treatment. *Social Work, 43*(6), 501–511.

Haynes, K. S. & Mickelson, J. S. (2010). *Affecting change* (7th ed.). Boston: Pearson.

Hayslip, Jr., B. & Kaminski, P. L. (2005). Grandparents raising their grandchildren. A review of the literature and suggestions for practice. *The Gerontologist, 45*(2), 262–269.

Head, J. W. (2008). *Losing the global development war.* Leiden, The Netherlands: Martinus Nijhoff Publishers.

Healy, L. (2008). *International social work: Professional practice in an interdependent world.* New York: Oxford University Press.

Healy, L. M. & Hokenstad, T. M. C. (2008). International social work. In T. Mizrahi & L. E. Davis, *Encyclopedia of social work* (20th ed.) (pp. 2:482–488). Washington, DC and New York: NASW Press and Oxford University Press.

Healy, L. M. & Link, R. J. (2012). *Handbook of international social work: Human rights, development, and the global profession.* New York: Oxford.

Herbert, R. J., Gagnon, A. J., Rennick, J. E., & O'Loughlin, J. L. (2009). A systematic review of questionnaires measuring health-related empowerment. *Research and Theory for Nursing Practice: An International Journal, 23*(2), 107–132.

Hillier, A. & Culhane, D. (2013). GIS applications and administrative data to support community change. In M. Weil's (Ed.) *The handbook of community practice* (pp. 827–844). Thousand Oaks, CA: Sage.

Hodge, D. R. (2005a). Social work and the House of Islam: Orienting practitioners to the beliefs and values of Muslims in the United States. *Social Work, 50*(2), 162–173.

Hodge, D. R. (2005b). Spiritual life maps: A client centered pictorial instrument for spiritual assessment, planning, and intervention. *Social Work, 50*(1), 77–87.

Hohman, M. (2012). *Motivational interviewing in social work practice.* New York: The Guilford Press.

Holland, T. (2008). Organizations and governance. In T. Mizrahi & L. E. Davis (Eds.) *Encyclopedia of Social*

Work (pp. 3:329–333). Washington DC: NASW Press and Oxford University Press.

Holland, T. P. & Kilpatrick, A. C. (2009). An ecological system—social constructionism approach to family practice. In A. C. Kilpatrick & T. P. Holland, *Working with families. An integrative model by level of need* (5th ed.) (pp. 15–31). Boston: Pearson.

Hong, P. Y. & Song, I. H. (2010). Glocalization of social work practice: Global and local responses to globalization. *International Social Work, 53,* 656–670.

Hopson, L. M. & Wodarski, J. S. (2009). Guidelines and uses of rapid assessment instruments in managed care settings. In A. R. Roberts, *Social workers' desk reference* (2nd ed.) (pp. 400–405). New York: Oxford Press.

Hudson, R. E. (2009). Empowerment model. In A. Gitterman & R. Salmon (Eds.), *Encyclopedia of social work with groups* (pp. 47–50). New York: Routledge.

Hull, G. H. & Mather, J. (2006). *Understanding generalist practice with families.* Belmont, CA: Thomson Brooks/Cole.

Human Rights Education Association. (n.d.). *Simplified version of the University Declaration of Human Rights.* Retrieved from http://www.hrea.org/index.php?base_id=104&language_id=1&erc_doc_id=5211&category_id=24&category_type=q3&group=

Ife, J. (2000). Localized needs and a globalized economy. *Social Work and Globalization (Special Issue), Canadian Social Work, 2*(1), 50–64.

"Indicators for the Achievement of the NASW Standards for Cultural Competence in Social Work Practice." (2008). In T. Mizrahi & L. E. Davis (Eds.), *Encyclopedia of social work. National Association of Social Workers and Oxford University Press, Inc. (e-reference edition).* Retrieved from http://www.oxford-naswsocialwork.com/entry?entry=t203.e427

International Association for the Advancement of Social Work with Groups, Inc. (IASWG) (2006). *Standards for social work practice with groups* (2nd ed.). Alexandria, VA: AASWG, Inc.

International Federation of Social Workers (IFSW). (2012). *Ethics in social work: Statement of principles.* Retrieved from http://ifsw.org/policies/statement-of-ethical-principles/

International Institute St. Louis. (2013). *Economic development.* Retrieved from http://www.iistl.org/edintro.html

Interorganizational Committee on Guidelines and Principles for Social Impact Assessment (IOCPG). (2003). Principles and guidelines for social impact assessment in the United States. *Impact Assessment and Project Appraisal, 21*(3), 231–250.

Jackson, K. F. & Samuels, G. M. (2011). Multiracial competence in social work: Recommendation for culturally attuned work with multiracial people. *Social Work, 56*(3), 235–245.

Janzen, C., Harris, O., Jordan, C., & Franklin, C. (2006). *Family treatment: Evidence-based practice with populations at risk.* Belmont, CA: Thomson Brooks/Cole.

Jayartne, S., Croxton, T. A., & Mattison, D. (2004). A national survey of violence in the practice of social work. *Families in Society, 85*(4), 445–453.

Jenson, J. M. & Howard, M. O. (2008). Evidence-based practice. In T. Mizrahi & L. E. Davis (Eds.), *Encyclopedia of social work* (20th ed.) (pp.2:158–165). Washington, DC and New York: NASW Press and Oxford University Press.

Johnson, M. & Austin, M. J. (2006). Evidence-based practice in the social services: Implications for organizational change. *Journal of Evidence-Based Social Work, 5*(1/2), 239–269.

Joplin Area CART. (n.d.). Homepage. Retrieved from http://joplinareacart.com/

Jordan, C. (2008). Assessment. In T. Mizrahi & L. E. Davis, *Encyclopedia of social work* (20th ed.) (pp 1:178–180). Washington, DC and New York: NASW Press and Oxford University Press.

Jordan, C. & Franklin, C. (2009). Treatment planning with families: An evidence-based approach. In A. R. Roberts, *Social workers' desk reference* (2nd ed.) (pp. 429–432). New York: Oxford Press.

Jung, M. (1996). Family-centered practice with single parent families. *Families in Society, 77*(9), 583–590.

Kagle, J. D. (2008). Recording. In T Mizrahi & L. E. Davis, *Encyclopedia of social work* (20th ed.) (pp. 3:497–498). Washington, DC and New York: NASW Press and Oxford University Press.

Kagle, J. D. & Kopels, S. (2008). *Social work records* (3rd ed.). Long Grove, IL: Waveland Press, Inc.

Kalyanpur, M. & Harry, B. (1999). *Culture in special education*. Baltimore: Paul H. Brookes.

Kasvin, N. & Tashayeva, A. (2004). Community organizing to address domestic violence in immigrant populations in the U.S.A. *Journal of Religion and Abuse*, 6(3/4), 109–112.

Kelley, P. (2008). Narratives. In T. Mizrahi & L. E. Davis (Eds.), *Encyclopedia of social work* (20th ed.) (pp. 3:291–292). Washington, DC and New York: NASW Press and Oxford University Press.

Kelley, P. (2009). Narrative therapy. In A. R. Roberts (Ed.), *Social workers' desk reference* (2nd ed.) (pp. 273–277). New York: Oxford Press.

Kelly, T. B. & Berman-Rossi, T. (1999). Advancing stages of group development theory: The case of institutionalized older persons. *Social Work with Groups*, 22(2/3), 119–138.

Kelly, M. S., Kim, J. S., & Franklin, C. (2008). *Solution-focused brief therapy in schools. A 360-degree view of research and practice*. New York: Oxford University Press.

Kettner, P. M. (2002). *Achieving excellence in the management of human service organizations*. Boston: Allyn & Bacon.

Kilpatrick, A. & Cleveland, P. (1993). *Unpublished course materials*. University of Georgia School of Social Work.

Kilpatrick, A. C. (2009). Levels of family need. In A. C. Kilpatrick & T. P. Holland, *Working with families*. An integrative model by level of need (5th ed.) (pp. 2–14). Boston: Pearson.

Kim, J. S. (2008a). Strengths perspective. In T. Mizrahi & L. E. Davis (Eds.), *Encyclopedia of social work* (20th ed.) (pp. 4:177–181). Washington, DC and New York: NASW Press and Oxford University Press.

Kim, J. S. (2008b). Examining the effectiveness of solution-focused brief therapy: A meta-analysis. *Research on Social Work Practice*, 18, 107–116.

Kim, H., Ji, J, & Kao, D. (2011). Burnout and physical health among social workers: A three-year longitudinal study. *Social Work*, 56(3), 258–268.

Kisthardt, W. E. (2013). Integrating the core competencies in strengths-based, person-centered practice. In D. Saleebey (Ed.), *The strengths perspective in social work practice* (6th ed.), pp. 53–78. Boston: Allyn & Bacon.

Kleinkauf, C. (1981, July). A guide to giving legislative testimony. *Social Work*, 297–303.

Kohli, H. K., Huber, R., & Faul, A. C. (2010). Historical and theoretical development of culturally competent social work practice. *Journal of Teaching in Social Work*, 30, 252–271.

Kong, E. (2007). The development of strategic management in the non-profit context: Intellectual capital in the social service non-profit organization. *International Journal of Management Reviews*, 10(3), 281–299.

Koop, J. J. (2009). Solution-focused family interventions. In A. C. Kilpatrick & T. P. Holland, *Working with families*. An integrative model by level of need (5th ed.) (pp. 147–169). Boston: Pearson.

Koren, P. E., DeChillo, N., & Friesen, B. J. (1992). Measuring empowerment in families whose children have emotional disabilities: A brief questionnaire. *Rehabilitation Psychology*, 37, 305–310.

Kreider, R. M. & Ellis, R. (2011). Living arrangements of children: 2009. *Household Economic Studies. Current Population Reports*, P70–126. Washington, DC: U.S. Census Bureau.

Kretzmann, J. P. & McKnight, J. L. (1993). *Building communities from the inside out: A path toward finding and mobilizing a community's assets*. Evanston, IL: Center for Urban Affairs and Policy Research.

Kretzmann, J. P. & McKnight, J. L. (2005). *Discovering community power: A guide to mobilizing local assets and your organization's capacity*. Evanston, IL: Asset-Based Community Development (ABCD) Institute.

Kurland, R. & Salmon, R. (1998). *Teaching a methods course in social work with groups*. Alexandria, VA: Council on Social Work Education.

Kurland, R., Salmon, R., Bitel, M., Goodman, H., Ludwig, K., Newmann, E.W., & Sullivan, N. (2004). The survival of social group work: *A call to action*. *Social Work with Groups*, 27(1), 3–16.

Kurland, R. (2007). Debunking the "blood theory" of social work with groups: Groups workers are made and not born. *Social Work with Groups*, *31*(1), 11–24.

Larson, K. & McGuiston, C. (2012). Building capacity to improve Latino health in rural North Carolina: A case study in community-University engagement. *Journal of Community Engagement and Scholarship*, *5*(1), 14–23.

Lawler, J. & Bilson, A. (2004). Towards a more reflexive research aware practice: The influence and potential of professional and team culture. *Social Work & Social Sciences Review*, *11*(1), 52–69.

Leahy, M. M., O'Dwyer, M., & Ryan, F. (2012). Witnessing stories: Definitional ceremonies in narrative therapy with adults who stutter. *Journal of Fluency Disorders*, *37*, 234–241.

Lee, J. A. B. & Berman-Rossi, T. (1999). Empowering adolescent girls in foster care: A short-term group record. In C. W. LeCroy (Ed.), *Case studies in social work practice* (2nd ed.). Pacific Grove, CA: Brooks/Cole.

Lee, J. A. B. (2001). *The empowerment approach to social work practice* (2nd ed.). New York: Columbia University Press.

Lee, M. Y. (2009). Using the miracle question and scaling technique in clinical practice. In A. R. Roberts, *Social workers' desk reference* (2nd ed.) (pp. 594–600). New York: Oxford Press.

Lee, M. Y. & Greene, G. J. (2009). Using social constructivism in social work practice. In A. R. Roberts (Ed.), *Social workers' desk reference* (2nd ed.) (pp. 294–299). New York: Oxford Press.

Lesser, J. G., O'Neill, M. R., Burke, K. W., Scanlon, P., Hollis, K., & Miller, R. (2004). Women supporting women: A mutual aid group fosters new connections among women in midlife. *Social Work with Groups*, *27*(1), 75–88.

Levy, R. (2011). Core themes in a support group for spouses of breast cancer patients. *Social work with groups*, *34*, 141–157.

Lewin, K. (1951). *Field theory in social science*. New York: Harper and Row.

Library of Congress. (1996). *Defense of Marriage Act*. Retrieved from: http://www.govtrack.us/congress/bills/104/hr3396#summary/libraryofcongress

Locke, B., Garrison, R., & Winship, J. (1998). *Generalist social work practice: Context, story, and partnerships*. Pacific Grove, CA: Brooks/Cole.

Logan, S. L. M., Rasheed, M. N., & Rasheed, J. M. (2008). Family. In T. Mizrahi & L. E. Davis (Eds.), *Encyclopedia of social work* (20th ed.) (pp. 2:175–182). Washington, DC and New York: NASW Press and Oxford University Press.

López, L. M. & Vargas, E. M. (2011). En dos culturas: Group work with Latino Immigrants and Refugees. In G. L. Greif & P. H. Ephross, *Group work with populations at risk* (3rd ed.) (pp. 144–145). New York: Oxford University Press.

Lotze, G. M., Bellin, M. H., & Oswald, D. P. (2010). Family-centered care for children with special health care needs: Are we moving forward? *Journal of Family Social Work*, *13*, 100–113.

Lovell, M. L., Helfgott, J. B., & Lawrence, C. (2002). Citizens, victims, and offenders restoring justice: A prison-based group work program bridging the divide. In S. Henry, J. East, & C. Schmitz (Eds.) *Social work with groups: Mining the gold* (pp. 75–88). New York: Haworth Press.

Lowery, C. T. & Mattaini, M. (2001). Shared power in social work: A Native American perspective of change. In H. E. Briggs and K. Corcoran (Eds.), *Social work practice: Treating common client problems* (pp. 109–124). Chicago: Lyceum Books.

Lum, D. (2004). *Social work practice and people of color: A process stage* approach (5th ed.). Belmont, CA: Thomson Brooks/Cole.

Lum, D. (2008). Culturally competent practice. In T. Mizrahi & L. E. Davis, *Encyclopedia of social work* (20th ed.) (pp. 2:497–502). Washington, DC and New York: NASW Press and Oxford University Press.

Lum, D. (2011). *Culturally competent practice: A framework for understanding diverse groups and justice issues*. Belmont, CA: Brooks/Cole Cengage Learning.

Lundahl, B. W., Kunz, C., Brownell, C., Tollefson, D., & Burke, B. L. (2010). A meta-analysis of motivational interviewing: Twenty-five years of empirical studies. *Research on Social Work Practice*, *20*(2), 137–160.

Lundy, C. (2011). *Social work, social justice, and human rights: A structural approach to practice*. North York, Ont.: University of Toronto Press.

Macgowan, M. J. (1997). A measure of engagement for social group work: The Groupwork engagement measure (GEM). *Journal of Social Service Research, 23*, 17–37.

Macgowan, M. J. (2000). Evaluation of a measure of engagement for group work. *Research on Social Work Practice, 10*, 348–361.

Macgowan, M. J. & Levenson, J. S. (2003). Psychometrics of the Group Engagement Measure with male sex offenders. *Small Group Research, 34*(2), 155–169.

Macgowan, M. J. & Newman, F. L. (2005). Factor structure of the Group Engagement Measure. *Social Work Research, 29*(2), 107–118.

Macgowan, M. J. (2008). Group dynamics. In T. Mizrahi & L.E. Davis, *Encyclopedia of social work* (20th ed.) (pp. 2:279–287). Washington, DC and New York: NASW Press and Oxford University Press.

Macgowan, M. J. (2009a). Evidence-based group work. In A. Gitterman & R. Salmon (Eds.), *Encyclopedia of social work with groups* (pp. 131–136). New York: Routledge.

Macgowan, M. J. (2009b). Measurement. In A. Gitterman & R. Salmon (Eds.), *Encyclopedia of social work with groups* (pp. 142–147). New York: Routledge.

Mackelprang, R. W. & Salsgiver, R. O. (2009). *Disability: A diversity model approach in human service practice*. Chicago, IL: Lyceum Books, Inc.

MacNeill, V. (2009). Forming partnerships with parents from a community development perspective: Lessons learnt from Sure Start. *Health and Social Care in the Community, 17*(6), 659–665.

Madigan, S. (2011). *Narrative therapy*. Washington DC: The American Psychological Association.

Manning, T. (2012). The art of successful persuasion: Seven skills you need to get your point across effectively. *Industrial and Commercial Training, 44*(3), 150–158.

Manor, O. (2008). Systemic approach. In A. Gitterman & R. Salmon (Eds.) *Encyclopedia of social work with groups* (pp. 99–101). New York: Routledge.

Maramaldi, P., Berkman, B., & Barusch, A. (2005). Assessment and the ubiquity of culture: Threats to validity in measures of health-related quality of life. *Health & Social Work, 30*(1), 27–36.

Marguerite Casey Foundation. (2012). *Organizational capacity assessment tool*. Retrieved from http://caseygrants.org/resources/org-capacity-assessment/

Mason, J. L. (1995). *Cultural competence self-assessment questionnaire: A manual for users*. Portland, OR: Portland State University, Research and Training Center on Family Support and Children's Mental Health. Retrieved from http://www.racialequity-tools.org/resourcefiles/mason.pdf

McCullough-Chavis, A. & Waites, C. (2008). Genograms with African American families: Considering cultural context. In C. Waites (Ed.), *Social work practice with African-American families: An intergenerational perspective* (pp. 35–54). New York: Routledge.

McGoldrick, M., Gerson, R., & Petry, S. (2008). *Genograms assessment and intervention* (3rd ed.). New York: W.W. Norton & Company.

McGoldrick, M. (2009). Using genograms to map family patterns. In A. R. Roberts, *Social workers' desk reference* (2nd ed.) (pp. 409–423). New York: Oxford Press.

McGowan, B. G. & Walsh, E. M. (2012). Writing in family and child welfare. In W. Green and B. L. Simon, *The Columbia guide to social work writing* (pp. 215–231). NY: Columbia University Press.

McKnight, J. L., & Block, P. (2010). *The abundant community: Awakening the power of families and neighborhoods*. San Francisco, CA: Berrett-Koehler Publishers.

McKnight, J. L. & Kretzmann, J. P. (1996). *Mapping community capacity*. Evanston, IL: Institute for Policy Research. Retrieved from http://www.racialequitytools.org/resourcefiles/mcknight.pdf

McNutt, J. & Floersch, J. (2008). Social work practice. In T. Mizrahi & L. E. Davis (Eds.), *Encyclopedia of social work. National Association of Social Workers and the Oxford University Press, Inc.(e-reference edition)*. Retrieved from http://www.oxford-naswsocialwork.com/entry?entry=t203.e375-s2

McWhirter, P. T., Robbins, R., Vaughn, K., Youngbull, N., Burks, D., Willmon-Haque, S., Schuetz, S., Brandes, J. A., & Nael, A. Z. O. (2011). Honoring the ways of American Indian women: A group therapy intervention. In A. A. Singh & C. F. Salazar, *Social justice in group work. Practical interventions for change* (pp. 73–81). New York: Routledge.

Meyer, C. (1993). *Assessment in social work practice*. New York: Columbia University Press.

Middleman, R. R. & Wood, G. G. (1990). *Skills for direct practice in social work*. New York: Columbia University Press.

Miley, K. K., O'Melia, M. W., & DuBois, B. L. (2013). *Generalist social work practice: An empowering approach*. Boston: Pearson Allyn & Bacon.

Miller, L. (2011). *Counselling skills for social work*. London, Thousand Oaks, CA: Sage.

Miller, W. R. & Rollnick, S. (2013). *Motivational interviewing: Helping people change* (3rd ed.). New York: The Guilford Press.

Minieri, J. & Getsos, P. (2007). *Tools for radical democracy*. San Francisco, CA: Jossey Bass.

Minuchin, S. (1974). *Families & family therapy*. Cambridge, MA: Harvard University Press.

Minuchin, P., Colapinto, J., & Minuchin, S. (2007). *Working with families of the poor* (2nd ed.). New York: The Guilford Press.

Mondros, J. & Staples, L. (2008). Community organization. In T. Mizrahi & L. E. Davis (Eds.), *Encyclopedia of social work* (pp.1:387–398). Washington DC and New York: NASW Press and Oxford University Press.

Morgan, A. (2000). *What is narrative therapy?* Adelaide, South Australia: Dulwich Centre Publications.

Moxley, D. (2008). Interdisciplinarity. In T. Mizrahi & L. E. Davis, *Encyclopedia of social work* (20th ed.) Washington, DC and New York: NASW Press and Oxford University Press. (e-reference edition). Accessed August 17, 2010 from: http:// www.oxford-naswsocialwork.com/entry?entry?t203.e200

Mulroy, E. (2008). Community needs assessment. In T. Mizrahi & L. E. Davis (Eds.), *Encyclopedia of social work* (pp. 1:385–387). Washington DC and New York: NASW Press and Oxford University Press.

Musick, K. & Meier, A. (2009). Are both parents always better than one? Parental conflict and young adult well-being. *Rural New York Minute, 28*, 1.

Nakhaima, J. M. & Dicks, B. H. (2012). A role for the religious community in family counseling. *Families in Society Practice and Policy Focus*.

National Association of Area Agencies on Aging. (2012). *National aging services network*. Retrieved from http://www.n4a.org/about-n4a/join/

National Association of Cognitive-Behavioral Therapists. (2013). *Cognitive behavioral therapy*. Retrieved from: http://www.nacbt.org/whatiscbt.aspx

National Alliance on Mental Illness (NAMI). (2013). *Cognitive behavioral therapy (CBT)?* Retrieved from http://www.nami.org/

National Association of Social Workers (NASW). (1996). *Code of ethics*. Washington, DC: NASW.

National Association of Social Workers (NASW). (2001). *NASW standards for cultural competence in social work practice*. Washington D.C.: NASW. Retrieved from http://www.naswdc.org/practice/standards/NAswculturalstandards.pdf

National Association of Social Workers. (2007b). *Indicators for the achievement of the NASW Standards for Cultural Competence in Social Work Practice*. Washington, DC: NASW.

National Association of Social Workers. (2007a). *Children and families*. Retrieved January 18, 2013 from: http://www.socialworkers.org/pressroom/features/issue/children.asp

National Association of Social Workers (NASW). (2008). *Code of ethics*. Washington, DC: NASW.

National Association of Social Workers (NASW). (2012). *Practice*. Retrieved from http://www.social-workers.org/practice/

National Association of Social Workers-California Chapter. (2012). *Online continuing education*. Retrieved from http://www.socialworkweb.com/nasw/

National Association of Social Workers. (2012–2014a). *Cultural and linguistic competence in the social work profession. Social work speaks: National Association* of Social Workers policy statements 2012–2014. Washington, DC: NASW Press.

National Association of Social Workers. (2012–2014b). *Professional self-care and social work. Social work speaks: National Association of Social Workers policy statements 2009–2012.* Washington, DC: NASW Press.

National Association of Social Workers. (2013a). *Draft NASW Guidelines for Social Worker Safety in the Workplace.* Washington, D.C.: NASW.

National Association of Social Workers. (2013b). *Standards for social work case management.* Washington, D.C.: NASW.

National Association of Social Workers (NASW). (n.d.) *NASW WebEd.* Retrieved from http://www.naswwebed.org/

National Conference of State Legislatures (NCSL). (2012). *Defining marriage: Defense of marriage acts and same-sex marriage laws.* Retrieved from: http://www.ncsl.org/issues-research/human-services/same-sex-marriage-overview.aspx

Netting, F. E., Kettner, P. M., & McMurtry, S. L., & Thomas, L. (2011). *Social work macro practice.* Boston: Pearson Allyn & Bacon.

Newhill, C. E. (1995). Client violence toward social workers: A practice and policy concern. *Social Work, 40,* 631–636.

New York Times. (2012, October 16). *Income inequality.* Retrieved from http://topics.nytimes.com/top/reference/timestopics/subjects/i/income/income_inequality/index.html

New Zealand Association of Social Workers (NZASW). (1993). *Code of ethics.* Aoteora, NZ: NZASW.

Nichols, M. P. (2011). *The essentials of family therapy* (5th ed.). Boston: Pearson.

Nissly, J. A., Barak, M. E. M., & Levin, A. (2005). Stress, social support, and worker's intentions to leave their jobs in public child welfare. *Administration in Social Work, 29*(1), 79–100.

North Central Regional Center for Rural Development, Iowa State University. (n.d.). *Community assessment.* Des Moines, IA: Author

Nowicki, J. & Arbuckle, L. (2009). The social worker as family counselor in a nonprofit community-based

agency. In A. R. Roberts, *Social workers' desk reference* (2nd ed.) (pp. 45–53). New York: Oxford Press.

O'Connor, M. K. & Netting, F. E. (2009). *Organization practice: A guide to understanding human service organizations.* Hoboken, NJ: Wiley.

Ohmer, M. L. & Korr, W. S. (2006). The effectiveness of community practice interventions: A review of the literature. *Research on Social Work Practice, 16*(2), 132–145.

Ohmer, M. L. (2008). Assessing and developing the evidence base of macro practice: Interventions with a community and neighborhood focus. *Journal of Evidence-based Social Work, 5*(3/4), 519–547.

Ohmer, M. L. & DeMasi, K. (2009). *Consensus organizing: A community development workbook.* Thousand Oaks, CA: Sage Publications, Inc.

Ohmer, M. L., Sobek, J. L., Teixeira, S. N., Wallace, J. M. & Shapiro, V. B. (2013). Community-based research: Rationale, methods, roles, and considerations for community practice. In M. Weil's (Ed.), *The handbook of community practice* (pp. 791–807). Thousand Oaks, CA: Sage.

Ohmer, M. L. & Brooks, F. (2013). The practice of community organizing. In M. Weil (Ed.), *The handbook of community practice* (pp. 233–248). Thousand Oaks, CA: Sage Publications

Oregon Legislature. (n.d.). *How to testify before a legislative committee.* Retrieved from http://www.leg.state.or.us/comm/testify.html

Packard, T. (2009). Leadership and performance in human services organizations. In R. J. Patti (Ed.), *The handbook of social welfare management* (pp. 143–164). Thousand Oaks, CA: Sage Publications.

Papell, C. P. & Rothman, B. (1962). Social group work models: Possession and heritage. *Journal of Education for Social Work, 2*(2), pp. 66–77.

Papero, D. V. (2009). Bowen family systems theory. In A.R. Roberts, *Social workers' desk reference* (2nd ed.) (pp. 447–452). New York: Oxford Press.

Parsons, R. J. (2008). Empowerment practice. In T. Mizrahi and L. E. Davis (Eds.), *Encyclopedia of social work* (20th ed.) (pp. 2:123–126). Washington, DC and New York: NASW Press and Oxford University Press.

Pawelski, J. G., Perrin, E. C., Foy, J. M., Allen, C. E., Crawford, J. E., Del Monte, M. Kaufman, M., Klein, J. D., Smith, K., Springer, S. Tanner, J. L., & Vickers, D. L. (2006). The effects of marriage, civil union, and domestic partnership laws on health and well-being of children. *Pediatrics*, *118*, 349–364.

Payne, M. & Askeland, G. A. (2008). *Globalizaiton and international social work: Postmodern change and challenge*. Burlington, VT: Ashgate Publishing Company.

Pfeffer, J. (1992). *Managing with power and politics: Influence in organizations*. Boston, MA: Harvard Business School Press.

Pippard, J. L. & Bjorklund, R. W. (2004). Identifying essential techniques for social work community practice. *Journal of Community Practice*, *11*(4), 101–116.

Plitt, D. L. & Shields, J. (2009). Development of the policy advocacy behavior scale: Initial reliability and validity. *Research on Social Work Practice*, *19*(1), 83–92.

Poole, D. L. (2009). Community partnerships for school-based services. In A. Roberts (Ed.), *Social workers desk reference* (pp. 907–912). New York: Oxford University Press.

Pope, N. D., Rollins, L., Chaumba, J., & Risler, E. (2011). Evidence-based practice knowledge and utilization among social workers. *Journal of Evidence-Based Social Work*, *8*, 349–368.

Potocky, M. (2008). Immigrants and refugees. In T. Mizrahi & L. E. Davis, *Encyclopedia of social work* (20th ed.) (pp. 3:441–445). Washington, DC and New York: NASW Press and Oxford University Press.

Putnam, R. D. & Feldstein, L. M. (2003). *Better together: Restoring the American community*. York: Simon & Schuster.

Ramanathan, C. S. & Link, R. J. (2004). *All our futures: Principles and resources for social work practice in a global era*. Belmont, CA: Brooks/Cole.

Randall, A. C. & DeAngelis, D. (2008). Licensing. In T. Mizrahi & L. E. Davis (Eds.), *Encyclopedia of social work* (20th ed.) (pp. 3:87–91). Washington, DC and New York: NASW Press and Oxford University Press.

Rapp, C. A. & Goscha, R. J. (2006). *The strengths model: Case management with people with psychiatric disabilities*. New York: Oxford University Press.

Rawls, J. (1971). *A theory of justice*. Cambridge, MA: Harvard University Press.

Reamer, F. G. (2003). Boundary issues in social work: Managing dual relationships. *Social Work*, *48*(1), 121–133.

Reamer, F. G. (2006). *Social work values and ethics* (3rd ed.) New York: Columbia University Press.

Reamer, F. G. (2008). Ethics and values. In T. Mizrahi & L. E. Davis (Eds.) *Encyclopedia of social work* (pp. 2:143–151). Washington DC: NASW Press.

Reamer, F. G. (2009). Ethical codes of practice in the US and UK: One profession, two standards. *Journal of Social Work Values & Ethics*, *6*(2), 4.

Redevelopment Opportunities for Women. (n.d.). *Economic programs*. Retrieved from http://www.row-stl.org/content/REAP.aspx

Reid, K. E. (1997). *Social work practice with groups: A clinical perspective* (2nd ed.). Pacific Grove, CA: Brooks/Cole.

Reid, K. E. (2009). Clinical social work with groups. In A. R. Roberts, *Social workers' desk reference* (2nd ed.) (pp. 432–436). New York: Oxford Press.

Reichert, E. (2011). *Social work and human rights: A foundation for policy and practice*. New York Columbia Press.

Reisch, M. (2013). Community practice challenges in the global economy. In M. Weil's (Ed.), *The handbook of community practice* (pp. 47–71). Thousand Oaks, CA: Sage.

Ringel, S. (2005). Group work with Asian-American immigrants: A cross-cultural perspective. In G. L. Greif & P. H. Ephross, *Group work with populations at risk* (2nd ed.) (pp. 181–194). New York: Oxford University Press.

Roberts, A. R. (2005). Bridging the past and present to the future of crisis intervention and crisis management. In *Crisis intervention handbook: Assessment, treatment and research* (3rd ed.) (pp. 3–34). New York: Oxford University Press.

Roberts, A. R. (2008). Crisis interventions. In T. Mizrahi & L. E. Davis, *Encyclopedia of social work* (20th ed.)

(pp. 1:484–491). Washington, DC and New York: NASW Press and Oxford University Press.

Roberts-DeGennaro, M. (2008). Case management. In T. Mizrahi & L. E. Davis, *Encyclopedia of social work* (20th ed.) (pp. 1:222–227). Washington, DC and New York: NASW Press and Oxford University Press.

Robinson, J. W. & Green, G. P. (2011) *Introduction to community Development: Theory, practice, and service-learning.* Thousand Oaks, CA: Sage.

Roche, S. E. & Wood, G. G. (2005). A narrative principle for feminist social work with survivors of male violence. *AFFILIA, 20*(4), 465–475.

Rogers, C. (1957). The necessary and sufficient conditions of therapeutic personality change. *Journal of Consulting Psychology, 22*, 95–103.

Rogers, E. E. (1975). *Organizational theory.* Boston, MA: Allyn & Bacon.

Roosevelt, E. (1958). Presentation of *In your hands: A guide for community action for the tenth anniversary of the Universal Declaration of Human Rights* [online]. Available: www.udhr.org/history/inyour.htm

Roscoe, K. D., Carson, A. M., & Madoc-Jones, L. (2011). Narrative social work: Conversations between theory and practice. *Journal of Social work Practice, 25*(1), 47–61.

Roscoe, K. D. & Madoc-Jones, L. (2009). Critical social work practice: A narrative approach. *International Journal of Narrative Practice, 1*, 9–18.

Rostoky, S. S. & Riggle, E. D. B. (2011). Marriage equality for same-sex couples: Counseling psychologists as social change agents. *The Counseling Psychologist, 39*(7), 956–972.

Rothman, J. (2008). Multi modes of community intervention. In J. Rothman, J. Erlich, & J. Tropman (Eds.), *Strategies of community intervention* (7th ed.) (pp.141–170). Peosta, Iowa: Eddie Bowers Publishing Co., Inc.

Rothman, J. C. (2009a). An overview of case management. In A. R. Roberts, *Social workers' desk reference* (2nd ed.) (pp. 751–755). New York: Oxford Press.

Rothman, J. C. (2009b). Developing therapeutic contracts with clients. In A. R. Roberts, *Social*

workers' desk reference (2nd ed.) (pp. 514–520). New York: Oxford Press.

Royse, D., Staton-Tindall, M., Badger, K., & Webster, J. M. (2009). *Needs assessment.* New York: Oxford University Press.

Rozakis, L. (1996). *New Roberts Rules of Order.* New York: Smithmark Reference.

Rubin, H. J. & Rubin, I. S. (2008). *Community organizing and development.* Boston: Pearson Allyn & Bacon.

Rutgers School of Social Work. (2007). *Continuing education.* Retrieved from http://socialwork.rutgers.edu/ContinuingEducation/ce.aspx

Sager, J. S. & Weil, M. (2013). Larger-scale social planning: Planning for services and communities. In M. Weil (Ed.), *The handbook of community practice* (pp. 299–325). Thousand Oaks, CA: Sage Publications

Sandfort, J. (2005). Casa de Esperanza. *Nonprofit Management & Leadership, 15*(3), 371–382.

Saint Louis University Pius Library (n.d.). *Research guides.* Retrieved from http://libguides.slu.edu/index.php

Saleebey, D. (2013). *The strengths perspective in social work practice.* Boston: Allyn & Bacon.

Scherrer, J. L. (2012). The United Nations Convention on the Rights of the Child as policy and strategy for social work action in child welfare in the United States. *Social Work, 57*(1), 11–22.

Schein, E. H. (1992). *Organizational culture and leadership.* San Francisco: Jossey-Bass.

Schiller, L. Y. (1995). Stages of development in women's groups: A relational model. In R. Kurland & R. Salmon (Eds.), *Group work practice in a troubled society: Problems and opportunities* (pp. 117–138). New York: The Haworth Press.

Schiller, L. Y. (1997). Rethinking stages of group development in women's groups: Implications for practice. *Social Work with Groups, 20*(3), 3–19.

Schiller, L. Y. (2007). Not for women only: Applying the rlational model of group development with vulnerable populations. *Social Work with Groups, 30*(2), 11–26.

Schiller, L. Y. (2003). Women's group development from a relational model and a new look at facilitator influence on group development. In. M. B. Cohen & A. Mullender, *Gender and group work* (pp. 16–31). New York: Routledge.

Schultz, D. (2004). Cultural competence in psycho-social and psychiatric care: A critical perspective with reference to research and clinical experiences in California, US and in Germany. In A. Metteri, T. Krôger, A. Pohjola, & P. Rauhala (Eds.), *Social work visions from around the globe* (pp. 231–247). Binghamton, NY: Haworth.

Schneider, R. L. & Lester, L. (2001). *Social work advocacy*. Belmont, CA: Brooks/Cole.

Schwartz, W. (1961). *The social worker in the group*. The Social Welfare Forum, 146–177.

Sebold, J. (2011). Families and couples: A practical guide for facilitating change. In G. J. Greene and M. Y. Lee, *Solution-oriented social work practice* (pp. 209–236). NY: Oxford University Press.

Seligman, M. & Darling, R. B. (2007). *Ordinary families, special children: A systems approach to childhood disability* (3rd ed.). New York: The Guilford Press.

Shakya, H. B., Usita, P. M., Eisenberg, C., Weston, J., & Liles, S. (2012). Family well-being concerns of grandparents in skipped generation families. *Journal of Gerontological Social Work*, 55, 39–54.

Shalay, N. & Brownlee, K. (2007). Narrative family therapy with blended families. *Journal of Family Psychotherapy*, 18(2), 17–30.

Shebib, B. (2003). *Choices*. New York: Allyn and Bacon.

Shelton, M. (2012, May 7). Tornado recovery offers Joplin students new lessons. *NPR*. Retrieved from http://www.npr.org/2012/05/07/151950143/tornado-recovery-offers-joplin-students-new-lessons

Sherraden, M. (1993). Community studies in the baccalaureate social work curriculum. *Journal of Teaching in Social Work*, 7(1), 75–88.

Shields, G. & Kiser, J. (2003). Violence and aggression directed toward human service workers: An exploratory study. *Families in Society*, 84(1), 13–20.

Shier, M. L. (2012). Work-related factors that impact social work practitioners' subjective well-being: Well-being in the workplace. *Journal of Social Work*, 12(6), 402–421.

Shin, J., Taylor, M. S., & Seo, M-G. (2012). Resources for change: The relationships of organizational inducements and psychological resilience to employees' attitudes and behaviors toward organizational change. *Academy of Management Journal*, 55(3), 727–748.

Shulman, L. (2009a). Developing successful therapeutic relationships. In A. R. Roberts (Ed.), *Social worker's desk reference* (pp. 573–577). New York, NY: Oxford University Press.

Shulman, L. (2009b). Group work phases of helping: Preliminary phase. In A. Gitterman & R. Salmon (Eds.), *Encyclopedia of social work with groups* (pp. 109–111). New York: Routledge.

Siebold, C. (2007). Everytime we say goodbye: Forced termination revisited, a commentary. *Clinical Social Work Journal*, 35(2), 91–95.

Simmons, C. S., Diaz, L., Jackson, V., & Takahashi, R. (2008). NASW Cultural Competence Indicators: A new tool for the social work profession. *Journal of Ethnic and Cultural Diversity in Social Work*, 17(1), 4–20.

Simon, B. (1990). Re-thinking empowerment. *Journal of Progressive Human Services*. 1(1), 29.

Sims, P. A. (2003). Working with metaphor. *American Journal of Psychotherapy*, 57(4), 528–536.

Singh, A. A. & Salazar, C. F. (2011). Conclusion: Six Considerations for Social Justice Group Work. In A. A. Singh & C. F. Salazar, *Social justice in group work*. Practical interventions for change. New York: Routledge.

Slade, E., McCarthy, J. F., Valenstein, M., Visnic, S., & Dixon, L. B. (2012). Cost savings from Assertive Community Treatment services in an era of declining psychiatric inpatient use. *Health Services Research*. doi: 10.1111/j.1475–6773.2012.01420.

Smith, B. D. (2005). Job retention in child welfare: Effects of perceived organizational support, supervisor support, and intrinsic job value. *Children and Youth Services Review*, 27(2), 153–169.

Sohng, W. S. L. (2008). Community-based participatory research. In T. Mizrahi & L. E. Davis (Eds.), *Encyclopedia of social work* (pp. 1:368–370). Washington DC and New York: NASW Press and Oxford University Press.

Sormanti, M. (2012). Writing for and about clinical practice. In W. Green and B. L. Simon, *The Columbia guide to social work writing* (pp. 114–132). NY: Columbia University Press.

Sowers, K. M. & Rowe, W. S. (2009). International perspectives on social work practice. In A. R. Roberts (Ed.), *Social workers' desk reference* (2nd ed.) (pp. 863–868). New York: Oxford University Press.

Specht, H. (1990). Social work and the popular psychotherapies. *Social Service Review, 64*, 345–357.

Speer, P. W. & Christens, B. D. (2011). Local community organizing and change: Altering policy in the housing and community development system in Kansas City. *Journal of Community & Applied Social Psychology, 22*(5), 414–427.

St. Anthony's Medical Center. (2010). *Family Intervention and Planning.* St. Louis, MO

Stempel, J. (2013). *Supreme Court to hear same-sex marriage cases in late March.* Retrieved from: http://www.reuters.com/article/2013/01/07/us-usa-court-gaymarriage-idUSBRE9060N820130107

Steger, M. B. (2009). *Globalization.* New York: Sterling Publishing.

Stevenson, M. (2010). Flexible and responsive research: Developing rights-based emancipatory disability research methodology in collaboration with young adults with Downs Syndrome. *Australian Social Work, 63*(1), 35–50.

Strand, V., Carten, A., Connolly, D., Gelman, S. R., & Vaughn, P. B. (2009). A cross system initiative sup- porting child welfare workforce professionalization and stabilization. A task group in action. In C. S. Cohen, M. H. Phillips, & M. Hanson (Eds.), *Strength and diversity in social work with groups* (pp. 41–53). New York: Routledge.

Streeter, C. (2008). Community: Practice interventions. In T. Mizrahi & L. E. Davis (Eds.), *Encyclopedia of social work* (pp. 1:355–368). Washington DC and New York: NASW Press and Oxford University Press.

Strom-Gottfried, K. J. (2000). Ensuring ethical practice: An examination of NASW Code violations, 1986–97. *Social Work, 45*(3), 251–261.

Strom-Gottfried, K. J. (2003). *Managing risk through ethical practice: Ethical dilemmas in rural social work.* Presentation at the National Association of Social Workers Vermont chapter, Essex, VT.

Strom-Gottfried, K. J. (2008). *The ethics of practice with minors.* Chicago, IL: Lyceum Books, Inc.

Swenson, C. R. (1998). Clinical social work's contribution to a social justice perspective. *Social Work, 43*(6), 527–537.

Tebb, S. C. (1995). An aid to empowerment: A caregiver well-being scale. *Health and Social Work, 20*(2), 87–92.

Tebb, S. C., Berg-Weger, M. & Rubio, D. M. (2013). The Caregiver Well-Being Scale: Developing a short-form rapid assessment instrument. In press, *Health and Social Work.*

Thomas, H. & Caplan T. (1999). Spinning the group process wheel: Effective facilitation techniques for motivating involuntary client groups. *Social Work with Groups, 21*(4), 3–21.

Thyer, B. (2008). Evidence-based macro practice: Addressing the challenges and opportunities. *Journal of Evidence-based Social Work, 5*(3/4), 453–472.

Thyer, B. A. (2009). Evidence-based practice, science, and social work. An overview. In A.R. Roberts, *Social workers' desk reference* (2nd ed.) (pp. 1115–1119). New York: Oxford Press.

Tolbert, P. S. & Hall, R. J. (2009). *Organizations structures, processes, and outcomes.* Upper Saddle River, New Jersey: Pearson Prentice Hall.

Tomasello, N. M., Manning, A. R., & Dulmus, C. N. (2010). Family-centered early intervention for infants and toddlers with disabilities. *Journal of Family Social Work, 13*, 163–172.

Toseland, R. W. & Horton, H. (2008). Group work. In T. Mizrahi & L. E. Davis, *Encyclopedia of social work* (20th ed.) (pp. 2:298–308). Washington, DC and New York: NASW Press and Oxford University Press.

Turner, H. (2011). Concepts for effective facilitation of open groups. *Social work with groups, 34*, 146–156.

UAW Global Organizing Institute. (2012). *Homepage.* Retrieved from http://www.uaw.org/page/uaw-global-organizing-institute–0

Uken, A., Lee, M.Y., & Sebold, J. (2013). The Plumas Project: Solution-focused treatment of domestic violence offenders. In P. De Jong & I.K. Berg, *Interviewing for solutions* (4th ed.) (pp. 333–345). Belmont, CA: Brooks/Cole.

United Nations. (1997). *Human rights at your fingertips.* United Nations Department Of Public Information, New York: Author. Retrieved from http://www.un.org/rights/50/game.htm#60

U.S. Census Bureau. (2010b). *Valentine's Day 2011: February 14.* Available at: https://www.census.gov/newsroom/releases/archives/facts_for_features_special_editions/cb11-ff02.html

U.S. Census Bureau. (2011a). *Current Population Survey, 2011 Annual Social and Economic Supplement.* Washington, D.C. Government Printing Office

U. S. Census Bureau. (2011b). *Census bureau releases estimates of same-sex married couples.* Retrieved from http://www.census.gov/newsroom/releases/archives/2010_census/cb11-cn181.html

U.S. Census Bureau. (2012a). *Current Population Survey, 2012 Annual Social and Economic Supplement.* Washington, D.C. Government Printing Office

U.S. Census Bureau. (2012b). *Most children younger than age 1 are minorities.* Retrieved from http://www.census.gov/newsroom/releases/archives/population/cb12–90.html

U.S. Department of Housing and Urban Development. (n.d.). *Connecting to success: Neighborhood networks asset mapping guide.* Retrieved from http://www.hud.gov/offices/hsg/mfh/nnw/resourcesforcenters/assetmapping.pdf

Van Den Bergh, N. & Crisp, C. (2004). Defining culturally competent practice with sexual minorities: Implications for social work education and practice. *Journal of Social Work Education, 40*(2), 221–238.

Van Hook, M. P. (2008). *Social work practice with families: A resiliency-based approach.* Chicago: Lyceum Books.

Van Soest, D. (2008). Oppression. In T. Mizrahi & L. E. Davis, *Encyclopedia of social work* (20th ed.) (pp. 3:322–324). Washington, DC and New York: NASW Press and Oxford University Press.

Van Treuren, R. R. (1993). Self-perception in family systems: A diagrammatic technique. In C. Meyer, *Assessment in social work practice* (p. 119). New York: Columbia University Press.

Van Wormer, K. (2009). Restorative justice as social justice for victims of gendered violence: A standpoint feminist perspective. *Social Work, 54*(2), 107–116.

Vinter, R. D. (1974). Program activities: An analysis of their effects on participant behavior. In P. Glassner, R. Sarri, & R. Vinter (Eds.), Individual change through small groups (pp. 233–243). New York: The Free Press.

Vodde, R. & Giddings, M. M. (1997). The propriety of affiliation with clients beyond the professional role: Nonsexual dual relationships. *Arete, 22*(1), 58–79.

Vodde, R. & Gallant, J. P. (2002). Bridging the gap between micro and macro practice: Large-scale change and a unified model of narrative deconstructive practice. *Journal of Teaching in Social Work, 38*(3), 439–458.

Wagner, E. F. (2008). Motivational interviewing. In T. Mizrahi & L. E. Davis, *Encyclopedia of social work* (20th ed.) (pp. 3:273–276). Washington, DC and New York: NASW Press and Oxford University Press.

Wahab, S. (2005). Motivational interviewing and social work practice. *Journal of Social Work, 5*(1), 45–60.

Walker, L. (2013). Solution-focused reentry and transition planning for imprisoned people. In P. De Jong & I. K. Berg, *Interviewing for solutions* (4th ed.) (pp. 318–328). Belmont, CA: Brooks/Cole.

Walz, T. & Ritchie, H. (2000). Gandhian principles in social work practice: Ethics revisited. *Social Work, 45*(3), 213–222.

Warde, B. (2012). The Cultural Genogram: Enhancing the cultural competency of social work students. *Social Work Education, 31*(5), 570–586.

Warren, R. L. (1978). *The community in America.* Chicago: Rand McNally.

Wayne-Metropolitan Community Action Agency (2007). *Wayne-Metropolitan Community Action Agency: Community Needs Assessment 2007.* Retrieved from http://infopeople.org/sites/all/files/past/2007/needs/needs_waynemetro_survey.pdf

Weick, A. (1999). Guilty knowledge. *Families in Society, 80*(4), 327–332.

Weick, A., Kreider, J., & Chamberlain, R. (2009). Key dimensions of the strengths perspective in case management, clinical practice, and community practice. In D. Saleebey (Ed.), *The strengths perspective in social work practice* (5th ed.) (pp. 108–121). Boston: Pearson.

Weil, M. & Gamble, D. N. (2009). Community practice model for the twenty-first century. In A. Roberts (Ed.), *Social workers' desk reference* (pp. 882–892). New York: Oxford University Press, Inc.

Weil, M. (2013). Community-based social planning. In M. Weil (Ed.), *The handbook of community practice* (pp. 265–298). Thousand Oaks, CA: Sage Publications.

Weil, M., Gamble, D. N., & Ohmer, M L. (2013). Evolution, models, and the changing context of community practice. In M. Weil (Ed.), *The handbook of community practice* (pp. 167–193). Thousand Oaks, CA: Sage Publications.

Weil, M., & Ohmer, M. L. (2013). Applying practice theories in community work. In M. Weil's (Ed.), *The handbook of community practice* (pp. 123–161). Thousand Oaks, CA: Sage.

Weil, M., Reisch, M., & Ohmer, M. L. (2013). Introduction: Contexts and challenges for 21st century communities. In M. Weil's (Ed.), *The handbook of community practice* (pp. 3–26). Thousand Oaks, CA: Sage.

West Virginia Board of Social Work Examiners. (n.d.). *Level D: LICSW: (Licensed Independent Social Worker) license.* Retrieved from http://www.wvsocialworkboard.org/licensinginfo/regular/licswlicense.htm

Wharton, T. C. (2008). Compassion fatigue: Being an ethical social worker. *The New Social Worker, 15*(1),4–7.

Wharton, T. C. & Bolland, K. A. (2012). Practitioner perspectives of evidence-based practice. *Families in Society: The Journal of Contemporary Social Services, 93*(3), 157–164.

Whitaker, T. & Arrington, P. (2008). *Social workers at work. NASW Membership Workforce Study.* Washington, DC: National Association of Social Workers.

Whitaker, T., & Wilson, M. (2010). *National Association of Social Workers 2009 compensation and benefit study: Summary of key compensation findings.* Washington DC: NASW.

Wheeler, W. & Thomas, A. M. (2011). Engaging youth in community development. In J. W. Robinson Jr. & G. P. Green (Eds.), *Introduction to community development: Theory, practice and service-learning* (pp. 209–227). Los Angeles, CA: Sage Publications.

Williams, N. R. (2009). Narrative family interventions. In A. C. Kilpatrick & T. P. Holland, *Working with families*. An integrative model by level of need (5th ed.) (pp. 199–223). Boston: Pearson.

Winship, K. & Lee, S. T. (2012). Using evidence-based accreditation standards to promote Continuous Quality Improvement: The experience of San Mateo County Human Services Agency. *Journal of Evidence-based Social Work, 9*(1–2), 68–86.

Wise, J. B. (2005). *Empowerment practice with families in distress.* New York: Columbia University Press.

Witkin, S. L. (2000). Ethics-R-Us. *Social Work, 45*(3), 197–212.

Wood, G. G. & Tully, C. T. (2006). *The structural approach to direct practice in social work: A social constructionist Perspective* (3rd ed.). New York: Columbia University Press.

Wood, G. G. & Roche, S. E. (2001). Representing selves, reconstructing lives: Feminist group work with women survivors of male violence. *Social Work with Groups, 23*(4), 5–23.

Work Group for Community Health and Development at the University of Kansas. (2012). *The Community Tool Box.* Retrieved from http://ctb.ku.edu/en/tablecontents/chapter_1033.aspx

World Health Organization. (2012). *Global democracy deficit.* Retrieved from http:// www.who.int/trade/glossary/story037/en/index/html

Yalom, I. D. & Leszcz, M. (2005). *The theory and practice of group psychotherapy* (5th ed.). New York: Basic Books.

Young, S. (2013). Solutions for bullying in primary schools. In P. De Jong & I. K. Berg, *Interviewing for solutions* (4th ed.) (pp. 308–318). Belmont, CA: Brooks/Cole.

Zandee-Amas, R. R. (2013). A good groups runs itself— and other myths. *The New Social Worker*, *20*(1), 10–11.

Zur, O., & Lazarus, A. A. (2002). Six arguments against dual relationships and their rebuttals. In A. A. Lararus & O. Zur (Eds.), *Dual relationships and psychotherapy* (p. 3–24). New York, NY: Springer.

CREDITS

Photo 1-A: © Lisa F. Young

Ex 1.1: Copyrighted material reprinted with permission from the National Association of Social Workers, Inc.

Ex 1.2: Healy, L. M. & Link, R. J. (2012). Handbook of international social work: Human rights, developmen, and the global profession. Used by permission of Oxford University Press.

Photo 1-B: © Lisa F. Young

Ex 1.4: University Declaration of Human Rights used by permission of Human Rights Education Association.

Photo 2-A: © marekuliasz

Ex 2.1: Copyrighted material reprinted with permission from the National Association of Social Workers, Inc.

Ex 2.3: Copyrighted material reprinted with permission from the National Association of Social Workers, Inc.

Ex 2.4: International Federation of Social Workers (IFSW). (2012). Ethics in social work: Statement of principles. Retrieved October 14, 2012 from http://ifsw.org/policies/statement-of-ethical-principles/. Used by permission.

Ex 2.5: New Zealand Association of Social Workers (NZASW). (1993). Code of ethics. Used by permission.

Photo 2-B: © Mark Bowden

QG 3: From Dolgoff/Loewenberg/Harrington. Ethical Decisions for Social Work Praictce, 8E. © 2009 Wadsworth, a part of Cengage Learning, Inc. Reproduced by permission. www.cengage.com/permissions

Ex 2.7: From The Propriety of Affiliation with Clients beyond the Professional Role: Nonsexual dual relationships. Arete 22(1) by R. Vodde and M.M. Giddings. © 1997. Reprinted with permission.

Photo 3-A: © Yuri Arcurs

Ex 3.3: From Skills for Direct Practice in Social Work by R. R. Middleman and G.G. Wood. Copyright © 1990. Columbia University Press.

Ex 4.1: From Assessing strengths, Identifying acts of resistance to violence and oppression, by K.M. Anderson, C.D. Cowger and C.A. Snively. In The Strengths perspective in social swork practice (the d.), by D. Saleebey (Ed.). Copyright Allyn & Bacon, 2009.

Ex 4.2: From Assessing strengths, Identifying acts of resistance to violence and oppression, by K.M. Anderson, C.D. Cowger and C.A. Snively. In The Strengths perspective in social swork practice (the d.), by D. Saleebey (Ed.). Copyright Allyn & Bacon, 2009.

Photo 4-A: © fatihhoca

QG 8: Adapted from The Strengths perspective in social work (5th ed.), by D. Saleebey. Copyright © 2009, Allyn & Bacon.

Photo 4-B: © Lisa F. Young

QG 10: Kagle, J. D. & Kopels, S. (2008). Social work records (3rd ed.). Long Grove, IL: Waveland Press, Inc. Used by permission.

Ex 4.8: Adapted from St. Anthony's Medical Center, St. Louis, Missouri. Reprinted with permission.

Ex 4.9: Adapted from St. Anthony's Medical Center, St. Louis, Missouri. Reprinted with permission.

Ex 4.10: Adapted from St. Anthony's Medical Center, St. Louis, Missouri. Reprinted with permission.

QG 11: From Dejong/Berg. Interviewing for Solutions, 4E. © 2013 Wadsworth, a part of Cengage Learning, Inc. Reproduced by permission. Www.cengage.com/permissions

Photo 5-A: © Alexander Raths

Ex 5.1: Adapted from The structural approach to direct practice in social work: A social constructionist perspective (3rd ed.), by G.G. Wood and C.T. Tully. Copyright 2006, Columbia University Press. Reprinted with permission.

Photo 5-B: © Adam Gregor

QG 16: Sormanti, M. (2012). Writing for and about clinical practice. In W. Green and B. L. Simon, The Columbia guide to social work writing (pp. 114-132). NY: Columbia University Press.

Photo 6-B: © Rob Hainer

Ex 6.4: From Culture in special education, by M. Kalyanpur and B. Harry. Copyright 1999, Paul H. Brookes. Reprinted with permission.

Ex 6.5: Van Hook, M.P. (2008). Social work practice with families: A resiliency-based approach. Chicago: Lyceum Books.

Ex 6.6: Van Hook, M.P. (2008). Social work practice with families: A resiliency-based approach. Chicago: Lyceum Books.

Ex 6.7: From Self-perception in family systems: A diagrammatic technique, by R.R. Van Treuren. In assessment in social work practice, by C. Meyer. Copyright 1993, Columbia University Press. Reprinted with permission.

QG 18: Warde, B. (2012). The Cultural Genogram: Enhancing the cultural competency of social work students. Social Work Education, 31(5), 570-586. Used by permission.

Ex 6.8: From The Culturagram, by E. P. Congress. In Social Workers' desk reference (2nd ed.), by A.R. Roberts. Copyright 2009, Oxford Press. Reprinted with permission.

QG 19: Adapted from St. Anthony's Medical Center, St. Louis, Missouri. Reprinted with permission.

Ex 7.1: From Hull/Mather. Undertssanding Generalist Practice with Families, 1E. © 2006 Wadsworth, a part of Cengage Learning, Inc. Reproduced by permission. www.cengage.com/permissions

Ex 7.2: Sebold, J. (2011). Families and couples: A practical guide for facilitating change. In G. J. Greene and M. Y. Lee, Solution-oriented social work practice (pp. 209-236). NY: Oxford University Press.

Photo 7-A: © Lisa F. Young

Ex 7.3: Adapted from St. Anthony's Medical Center, St. Louis, Missouri. Reprinted with permission.

Ex 7.4: Adapted from St. Anthony's Medical Center, St. Louis, Missouri; Missouri Department of Social Services.

QG 22: From Enabling and empowering families: principles and guidelines for practice, by CX.J. dunst, C.M. Trivette and A.G. Deal. Copyright 2003, Brookline Books. Reprinted with permission.

QG 23: From Enabling and empowering families: principles and guidelines for practice, by CX.J. dunst, C.M. Trivette and A.G. Deal. Copyright 2003, Brookline Books. Reprinted with permission.

Photo 7-B: © Lisa F. Young

Photo 8-A: © Blaj Gabriel

Ex 8.6: From Teaching a methods course in social work with groups, by R. Kurland and R. Salmon. Copyright 1998, Council on Social Work Education.

Ex 8.7: Kurland, R. & Salmon, R. (1998). Teaching a methods course in social work with groups. Alexandria, VA: Council on Social Work Education.

Photo 8-B: © Yuri Arcurs

Ex 8.8: Adapted from St. Anthony's Medical Center, St. Louis, Missouri. Reprinted with permission.

Ex 8.9: Adapted from Women's Support and Community Services, St. Louis, Missouri.

Photo 9-A: © Glynnis Jones

QG 27: From Group Composition, diversity, the skills of the social worker, and group development, by T. Berman-Rossi and T.B. Kelly. Presented at the Council for Social Work Education Annual Meeting, Atlanta, February. Copyright 2003. Reprinted with permission.

Photo 9-B: © Blaj Gabriel

Ex 9.1: Comer, E. & Meier, A. (2011). Using evidence-based practice and intervention research with treatment groups for populations at risk. In G.L. Greif & P.H. Ephross (Eds.) Group work with populations at risk (3rd ed.) (pp. 459–488). New York: Oxford University Press

Ex 9.2: Greif, G. & Ephross, P.H. (Eds.). (2011). Group work with populations at risk (3rd ed.). New York: Oxford University Press.

QG 28: Berg, R.D., Landreth, G.L., & Fall, K.A. (2013). Group counseling concepts and procedures (5th ed.). New York: Routledge.

Photo 9-C: © vm

Photo 9-D: © Yuri Arcurs

QG 30: Berg, R.D., Landreth, G.L., & Fall, K.A. (2013). Group counseling concepts and procedures (5th ed.). New York: Routledge.

Ex 9.7: Berg, R.D., Landreth, G.L., & Fall, K.A. (2013). Group counseling concepts and procedures (5th ed.). New York: Routledge.

Photo 10-A: © vm

Ex 10.2: Adapted from Center of Organizational and Social Research, Saint Louis University.

Ex 10.3: Royse, D., Staton-Tindall, M., Badger, K., & Webster, J.M. (2009). Needs assessment. New York: Oxford University Press.

Photo 10-B: © CREATISTA

Ex 10.4: City of Seattle, Washington. (2012). A Community Assessment of Need for Housing and Services for Homeless Individuals and Families in the Lake City Neighborhod. Retrieved from http://seattle.gov/realestate/pdfs/Needs_Assessment_data_report.pdf

Ex 10.7: Kretzmann, J. P. & McKnight, J. L. (1993). Building communities from the inside out: A part toward finding and mobilizing a community's assets. Evanston, IL: Center for Urban Affairs and Policy Research. Used by permission.

Ex 10.8: McKnight, J. L., & Block, P. (2010). The abundant community: Awakening the power of families and neighborhoods. San Francisco, CA: Berrett-Koehler Publishers

Ex 10.9: McKnight, J. L., & Block, P. (2010). The abundant community: Awakening the power of families and neighborhoods. San Francisco, CA: Berrett-Koehler Publishers

Photo 10-C: © THEGIFT777

Ex 11.1: Netting, F. E., Kettner, P. M., McMurtry, S. L., & Thomas, L. (2011). Social work macro practice. Boston: Pearson Allyn & Bacon.

Ex 11.3: Source: 350.org

Photo 11-A: © 350.org

Photo 11-B: © 350.org

QG 35: Adapted from Organizing for social change, by K. Bobo, J. Kendall, and S. Max. Copyright 2010, The Forum Press. Reprinted with permission.

QG 36: Adapted from New Robert's Rules of Order by L. Rozakis. Copyright 1996, Smithmark Reference.

QG 37: Adapted from Consensus through conversation, by L. Dressler. Copyright 2006, Berrett-Koehler Publishers, Inc. All rights reserved. www.bkconnection.com

Ex 11.5: Kretzmann, J. P. & McKnight, J. L. (2005). Discovering community power: A guide to mobilizing local assets and your organization's capacity. Evanston, IL: Asset-Based Community Development (ABCD) Institute. Used by permission.

Ex 11.7: Ohmer, M.L. & DeMasi, K. (2009). Consensus organizing: A community development workbook. Thousand Oaks, CA: Sage Publications, Inc.

Ex 11.9: Ohmer, M.L. & Korr, W.S. (2006). The effectiveness of community practice interventions: A review of the literature. Research on Social Work Practice, 16(2),132–145. Used by permission.

Ex 11.10: From Localized needs and a globalized economy, by J. Ife in Social work and globalization (Special Issue), Canadian Social Work, 2(1). Copyright 2000.

Ex 11.11: From International social work: Issues, strategies and programs by D. Cox and M. Pawar. Copyright 2006, Sage Publications, Inc. Reprinted with permission.

Ex 11.14: Speer, P. W. & Christens, B. D. (2011). Local community organizing and change: Altering policy in the housing and community development system in Kansas City. Journal of Community & Applied Social Psychology, 22(5), 414-427. Used by permission.

Photo 12-A: © Digital Vision

Ex 12.1: National Association of Area Agencies on Aging. (2012). National aging services network. Retrieved from http://www.n4a.org/about-n4a/join/

Photo 12-B: © Chris Fertnig

Photo 12-C: © CandyBox Images

Ex 12.5: Gambrill, E. (2012). Social work practice: A critical thinkers guide. New York: Oxford University Press.

QG 39: From Locke/Garrison/Winship. Generalist Social Work Practice, 1E. © 1998 Wadsworth, a part of Cengage Learning, Inc. Reproduced by permission. www. Cengage.com/permissions

Ex 12.7: Larson, K., & McGuiston, C. (2012). Building capacity to improve Latino health in rural North Carolina: A case study in community-University engagement. Journal of Community Engagement and Scholarship, 5(1), 14-23.

QG 42: Plitt, D. L. & Shields, J. (2009). Development of the policy advocacy behavior scale: Initialreliability and validity. Research on Social Work Practice, 19(1), 83-92. Reprinted by permission.

Photo 13-A: © gosphotodesign

Ex 13.1: Evans, S. D., Hanlin, C. E., & Prillehensky, I. (2007). Blending Ameliorative and transformative approaches in human service organizations: A case study. Journal of Community Psychology, 35(3), 329-346. Used by permission.

Ex 13.2: Netting, F. E., Kettner, P. M., McMurtry, S. L., & Thomas, M. L. (2012). Social work macro practice. Boston: Pearson Allyn & Bacon.

Ex 13.4: Netting, F. E., Kettner, P. M., McMurtry, S. L., & Thomas, M. L. (2012). Social work macro practice. Boston: Pearson Allyn & Bacon.

QG 43: Manning, T. (2012). The art of successful persuasion: Seven skills you need to get your point across effectively.

Ex 13.6: O'Connor, M. K., & Netting, F. E. (2009). Organization practice: A guide to understanding human service organizations. Hoboken, NJ: John Wiley and Sons.

Ex 13.7: Winship, K. & Lee, S. T. (2012). Using evidence-based accreditation standards to promote Continuous Quality Improvement: The experience of San Mateo County Human Services Agency. Journal of Evidence-based Social Work, 9(1-2), 68-86.

GLOSSARY/INDEX

impact evaluation reviews the impact of the efforts, such as individual or community-level behavioral changes, 557
impartiality, 58
implementation skills, 545
"incestuous families", 246
income-generation programs, 483
independence, 327, 330
indigenous leadership occurs when a member or members of a social group exert leadership from within that group, 385, 468
indirect questions are questions phrased as sentences, rather than questions, 78
Individual Development Accounts (IDAs), 468–469
individual exclusion, 94
individual work is the professional application of social work theory and methods to the treatment and prevention of psychosocial dysfunction, disability, or impairment, including emotional and mental disorder, 7
assessment and planning, 103–172
engagement, 67–102
evaluation, 97, 208–209, 217–229, 232
intervention, 173–209, 231–232
termination, 209–217, 232
individualism, 327, 330, 413
inequality
community practice, 448
violence and, 154
inertia according to systems theory, is the tendency of organizations to seek to maintain the status quo, or stability, by working actively against change, 539
inferred empathy, 78–79
informal resources are those resources that exist naturally in individuals, families, organizations, and communities that may benefit people seeking services; contrast to formal services, 137, 138, 193, 213
information provision, 185–186, 211, 384
information sharing, 449
inputs are those resources necessary to implement a change effort or a program, 557, 558
institutional settings
group work, 340–341, 376
termination, 217
intangible resources are non-concrete resources, such as individual counseling and education groups, 88
integrity, 42

interdependence occurs when people depend on one another for the goods, services, relationships, social, and spiritual dimensions required to function, 413
interdisciplinary practice, 200, 256
interfaith alliances, 518
intergenerational patterns refer to the assertion that families transmit their patterns of relationship from one generation to the next, 244, 246–247
internalization of oppression is the process by which individuals come to believe that the external judgments are valid, thus resulting in a devaluing of one's self, 141, 169
International Association for Social Work with Groups (IASWG), 332, 383
international community development, 481–485
international families, 260–262
International Family Strengths Model, 261
International Federation of Social Workers (IFSW), 42, 43–44
international financial institutions, 21
international social work describes work with international organizations using social work methods or personnel, social work co-operation among countries, and transfer among countries of methods or knowledge about social work, 7
interpersonal power is the personal attribute characterized by the ability to build strong relationships, develop rapport, and persuade people, 88
interpretation, 187–188, 267
interpreters, 81
interprofessional collaboration refers to situations when professionals from different disciplines integrate their professional knowledge to work together toward a common goal, 199–201, 506–507
interprofessional team is an organized group of people, each trained in different professional disciplines, working together to resolve a common problem or achieve common goals, 256, 506–507
intervention is the joint activity of the client system and the social worker that will enable the client and the practitioner to accomplish the goals decided upon in the assessment, 10
case summary, 227–228
client's environment, 182–189
cognitive behavioral therapy, 180–181
community practice, 455–485, 491–492

critical social construction, 33
dual relationships, 59
empowerment practice, 201
family social work, 290
group work, 369
involuntary clients, 153–154
minimization of distance, 19
organizations, 502–504
shared, 18–19, 88, 166
sources of, 87–89
power and control refers to a developmental stage in social group work practice in which members vie for control among themselves and between members and the worker, 375, 376
power dependency theory, 420, 459
power politics model describes a theoretical perspective used to depict organizational change that emphasizes competition for resources, personal advancement, and inter- and intra-power struggles and asserts that the primary path to change requires strategic access to the persons who have the greatest power and primary influence, 535–536
practice is the way in which organizations implement basic functions, 542, 543, 551
practice, social work, 2–3
family social work, 299
group work, 351–353
host settings, 505–507
practice framework, 8–11
practice wisdom, 5, 229–230, 255
preaffiliation refers to a developmental stage in group practice, according to the Boston Model, in which members are ambivalent about joining a group, 374–375, 377
preference questions, 123, 293
preferred reality is a term reflecting the client's desire for creating a different reality that is consistent with her or his dreams and goals, 125, 152, 312
prejudice, 93
primary education, 483
privacy, 55, 59, 96–97, 99
private organizations, 525
privilege, in the context of culture and diversity, is a phenomenon in which unearned advantages are enjoyed by members of a particular dominant group simply by virtue of membership in it, 33, 92, 272, 369, 372–373, 451
problem solving, 387–388, 474, 545

problem-saturated story is a term used in narrative and solution-focused practice to refer to one-dimensional perspectives on a truth that client systems created about themselves that may or may not be based in fact, 123, 264–265, 293, 294
procedures are the processes through which the agency members interact with client systems that impact the experience of the client system, 513, 520, 552, 553
process evaluation focuses on the degree to which an effort functioned optimally, 556–557
professional boundaries, 59, 62, 95, 214
professionalism, 14, 461–462
program is a prearranged set of activities designed to achieve a set of goals and objectives, 541, 543, 551
community needs assessment, 428
structure of programs and services, 510, 519–520
Project Restore (PR), 484
project teams consist of a group of persons who collectively work on organizational challenges or opportunities through committee or task force structures, 502, 504
projects are prearranged sets of activities designed to achieve a set of goals and objectives, and are smaller and more flexible than programs; they can be adapted to changing needs relatively easily, and are not permanent, 541–542, 551
pseudopaternalism, 61
psychoanalytic theory is a classic theory, primarily associated with Sigmund Freud, that maintains that unconscious processes direct human behavior, 25, 107
psychoeducational group is a group focused on educating members regarding psychological processes or principles and supporting their common experiences, 3, 335–336, 379
psychosocial history, 68, 106
psychotherapy, 14
public, working with the, 197
public organizations, 497, 524–525
public participation, 43
public relations is the practice of managing communication between an organization and the public, 511, 520

qualifications, 11
qualitative evaluation, 218, 223–229

remedial groups are developed for the purpose of changing behavior, restoring function, or promoting coping strategies of the individual members who join the group voluntarily or involuntarily, 336, 339, 379, 405

remedial model is a classic model for group work practice in which the emphasis is on healing or changing individual members who have deviated from societal expectations, 331, 332

reparenting, 26

representation of the self is a social constructionist term to describe a view of her or his own worth and identity, 380–381

resiliency is the human capacity to deal with crises, stressors, and normal experiences in an emotionally and physically healthy way, 83, 98, 109
  community practice, 421
  family social work, 265–266, 288

resistance, 105, 391

resocialization is a term used to describe an intervention to assist people in dealing with their feelings and inner perceptions that are primarily related to the "self"; thought to be more relevant to psychology than social work, 14

resource mobilization theory, 420, 459

resources
  advocacy, 197
  agency, 88
  assessment of, 137–139, 169
  brokering role, 193
  case summary, 226
  client's environment, 184
  community, 414, 419, 420, 438, 456, 457
  developing, 94
  dual relationships, 59
  empowerment approach, 289, 290, 291
  group work, 379, 387
  IFSW *Statement of Principles*, 45
  organizations, 495, 516, 520
  single parent families, 254
  solution-focused approach, 295
  strengths perspective, 30, 109, 112
  termination planning, 213

respect, 6

restorative justice groups focus on crime as an interpersonal conflict that has repercussions for the victim, offender, and community at large; they emphasize the harm done to the relationships of those involved rather than the violation of the law, 380, 382–383

Richie, H., 62

Richmond, Mary, 14, 104–105, 285

risk, suicide, 148, 158–163

risk management is the identification, assessment, and prioritization of risks, 41, 62–63, 64

risk taking, 46–47

Riverton Against Youth Drinking (RAYD), 337, 403

Riverton Children's Grief Support Group, 338, 404

Riverton Mental Health Center Group for Persons with Dual Diagnosis, 339, 405

Roberts, A. R., 156–158

Robert's Rules of Order, 468, 471, 476. *See also* **parliamentary procedure**

Rogers, Carl, 69–70

role-playing, 304

Rollnick, S., 205, 303, 383

Roman Catholic Church, 54

Roosevelt, Eleanor, 173

Rothman, J., 457, 492

rules and procedures are the set of laws and processes that provide the structure for the day-to-day functioning of the program, 552, 553

rural areas
  brokering role, 192–193
  dual relationships, 58, 214

safety issues, 154–156

Salazar, C. F., 369

Saleebey, D., 29, 109, 121–122, 131, 175, 421

Salmon, R., 346

sanctions, 50

Sanger, Margaret, 464

Santa Clara County Social Service Agency (SSA), 538

scaffolding, 293–294

scaling questions
  family social work, 269, 295, 311–312
  group work, 373, 395
  individual work, 125, 178–179

scanning is a social work group practice skill in which the social worker uses visual contact with all group members to monitor their affect, participation, and non-verbal communication, 385

Schiller, L. Y., 376

school drop-outs, 497

school-based services, 195, 469, 505, 507
  case management, 191
  LGBTQ students, 523–524, 531–532, 537–539, 541–542, 544, 546–549, 552–553

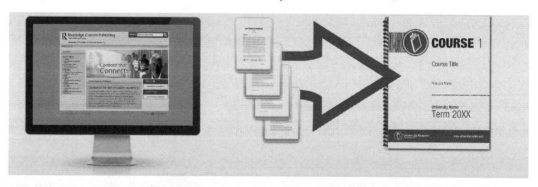

NEW DIRECTIONS IN SOCIAL WORK

SERIES EDITOR: ALICE LIEBERMAN, UNIVERSITY OF KANSAS

New Directions in Social Work is an innovative, integrated series offering a uniquely distinctive teaching strategy for generalist courses in the social work curriculum, at both undergraduate and graduate levels. The series integrates five texts with custom websites housing interactive cases, companion readings, and a wealth of resources to enrich the teaching and learning experience.

Research for Effective Social Work Practice, Third Edition

Judy L. Krysik, Arizona State University and Jerry Finn, University of Washington, Tacoma

HB: 978-0-415-52100-0
PB: 978-0-415-51986-1
eBook: 978-0-203-07789-4

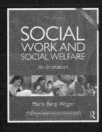

Social Work and Social Welfare, Third Edition

Anissa Taun Rogers, St. Louis University

HB: 978-0-415-52080-5
PB: 978-0-415-50160-6
eBook: 978-0-203-11931-0

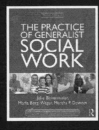

The Practice of Generalist Social Work, Third Edition

Julie Birkenmaier, Marla Berg-Weger, both at St. Louis University, and Martha P. Dewees, University of Vermont

HB: 978–0–415–51988–5
PB: 978–0–415–51989–2
eBook: 978–0–203–07098–7

Social Policy for Effective Practice: A Strengths Approach, Third Edition

Rosemary Chapin, University of Kansas

HB: 978–0–415–51991–5
PB: 978–0–415–51992–2
eBook: 978–0–203–79476–0

Made in the USA
Lexington, KY
26 January 2017